Water Matters

Water Matters

Franklin M. Fisher

QueenBeeEdit

Bloomfield, Connecticut, U.S.A.

Water Matters by Franklin M. Fisher

© 2020 Ellen Paradise Fisher

First edition, January 2020.

ISBN: 978-0-578-63281-0 (print)

Library of Congress Cataloging-in-Publication Data
Name: Fisher, Franklin M., author.
Title: Water Matters / Franklin M. Fisher.
Description: Connecticut: QueenBeeEdit Books, [2020].
Identifiers: LCCN 2020901079 (print). ISBN: 978-0-578-63281-0 (paperback).
Subjects: 1. BUSINESS & ECONOMICS / International / General. 2. POLITICAL SCIENCE / International Relations / Diplomacy. 3. TECHNOLOGY & ENGINEERING / Environmental / Water Supply.
Classification: LCC HD333.F57 2020 (print).

www.queenbeeedit.com

Cover art by Colleen Humphreys.
Cover design by Stephanie C. Fox.

In Memoriam

Franklin M. Fisher
December 13, 1934 – April 29, 2019

וּשְׁאַבְתֶּם-מַיִם בְּשָׂשׂוֹן מִמַּעַיְנֵי הַיְשׁוּעָה

Isaiah 12:3

This Memoir is dedicated fondly to my grandchildren, a.k.a.
The Eight:

Wendy

Jamie

Teddy

Val

Eve

Harry

Jesse

Beth

Contents

Preface

My youngest daughter, Naomi, was in college in 1991 when Iraq's Scud missiles began to fall on Israel during the Gulf War. It was the first time she remembers Israel being in physical danger, and she was quite upset. Some friends asked her why she was so distressed. They wanted to know whether she had family in Israel.

"I was going to tell them 'no'," she told me, "and then I thought of Robert Frost's line "home is the place where when you have to go there, they have to take you in.""[1]

A book's preface is meant to explain how a book came into being, or how the idea for a book was developed. That sort of preface would be very simple: I would simply write that the book came into being because I kept a "journal" as I worked on the water project that is the backbone of this book, and I turned it into something like a memoir. The idea of keeping a diary came from my colleague and friend Peter Temin, who is an economic historian and has complained that while physical scientists often write books about the development of their ideas, social scientists rarely, if ever, do. This book is an attempt to begin to fill that need.

The early chapters of this memoir explain *where* the basic economic idea of what became the water project—that backbone—had its genesis. In this preface, I try to explain why I wanted to do this work.

The story of my daughter and the Scud missiles is one I tended to talk about in my capacity as a leader of the New Israel Fund, but there was an occasion when I made a point of telling it to an Israeli official in the Finance Ministry. I was trying to win support for the project within the Israeli government, and I wanted the official to know how I felt about Israel.

I went on to remark that I hoped we would be able to perpetuate in later generations that same feeling Robert Frost had expressed, but noted that it depended very much on what kind of country Israel was. The official gave somewhat of an "automatic" reply one would hear from Israelis of a certain generation.

"My father moved here," he said, "so he wouldn't have to worry about things like that."

I thought that was quite true. But I also replied that it would really matter to both Israel and to Jews all over the world whether the nature of the Israeli state was such as to make people like my grandchildren have strong feelings for it. It would not only be about what might happen in the peace process, but what happens in the areas of human rights, democracy, religious pluralism, and regional cooperation and development.

The official seemed to get the message from my daughter's story. It was about how my family and I felt about Israel. And so, he began to discuss the possibility of arranging for me to meet with a key player very much higher up in the Israeli government.

The work on the economics of water in the Middle East is unfinished, but the techniques are well developed and, when the region's nations are ready to negotiate seriously about the sharing of water, it is my hope that the tools to make a just settlement of this issue will be waiting for them.

Acknowledgments

In a project of this length and complexity, many people participate in ways large and small, and even some who played important roles get lost over time. People change positions and participants move in and out. This project began more than 30 years ago and continues in all three countries—Israel, Palestine, and Jordan—in different ways with a changing cast of characters (we have identified at least 350). By the point at which it was time to write this acknowledgements section, Franklin was unable to assist in its compilation and I—his wife—have completed the task. If I have failed to acknowledge or thank any participant, it is only from ignorance and through no malice.

Several very important participants in *Water Matters* must be remembered and thanked. Most obvious is the late Gideon Fishelson, whose insight into water as a valuable resource began this entire endeavor. Also present at the outset was Myron (Mike) Fiering, who was to co-chair the project and died too soon and too young. While he could, Robert Dorfman, grand old man of the economics profession, worked diligently to keep the technical work up to par. And Howard Raiffa lent valuable insight.

Of course, the person without whom the water project would never have seen the light of day is Leonard Hausmann, chairman of ISEPME and Frank's indefatigable cheerleader to this day. Through thick and thin, he pushed and pulled to try to make good things happen.

And thus we get to the people who must be thanked not only for their faith, but their help. First is Theresa Benevento, Frank's secretary of 25 years, whose flawless transcription of the initial 400 pages of the manuscript (including Arabic and Hebrew names and places) made this book possible.

Both Frank and I are grateful to her for any successes he achieved while she was at MIT. As the book neared completion, Monica Collins did some editing, and our son Abraham also edited and gave advice. Frank is proud of him as always. The person who pulled the manuscript together and almost completed the work was Scott Cooper. Finally, Stephanie C. Fox was the magician who actually brought this memoir over the finish line.

If one looks at the co-authors for the earlier book *Liquid Assets*, a long list appears after Frank's name. As you read *Water Matters*, you will see that most of these wonderful colleagues continued in their participation as the project grew into MYWAS and AGSIM and became more technical and useful. First and foremost is Annette Huber, who began as a technical assistant while a graduate student at Harvard, remains a dear friend, and still a consultant to those in Israel, Palestine, and Jordan who use the project methodology to develop plans for infrastructure and sharing of resources. Beyond *Liquid Assets*, she was a skilled colleague and frequent joint author of water-related papers. Annette was recruited to ISEPME by Harshadeep (Harsh), another brilliant assistant who grew up and moved to the World Bank, where he works on the problems of water in Pakistan and India.

Other ISEPME staff important over the years include Tamar Miller, Anni Karasik Thurow, and Yasmina Mudarres. Theo Panayotou of HIID was indispensible in his help with field work. There were a number of students who helped with computer work and research, such as Aviv Nevo, Daniel Passerman, Yuval Nachtom, and Bilal Zia. Atif Kiburski gave great support to the Palestinian team.

Among many Dutch collaborators and government officials, first and foremost was Louise Anten, the supporter and advisor who was a friend and confidant, even when her change of title took the project out of her purview. Other Dutch friends and government supporters include Gerben DeJong, the indefatigable Hans Wesseling, Minister Jan Pronk, and Ambassador Como van Hellenberg Hubar.

In the later years, Pavel Seifter was instrumental in arranging Czech support for the Palestinians.

Numerous academic and personal friends provided useful contacts and support and cheered Frank along when he lost hope for the project and the Middle East. Eitan and Ruth Sheshinsky, who run the best B&B in Jerusalem, James Poterba, Lady Julia Neuberger, and Alon and Rachel Liel all provided friendship, useful contacts, and much more. Then there was Yair Hirschfeld, diplomat *extraordinaire*. Eliezer Yaari and Menachem Yaari gave helpful advice. Negotiating guru Robert Mnookin helped at an important moment, as did retired (but not inactive) Yoav Kislev, who in the final days of Frank's involvement, agreed to organize an Israeli team.

Among the most important participants were the members of the teams who actually did the work for so long. In Jordan, first and foremost was Munther Haddadin, indispensable water expert, year in and year out. Minister Jawad Annani was a steady supporter and the man who knew everyone. Hazim el-Nasser, Salem Hamati, Faisal Hassan, Rima Khalaf, and Hanni Mulki helped over the years. More Jordanians included Iyud Abu-Moghli, Bassem Awadallah, the Fakhourys, Ali Ghezawi, and Elias Salameh.

The mainstays of the Israeli team were Uri Shamir, considered god to young water mavens, Zvi Eckstein, Shaul Arlosoroff, the energetic Hillel Shuval, and Ilan Amir. Among the many Israelis who helped, or at least listened, over the years were Yehuda Bachmat, Haim Ben-Shahar ("Habash"), Ephraim Kleiman, Ambassador Itzhak Levanon, Yuval Nachtom, Yossi Vardi, and Dan Bitan. Also involved were Shimon Tal, Shalom Turgeman, and, of course, Rafi Cohen ("Moshe") and his family always provided the transportation. Sharon Friedman of the Jewish National Fund helped, as did Gidon Bromberg of FOEME.

Nabil Sha'ath was the project's biggest booster among the Palestinians. Karen Assaf was probably the most loyal of all the Palestinian team members. Others in the project at one time or other were Khairy El-Jamal, Riyad El-Khoudary, Marwan Haddad, Anan Jayousi, Said Abu-Jallalah, Ali Sha'ath, Nabil Sharif, Mohammad Nashishibi, and Samid Al-Abed. Also, Issa Khatar, Khalid Qahman, Dr. Samir, and Luai Sha'at. Last but not least was Jad Isaac, always true to his beliefs, of whom Frank once said that his friendship made the years he spent on the project worthwhile.

To all these and many others who made this endeavor worthwhile, Frank owes his deepest gratitude.

Ellen Paradise Fisher
Cambridge, Massachusetts
February 2019

I. 1989 through 1994

1

The Beginning: 1989-1993

I became involved with Harvard University's Institute for Social and Economic Policy in the Middle East (ISEPME) shortly after it was founded in 1988. Under the leadership of Leonard Hausman, director, and Anni Karasik, associate director, the organization had begun to hold large conferences of economists—Palestinians, other Arabs, Israelis, and Americans—to discuss the economics of peace in the Middle East. Hausman wanted to involve Harvard and MIT economists interested in the region. I qualified as an economist, and I was seriously interested in the Middle East, but at the time that interest was not professional. Still, my close friend Eytan Sheshinski of The Hebrew University, who had earned his Ph.D. in economics at MIT, recommended me to Hausman.

Those early conferences were quite general. The main progress we made was managing to get past the difficulty of all being in the same room and talking to each other. However, no serious intellectual content was being produced. My involvement up until that time could be characterized more as general statesman-like helpfulness—my children do not believe I possess the qualities of a statesman, and so I must have been sufficiently bored to exhibit what they hadn't seen. I wasn't offering much in the way of substantive economics, but I did provide some assistance in those early conferences when the sometimes deep political waters needed navigating.

The 1990 London conference was one of those occasions. Attending were Americans, Israelis, Jordanians, Palestinans (including members of the

Palestine Liberation Organization, PLO), and other Arabs. Suddenly two more PLOers appeared, seized the podium, and demanded that the conference call for the establishment of a Palestinian State. My friends among the Israelis present—all committed to peace and the end of the occupation—became more than a little uncomfortable. I realized I would have to speak. So, I explained calmly that such a declaration would make it impossible for activists who support the Israeli peace movement to continue participating in these conferences, myself included. The interlopers disappeared and never returned.

Apparently, my words made a big impression. Even the Arab participants, including the other Palestinians, were pleased. This included, especially, Nabil Sha'ath, an important figure in the Palestinian government. A U.S.-trained economist, he has been the Palestinian chief negotiator, a cabinet minister, planning minister, and even acting prime minister of the Palestinian National Authority. As I learned later, ISEPME's leaders regarded my statement as a major moment.[*]

The London conference was also a turning point, because then-Dean Robert Putnam of Harvard's Kennedy School, which housed ISEPME, began to insist on intellectual results, not just rounds of self-congratulation among participants who had succeeded in sitting together in the same room. Lenny Hausman began enlisting various teams to work on specific tasks. A steering committee of Harvard and MIT economists, chaired by the late Thomas Schelling, a professor of political economy at the Kennedy School, was established, and work was divided into several areas; most became focused around the project on the economics of transition, which resulted in a serious publication in early 1994.[1] Other areas included refugees, data, health, and similar topics. Lenny strongly urged me to chair one of the teams, and I agreed to take on *water*.

I cannot say I accepted the task with great enthusiasm. I knew nothing about the economics of water. My area of interest and study throughout my

---

[*] After the London Conference, my wife Ellen and I traveled to Israel and visited with Mariam Maari, a Haifa University faculty member. A self-styled "Palestinian Israeli," she is a strong presence. When I described the Arab reaction at the conference, she said, with some surprise, "That was you?" She almost certainly learned about the episode from Sha'ath, with whom she was very close. (It was illegal then for Israeli citizens to speak with members of the PLO.)

academic career, though, had been microeconomics, and water fit into the crucial microeconomics question of how to allocate scarce resources. I felt I had to do something, and water seemed plausible. But I did nothing for the next two and a half years, basically letting the project coast. It was always at the bottom, instead of the top, of my tray.

Finally, in autumn 1993, I turned my full attention to water.

The Water Project Before Fall 1993

I remember vividly the first time I truly understood that water could be treated as a natural resource, within the economic theory of resource allocation (and conservation), and with a price attached. It was at the 1990 London conference, when Gideon Fishelson, Tel Aviv University economist, spoke and pointed out that water is a scarce resource, and scarce resources have value, but that such value could not rationally be above the cost of replacing the water. He then explained that the cost of desalinating water on Israel and Gaza's Mediterranean coast was on the order of $1 per cubic meter, which is about 264 gallons, and that the entire value of the water in dispute in the region could not be very great (the desalination cost today is about half as much). The fact that this made me sit up with a start shows how powerfully we believe that water presents an intractable problem, and how unlikely it is to think of monetizing water disputes. There I was, an experienced microeconomist, trained to think of prices as associated with scarcity—and if I was surprised by Fishelson pointing out something I should have found entirely natural (and indeed did once he said it), how much more of a jolt in thinking it must be for non-economists to come to that idea.

That jolt in thinking, as well as the idea itself, became basic to the Water Economics Project that began, rather awkwardly, in the winter of 1992.

I was on a steering committee with Robert Pindyck (professor, MIT Sloan School of Management), Robert Dorfman (Harvard professor of political economy), and Myron Fiering (professor of engineering and applied mathematics at Harvard). We were to meet and plan a conference for the coming summer. The committee, though, did not take an active role. Pindyck may have attended the first meeting but then never reappeared. Dorfman came, but I don't remember whether Fiering did. Still, though, a conference was scheduled; it would be a preliminary meeting to which participants from the region were invited and encouraged to think in terms of the economics of water, especially the valuation of water through demand curves. The aim was to stimulate work on the demand side.

The mid-summer conference was a flop. Only Fishelson and Fiering seemed to have understood what I was talking about in discussing the demand side for water. The Arab participants were all serious and pleasant enough, but seemed unable to deal with the idea that water could be given a monetary value. They insisted instead on discussing who had the rights to the water. In retrospect, I see that I failed to understand how to deal with that issue—but in any event, the result was depressing.

The conference adjourned with mild agreement that participants would attempt to answer questions about what the social and economic consequences would be of having more or less water.

Shortly thereafter, Fiering and I met and, at his suggestion, formulated a letter asking participants to focus on the consequences of having *more* water. Fiering had suggested that this might arouse less emotion than the possibility of having to deal with less.

The results were disappointing. Fishelson turned in a paper—informative, if preliminary—on Israeli demand for water. The Palestinians submitted nothing. The Jordanians, Elias Salameh and Ali Ghezawi, produced an intelligent paper, but it did not mention price and was really pretty useless from the project's point of view They are not economists, and it showed.

In October 1992, Myron Fiering died suddenly. It was a great tragedy, and a considerable loss to the project. He was the only one in Cambridge who seemed interested in taking an active role along with me.

Matters stood in a depressing state until the following spring, when I advised Hausman and Karasik at ISEPME to convene a different group for a second try. I believed it was important for them to recruit economists. I wasn't hopeful, though, and I wasn't putting any time into the project—which simply stalled.

The First Team Comes Together

Looking back, the failure of the project's first stage was probably a good thing. The basic idea of national policy demand curves had certainly not occurred to me at the time of the first conference. Even if I had that idea at the time, the participants at the first conference would not have been ready to receive it. The hopeful atmosphere of the September 1993 Oslo peace accord between Israel and the PLO, which certainly mattered later, was not yet in place. The right central team for the project was not yet there.

For the second conference, we began to see a team that would somewhat serendipitously bring the right skills and contacts at the right time. Two events in Spring 1993 helped to bring that team together. The first was an inquiry from Hillel Shuval, an environmental engineering professor and water expert at The Hebrew University. We'd known each other for some years. During an earlier visit to Boston, he became active with Boston Friends of Peace Now, an organization with which I was also becoming involved.

Shuval wrote, inquiring about the project. He mentioned he would be visiting Harvard for the 1993–1994 academic year. He was in touch with Lenny Hausman, but he needed an office. In Spring 1993, he came to Cambridge and presented a paper to Hausman, Dorfman, and me in which he predicted that there would be no water for agriculture in the region thirty years out. He showed considerable interest in the project and, after some negotiation, we agreed to finance, in part, his time at Harvard. He wanted to spend much of that academic year involved with the water problem.

Hillel and I wrote a proposal for a serious water project. It was not a surprise that his first draft didn't include much about the demand for water. He wasn't an economist. I supplied the missing pieces.

That same spring, word came that some people at Tel Aviv University— in particular, economist Zvi Eckstein—were working on a water allocation model for the region (I refer to it as the "Eckstein-Fishelson" model from here on). It was reported to me as available only in Hebrew. I suggested they talk to Shuval and expressed considerable interest. The model was to play a very important role in what later transpired.

I learned that Eckstein would be visiting Boston University in the fall, and Fishelson would be visiting New York University. The Israelis wanted to bring Yuval Nachtom with them; he was a student who had helped develop their model and program. We agreed to pay for Nachtom and provide some other money. Thus, the Eckstein-Fishelson model came into the water project.

It was an important development. Eckstein and his associates, using somewhat sketchy data, had produced a model of supply and demand for water for Israel and Palestine that accounted for conveyance costs between parts of the region. In that model, prices (which differed over locations only by conveyance costs) were adjusted until total supply and demand in the region were balanced. Those prices were then the efficiency prices given the supply and demand curves. Note, however, that the adjustment of prices could be done only by assuming that all conveyance of water is in a single direction (assumed to be north to south), so that only one price needs to be adjusted. Unfortunately, that assumption was incorrect, as we quickly discovered. Nevertheless, the Eckstein-Fishelson model became the prototype of the project, the template of the computer model we had already contemplated building—because it would be needed if we were going to deal with the supply and demand for water.

Economists know how to do that.

Although I did not realize it then, a good team had come together—with one notable exception. The project's success still required serious participation on the Arab side—Palestinians and Jordanians. We scheduled a conference for late October and, importantly, convened shortly after the handshake on the White House lawn between Israeli Prime Minister Yitzhak Rabin and Yasser Arafat, the head of the PLO, which initiated the Oslo Accords and ushered in an all-too-brief spirit of reconciliation.

This time, conference participation was somewhat different. In particular, Jad Isaac was the principal Palestinian representative. A very smart man from Bethlehem, Isaac was a water expert and agronomist with high-level connections in the PLO. Hausman had evidently succeeded in getting the PLO to pay some attention to the project.

Other Arab participants at the October conference were also important. They included Elias Salameh from Jordan, who had attended the first conference; Issam Al Shawwa from Gaza; Mussa Neimah from Lebanon; and Sharif El-Musa, a Palestinian working in Washington who was about to join Jad Isaac in Bethlehem and who had produced a very intelligent paper that Bob Dorfman had called to my attention.

National Policy Demand Curves and Property Rights Disputes

Zvi Eckstein, Yuval Nachtom, Hillel Shuval, Lenny Hausman, and I met a couple of weeks before the second water conference, which was held on October 21–22, 1993, to prepare for it. Our principal task was Zvi's presentation of the Eckstein-Fishelson model. Hillel emphasized that the presentation was likely to set off various political landmines. So, we decided I should give an introduction to try to bring the participants around to thinking about water as a monetized commodity. I suggested I talk about crude oil and the Rule of Capture, my standard subject for more than 35 years for discussing externalities in natural resources. The Rule of Capture, common law from England that has been adopted in many U.S. jurisdictions, essentially establishes a rule of non-liability and ownership of captured natural resources. It illustrates that the distribution of property rights is analytically separate from the efficient operation of the system, known as the Coase Theorem.*

Hillel urged the importance of a careful presentation, and I sat down to write out very meticulous notes. Those notes turned out to be quite important in the history of the project. They were distributed at the meeting, and they later became the basis for my paper on what the project was about. Writing them brought home to me two basic ideas of great importance.

The first of these was the importance of the separation of the property rights dispute from the question of efficient operation. It was not a new idea in economics, but its centrality to our project became very clear.

The second idea was new. The Eckstein-Fishelson model—the Eckstein paper—really did not cope very well with the question of what happens when one country departs from free market water allocation and adopts a water policy. It pointed out that the effects of a water subsidy would lead to other countries raising the price of water. While true, this was insufficient. In the first place, other countries might be net consumers of water and therefore not benefit. Further, such an argument re-entangled the questions of property rights and water management.

Later, while walking along the Charles River, my wife Ellen impressed on me the fact that we needed a satisfactory solution to questions involving what happens when countries adopted subsidization policies for social or political reasons. In essence, these were questions concerning the public rather than

---

* Named for Ronald Coase, for whom it earned the 1991 Nobel Prize in Economics, the Coase Theorem states that if trade in an externality is possible and there are sufficiently low transaction costs, bargaining will lead to a Pareto efficient outcome—that is, no one will be made better off without making someone else worse off—regardless of the initial allocation of property.

the purely private benefits of water. (For simplicity, the term "social benefits of water" should be understood to mean non-private benefits.) While writing the notes for my introduction, I realized this could be done by expressing non-private, social benefits of water in terms of a "national policy demand curve," a deformation in the national demand curve for water relative to what would occur in the free market. It would be a deformation brought about by national policy. In effect, just as a private demand curve gives the marginal value of water for private use, the national policy demand curve would include the marginal social benefit of water—the benefit beyond private use. This fit very nicely into a discussion of the economics of natural resources, which, together with the Rule of Capture discussion, was to form a standard part of the introduction.

It became clear to me that the social benefits of water expressed by any coherent national policy could be expressed as a demand curve. It was only later, however, in discussion with Zvi Eckstein, that I understood exactly how it could be done. One does not have to measure the social benefit of water directly. Rather, figuring out what a national policy does to the demand curve for water implicitly measures those benefits. After the conference, I realized there would have to be some iteration between policies and results. Since policy makers don't think in these terms, they may choose policies that lead to demand curves they don't actually like. (Later in the project, it became clear that there were easier ways to handle such problems; these are discussed in subsequent chapters.)

The big advantage to the idea was that countries, once a national policy demand curve was specified, could not reasonably refuse to trade water at the corresponding prices. Hence, prices calculated incorporating national policy demand curves would in fact incorporate a "true" value of water including national policy value, the social benefits. Negotiation over property rights in dispute could then be accommodated and facilitated by valuing the property rights at the equilibrium prices. Since the cost of desalination puts a ceiling on equilibrium prices, the value of the property rights in dispute could not be astronomical. Further, negotiation over policy could take place using consumer and producer surplus generated with supply curves and national demand curves. (Later, the idea of "equilibrium prices" disappeared from the model.)

Again, there were three basic ideas for the model: 1) water is a scarce resource and therefore must have a price, and that price cannot be above replacement cost; 2) the property rights disputes are analytically separate from questions about national policy and the management of water; 3) the model should incorporate national policy demand curves.

The Second Water Conference

I was nervous as the second water conference began on October 21, 1993. Both Hausman and Eckstein had obligations for the first part of the first day, so it was up to me to introduce the conference pretty much on my own. Shuval and I had put the program together and decided it would be politic to put the Eckstein-Fishelson model on the agenda relatively late in the conference, in the middle of the second day. My introduction would be followed by reviews of the papers by Shuval, Salameh, Ghezawi, and finally Eckstein.

My presentation took the entire first morning. I took questions as I went along, and concentrated quite hard on trying to satisfy everyone's political sensibilities. I suppose years of exposition as a teacher and as an expert witness in litigation paid off. But I could not keep from thinking of an argument I had with my mother, some time before, about a health problem my sister may or may not have had. I got my mother to consider the probabilities of two scenarios: one with a real problem and one in which the appropriate tests had given incorrect results. We didn't know the true probabilities, but I thought we had agreed that the good outcome was about a million times more probable than the bad one. Later, when I told my mother I thought the emergency was over, she became quite indignant. "How can you apply these cold statistical and mathematical arguments to your own sister?"

I supposed my mother might have asked now, "How can you talk about the value of a cup of water in the desert?!" I feared I would be met with the question of how I could apply such economics and mathematical techniques to something as important as water.

Somewhat to my surprise and great pleasure, my lengthy introduction went off far better than anyone could have expected. Although the participants occasionally lost their focus as the conference progressed, I apparently had persuaded them of this way of thinking about water. Jad Isaac had plainly grasped what was happening most clearly—and he was

11

unquestionably the most important person to convince. Lenny and especially Hillel were very excited—even if I found myself limp and, by the end of the day, nearly trembling.

Returning home before the event dinner, I had a momentous realization. When Ellen asked me how things had gone, I said "You know, I may just have done something of historic importance.

At that evening's dinner, I was seated next to Jad. He turned to me and said, "I like you." It was pleasantly odd but touchingly significant. I was a little taken aback, but pleased by his comment.

"I like you, too," I responded.

Jad made it clear that he wasn't merely being polite. He complimented me on my sense of humor and indicated he was interested in working with me. In retrospect, it became clear he had decided I was someone he could trust and was committing himself to work on a project he regarded as quite important.

The remainder of the conference went off well, although there was some argument over the particular data in formulations in the Eckstein-Fishelson model. We had to keep reminding the participants that it was only a prototype and that they would be free to provide their own data. Further, we were not forcing an Israeli model on them, nor was our model going to take any position politically. For instance, we weren't defining districts within countries or deciding who got East Jerusalem.

Unfortunately, I had not given much planning or thought to the session up next, and the meeting proceeded much too hurriedly. It was decided that the Palestinians and Jordanians would each form a team, headed by Jad Isaac and Elias Salameh, respectively. Aided by the Harvard Institute for International Development (HIID) and, in particular, by Theo Panayotou, a Greek Cypriot by way of Canada at HIID who participated in the conference, each team was to produce supply and demand curve estimates for its country to fit into the template of the Eckstein-Fishelson model estimates for (roughly) 1990, 2010, and 2020. ISEPME would fund their efforts.

Because of the real importance of everyone understanding the model and knowing that nothing was being done in secret, we urged the Arab participants to select as a technical consultant someone they felt they could trust. They chose Atif Kubursi of McMaster University, a Lebanese economist who had attended some of the earlier general conferences. He joined the project in early February 1994.

I had some concern about the work plan, mostly concerning the Jordanian team. It didn't seem to me that Elias Salameh, who was not an economist, really understood the project in the same terms as, say, Jad Isaac. I knew he could do the supply part of the project well, but I had doubts about the demand end of things. Fortunately, my doubts ultimately were unfounded. Salameh brought on a very able economist, Maher Abu-Taleb, to assist him.

The participants deemed it desirable to have at least a preliminary set of results ready before the next meeting of the "Multilaterals."* Water was an important topic for these sessions, scheduled to take place probably in March. We arranged for the project teams to meet in Cyprus in late February or early March, and the conference adjourned.

The First Middle East Trip

There was a sense of increasing excitement over the last two months of 1993 as we began to realize what we might achieve. Scholarly work on the project proceeded mostly in the three countries. I wrote up my remarks from the conference in a paper and added in a section called "Questions and Answers."

When I spoke to the ISEPME board of directors at Harvard on November 9, the members were quite interested—particularly Zein Mayassi, a Palestinian engineer of considerable wealth and high PLO connections who lived in London. We agreed I would go and speak with him at some future date.

Meanwhile, largely due to Hillel, the news began to spread through the "water world." Shuval and I spoke with Jerry Delli Prescoli of the U.S. Corps of Engineers, one of the U.S. delegates to the multilateral negotiations on water, who gave a seminar at the Institute. We discussed the possibility the Americans might sponsor us for an intersessionary event at the multilaterals, which had by then been rescheduled for April. Specifically, Hillel and I

---

* Born at the Madrid Conference in October 1991, the Middle East Multilaterals were conceived as a parallel track to bilateral negotiations between Israel and Arab countries that would address economic, political, and technical relationships between the regional participants. It involved not only Middle East nations but also the United States, Russia, Canada, several Asian countries, the European Union, the World Bank, and other institutions.

proposed to put on a show that involved using interactive software that would allow for hands-on "gamification" of negotiations. Simply getting the parties to think about negotiations in this way would be a great gain.

We had agreed at the conference that at least Theo and I would visit the Middle East and talk to the project teams, and as 1993 came to a close, it became clear to me that our visit would have to accomplish two other principal—and possibly more important—objectives. First, we needed to learn about likely national policies of the different countries so that variations among them could be included in the model as national policy demand curves. Second, it would be important to sell our approach to higher-ups. We began to discuss the question of going to Tunis to see Arafat and others.

The trip to Tunis did not then take place, however. Instead, we went to Jordan, Palestine, and Israel. Lenny Hausman went to Damascus, Syria, where there was also some interest.

2

The First Trip to the Middle East

On the advice of Bishara Bahbah, the Palestinian assistant at ISEPME, we decided that the time was not right to go to Tunis. That decision was made, in part, because of the somewhat unhappy state of the peace negotiations. So, instead, my wife Ellen and I left Boston on January 9, stopping in London on our way to Jordan.

We spent several days in London. I met with Palestine engineer Zein Mayassi, first in his home, where his family entertained us at a perfectly glorious dinner. He asked me why Bishara had advised against going to Tunis, insisting that it probably was the right time. "The Chairman," he said, referring to Yasser Arafat, "is in a good mood this week."

We then met at his office, where we discussed the water project. By then, he had given more thought to Tunis. Water should follow politics, he told me, and suggested he might go with us to Tunis. He was enthusiastic about the water project, although I was not sure he understood it would go beyond having a regional development corporation with a market.

From London, we flew to Amman. Theo Panayotou joined us on the plane, and Lenny Hausman—who had extensive contacts in Amman—joined us later.

A Visit to Jordan

I had been to Israel several times, but Jordan was my first foray to any other Middle Eastern country. It was a somewhat complicated experience. The internal movements and tensions in Jordanian society—such as between East Bankers and Palestinians—were plain to see, and I wasn't sure I fully understood everything that was happening. But I resolved do my best.

We moved among people—politicians and businessmen—of incredible wealth in a country whose per-capita gross domestic product was said to be about $1,100. Najeeb Fakhoury, the Institute's representative in Jordan, operated a travel agency, but clearly did much more. He was also a man of considerable wealth and, apparently, some influence. His family was particularly nice to us, guiding us around and keeping us entertained.

Before any official project business, we visited Petra, a historic city in southern Jordan that is an archeological wonder famous for its rock-cut architecture and water conduit system. It was as gorgeous and as impressive as I had always believed. We went to the top of a mountain where there were shrines to the gods of water and security. It was as far as we could get from the nearest shelter, and then the country's entire annual rainfall descended on us.

The god of water did his job. I pointed out to Theo that all that was missing was a shrine to the god of economic development.

Returning to Amman from Petra, we held an all-day meeting with the project team on January 16 at the Jordanian Royal Scientific Society. Some twenty-five people, all involved with water in one way or another, attended. It was largely an opportunity for Theo and me to sell the project concept, to which the morning was devoted, and for the Jordanian team to report on what they were doing.

There had been some political problems leading up to the meeting. Elias Salameh, who had attended the first conference, had recruited an economist, Maher Abu-Taleb, to assist; he headed a consulting group. Elias, Maher, Maher's group, and representatives from HIID had done a substantial amount of work in advance of our arrival, with the product promising to be much better than I had expected at the time of the October conference. Someone the Crown Prince's office had told us to ask for advice had suggested Elias. However, Elias had apparently begun to fall out of favor as far back as 1992. The November prior to our arrival, Lenny had received a message saying we should proceed no further on the project without permission from a certain Munther Haddadin. So, Lenny had traveled to Jordan and met with Munther and Elias. Munther, reported Lenny, was at first quite hostile and quite difficult, but Elias had defended the project very well, showing a higher degree of understanding than I had given him credit for. In the end, they apparently talked Munther around and we were permitted to proceed. In particular, we were permitted to approve Elias's budget and resume correspondence. Things were set by the time we arrived in Amman.

Munther would eventually become a leader on the project and also a dear friend. He held a number of water-related positions, including head of the Jordan Valley Water Authority and later chief water negotiator, and eventually water minister. That day, he was slated to chair the afternoon session and, in effect, was part chairman at the morning session. He entered the room dressed in a business suit just like everyone else but also wearing the Jordanian keffiyeh—red rather than the Palestinian black one—wrapped dramatically around his very large head. It was an impressive sight, and quite a performance. I was later told of an even more impressive performance, when he met with members of the GDZ consulting team from Germany dressed in full Arab clothing and not just the keffiyeh.

Munther sat down and said, loudly, "Does no one else wear the headdress?" No one else did.

Munther was exquisitely polite during my presentation, but also could be charitably described as totally skeptical. He interjected several times: "We are a desert people, and you with your Western ways cannot possibly understand the value of water to us." Also, "If there were three of us lost in the desert east of here, what then would be the value of a cup of water?" This did not make me cheerful, particularly—as I later found out—because Munther was not only the Jordanians' chief water person, but was even known as "Mr. Jordan Valley Water."

Unfortunately, I didn't realize until some years later how I should have answered him. The value of water in the place in the desert would be very high, I could have noted, but the value of water in the rivers or aquifers would not be affected. What was scarce in his example, I would have said, was not water, but rather a conveyance system to bring water to the stranded travelers he described. That phenomenon showed up very clearly in runs of our original water allocation system (WAS) model and remains important.

As the day wore on, Munther did become increasingly receptive. When the conference ended, he invited Lenny, Theo, and me to his home—a largely

Westernized home where he lives with his wife, who is from South Dakota—for drinks. There, he took off the headdress. He is almost entirely bald, and much less impressive looking without the costume. It turned out his daughter was applying to Harvard, and he asked Lenny to write a letter on her behalf. Munther also said he would support us at the multilaterals.

That night, the Fakhourys gave a big party for us. There was a similar (even larger) party the following night at the home of Mr. Saifi, an extremely wealthy businessman. Many of the big-time players (businessmen and politicians, including Munther) came to both parties. It was clear that the businessmen, particularly the Palestinians among them, understood very well what the project was about (when it was explained to them) and were heavily in favor both of it and of peaceful economic relations with Israel. I was not quite so sure about the politicians.

We spent Monday, January 17, calling on various politicians in their offices, including the ex-Speaker of the House and a Deputy Prime Minister. At the office of the Water Commissioner, we met both Munther and Ali Ghezawi, who had attended the first Water Conference but had dropped out of the project after entering the government. A meeting with the Crown Prince did not come off because of the arrival of Ron Brown, the U.S. Secretary of Commerce, on the same day, but a royal meeting was planned for a future date—although at the time I didn't think that meant it would occur well over a decade later.

Everyone we saw was at least interested—with one exception. The Deputy Prime Minister was perfectly polite, but pointed out he had come from the area of education and knew nothing whatsoever about water. After listening to us, he spoke about the difficulties of making peace with Israel—a process of which he was not entirely in favor. He explained the challenge of changing feelings created over fifty to one hundred years. Such changes, he said, take time. This was in contrast to everyone else we had spoken to, including people at Petra, all of whom were chomping at the bit for peaceful economic relations.

I had the feeling I was in the presence of an old-time segregationist in the American South explaining how one cannot litigate changes in attitudes. Moreover, the Deputy Prime Minister's view of history was not exactly the same as mine. Under the circumstances, I did not feel that I ought to argue with him, but, when he told us that it would have been better if the Jews had "accepted Kenya when it was offered," I thought I could at least correct him by pointing out that it was "Uganda, actually."

Finally, he said that what we were doing made no sense because "water is beyond price." I thought for a moment and pointed out the severe water shortage in Amman. Why, I asked, did Jordan not desalinate water at Aqaba (far to the south) and pump it (a very long uphill way) to Amman.

"Oh no!" he replied. "That would be too expensive."

I could not make him understand he had just contradicted himself and stated that water is *not* beyond price.

Jordan has an immense water problem. We did, during the trip, learn much about the specifics of the water situation and about Jordanian attitudes towards water policy.

As Ellen pointed out, Jordan is not a geographically viable entity. Winston Churchill created it out of the British mandate in the 1920s. And the problems involved show up in water. Most of the country, especially the eastern part, is desert. There is some agriculture in the highlands near Amman, and a lot more in the Jordan Valley—but Jordan River water is brackish and not used. The only place with a seacoast is Aqaba, far to the south on the Red Sea. Hence, desalination is not really practical unless Jordan is part of a larger regional cooperation. Further, there were disputes with Saudi Arabia and Syria concerning fossil aquifers, structures that contain water but into which water no longer flows. Finally, well over 50 percent of the water brought into Amman was unaccounted for, meaning it was either lost through leakage or stolen through unmetered use (a situation that persists).

Jordanian thinking about water seemed to be almost exclusively about supply. Perhaps that is natural in so water-poor a country. The policy was to get all their rights to the water and use it—not a rational way of thinking about water policy. Nevertheless, it was very difficult to get the Jordanians to think in any other way. If we spoke of the *value* of water, the response tended to be to describe it as *beyond* value—as in the example from the Deputy Prime Minister. Nor did those we spoke with connect it with their willingness to allow that kind of leakage without making the capital expenditures necessary to correct the situation. There was a lot of education needed here.

One way to describe something the project had to offer the Jordanians concerned their claim that the Israelis pumping so much water out of the Sea of Galilee into the national water carrier is what made the Jordan River brackish—a claim that may well be perfectly correct. In effect, the Jordanians were saying that the Israelis were doing them an incalculable harm. My position was that while the project would take no position on whether the Israelis were *entitled* to do that pumping, we could certainly show how to *calculate* any harm.

There was (as least) one exception to the Jordanian concentration on supply, and it was a major one: the policy toward water in the Jordan Valley, where it was supplied to agriculture at a flat price of six fils per cubic meter (less than a U.S. penny) and was also rationed. It was a very big subsidy indeed.

Discussion in Jordan turned up something interesting about the subsidy. A principal reason for it was social stability: farm workers would become unemployed were water not heavily subsidized in the Jordan Valley. It turned out, however, that more than half the agricultural laborers in the Jordan Valley at that time were neither Jordanian nor even Palestinian; rather, they were Egyptians, Pakistanis, and other foreign workers. The mystery of that water policy was solved when we realized that a large section of the power elite (including Munther Haddadin) were large landholders in the Jordan Valley.

On to Palestine

Very early on the morning of Tuesday, January 18, after spending the previous day calling on Jordanian officials, we left Amman and were driven to the Allenby Bridge on the Jordan River. We crossed the bridge with very little fuss, but were held up getting a taxi on the other side, as arrangements— which Lenny claimed had never previously been necessary—had not been made in advance. As a result, we found ourselves going through Israeli immigration behind an entire busload of Indonesian pilgrims all dressed in similar colorful jackets.

I pointed out that if we were with an Israeli, he or she would certainly have known one of the soldiers from an old kibbutz. I offered to go inside and shout, "Does anyone here know Sheshinski?" I was reasonably certain of a favorable reply.

Lenny went inside instead and found someone he did know. After some discussion, we were on our way.

The distance between Amman and Jerusalem is so short that, having left Amman a little before 7:00 a.m. and even with about a 45-minute delay at the bridge, we still managed to drop Ellen at the government guest house and find Jad Isaac's Institute in Bethlehem by 11:00. We were right on time.

We spent the next two days largely with Jad, his team, and other Palestinians he brought in to meet us. That first night, we had dinner at the Sheshinskis, mostly with old friends of mine; it was there I first met Professor Uri Shamir, who had come from Haifa for the occasion. Uri was (and is) a water expert from the Technion, Israel's technical university, and an important Israeli water negotiator, smart and very knowledgeable. My colleague Annette Huber-Lee referred to him once as "one of the gods of water."

Uri had read about the Eckstein-Fishelson model in the newspaper and thought it quite naïve. I gave him material to read. He took it with him. I talked to him twice after returning to the United States, and he sent me extensive comments. I didn't think the comments were correct, and wrote him back. Uri showed substantial enough interest in the project to turn over the materials to a group at the World Bank and to recommend our possible involvement in reviewing one of the World Bank's projects. Later, he became a major leader of the water project.

The meetings with the Palestinians were far more informal than those with the Jordanians. The Palestinian team had essentially finished its work on a first draft. They were all extremely enthusiastic about the project—Jad in particular, who was leading the charge. He correctly saw our model as an opportunity to make rational internal water policy. I hoped we could deliver.

It seemed Jad's importance was rising in a prospective Palestinian state. He was named head of the Environmental Council while we were there, but he was not the chief water negotiator. Who played that role was unclear to me. Sharif El-Mussa was another water negotiator, but left the project later—apparently over differences with Jad.

Relations with Jad were very good. We liked each other from when we first met. Our time together was relaxed, perhaps because he felt comfortable with us on his own turf. When Theo expressed a desire to buy something for his family, Jad drove us to a jewelry-souvenir store run by a cousin, who was quite fond of him. The cousin offered us a 50 percent discount. Theo went on bargaining, which was embarrassing.

21

Jad also met us twice for drinks in Jerusalem—but it had to be East Jerusalem. On one occasion, he said he was technically violating curfew by being there after dark, but that it was unlikely there'd be trouble given his status. I said we would protest should trouble ensue, which led to a short discussion about my motivation for involvement in the water project in which Jad learned that I was on the board of the New Israel Fund.* It surprised him.

"They were very helpful to us in Beit Sahour," he said, referencing the Palestinian town east of Bethlehem that had been the scene of a tax uprising against the Israeli occupation. "But I never connected you with the New Israel Fund."

"Jad," I responded, "do you think I'm doing this because I care desperately about *water*?"

I also told Jad about my visit with Mariam Maari and the time I invited her as a guest to a dinner of the Society of Fellows at Harvard University. She and I were speaking with someone who picked up that she was Israeli, but not that she was an Israeli Muslim. The conversation turned to religious beliefs, and Mariam or I would alternate between saying "I believe" this or that and "so do I."

"I can't distinguish between the two of you," the gentleman said to Mariam and me. He assumed we were both Jews.

Jad introduced us to David Grey, an English ground water expert with the World Bank, seconded to the United Nations Development Program (UNDP), who was in Israel coordinating research on water. We spoke with him briefly one afternoon and gave him materials to read. We then joined him for drinks at the American Colony hotel that evening—by which time he had read the papers and become quite enthusiastic. By then it was too late to go to Beit Sahour.

Our work, he told me, is "totally without precedent. Do you realize that?"

---

* The New Israel Fund (NIF), founded in 1979, has grown to be a very important guardian of democratic values in Israel. Its Board is made up of Americans, Israelis (including Palestinian Israelis), and some Europeans. At the time, I was treasurer; later, I became president, then treasurer again, and then the "Institutional Memory." I joined the Board in 1983 when it was formally constituted, making it its longest-serving member. I have remained so long that I have adapted an old Israeli political joke in which the Pharaoh of the Exodus is brought back to life after about 3,500 years and eventually asks whether I am still on the Board of the New Israel Fund.

"One has managed to generate a certain amount of internal enthusiasm," was my reply.

There was a momentary silence.

"Internal enthusiasm?" Grey had directed this to Ellen. "How do you stand staying married to this guy?"

I thought out-understating an Englishman was a significant accomplishment.

At Jad's request, we invited Grey to the Cyprus meeting scheduled for the beginning of March.

Jad and I also discussed the question of going to Tunis. When I mentioned that we had consulted Bishara Bahbah, Jad immediately interjected, "And he told you not to go yet."

Then when I said we had also consulted Zein Mayassi, Jad said, "And he told you to go right away."

"If there is one thing I understand even less than water," I began to say to Jad, but he finished my sentence for me.

"It is our internal politics," he noted, correctly.

We agreed that we would certainly consult him before proceeding further in this regard.

I may not have understood the internal politics, but at least at the level of the people we were seeing, discussion of water policy was almost entirely rational. The Palestinians (including Jad) saw themselves as starting with a new government and therefore with an opportunity to avoid getting caught in the long-term agricultural subsidy trap—even if they might have to subsidize agriculture briefly to accommodate the return of refugees, a possibility for which they were planning but hoped against.

I was particularly impressed with the people from Gaza. They spoke very sensibly, not only about the enormous need for fresh water in Gaza but also about the ways in which they and the Israelis might interact. One proposal was that Israel should sell fresh water to Gaza, while Gaza would treat the wastewater and then resell it to Israel for use for agriculture in the Negev desert, where there is no underground aquifer to pollute.

The proposal made a lot of sense to me, and David Grey, in particular, thought it was an important partial solution to a general problem: since none of the nations in the area trust each other, none may wish to be dependent on water received from their neighbors. In this circumstance, since autarky (being self-sufficient) is terribly costly, the only solution may be to make them totally interdependent with water sales going back and forth (what Grey called "the circular tap"). One possibility suggested was for the Israelis to sell water to the Palestinians at Gaza and buy water from them in the northern part of the West Bank.

Later in the project, runs of the WAS model yielded results that implied the general idea was sound; we thought Israel could make a confidence-building gesture by offering to build the required treatment plant for Gaza.

In terms of the model, David Grey's principal concern—one Jad echoed—was that we hadn't correctly modeled that pumping of water in one location raises the cost of pumping it in other locations on the same aquifer. Instead, we had simply restricted the amount of pumping in each location. I committed to try to do something about this improper treatment of hydrology, the properties and effects of water, after the Cyprus meeting. We needed to put a hydrologist and an economist together to model the surface involved. It seemed obviously necessary—if only to convince hydrologists we had a sensible model. I did not believe it would seriously influence the output.

It also became clear that we needed the model to handle wastewater and brackish water adequately, and fairly soon. At the time, the model assumed an existing wastewater treatment plant would be another relatively low-cost source of water. It looked as though agricultural demand in the districts with existing treatment plants was more than enough to use up the wastewater, so we were not in danger of implying that people would drink it. Still, our handling of the problem could be regarded as inelegant, at best.

Addressing that would also be something for the next phase.

Into Israel

Our first meeting in Israel was on January 20, at the Tel Aviv office of Israeli Water Commissioner Gideon Tzur. Gideon Fishelson and Anni Karasik joined us, along with an "interpreter" who—as they told me afterwards—did not properly interpret. He appeared to have been there as something of a watchdog. Still, we came away from the meeting with what was essentially a promise of full support. Tzur offered a working group to provide us with any information we needed; we referred him to Fishelson (who he already knew).

Israeli water policy appeared to be based first on historic allocations of water to agriculture. Each farm or cooperative was entitled to a certain amount, and each allotment was divided into three parts, all available at a different price. In the year before our visit, the high-priced parts had not been purchased. When I inquired about how the authorities set the sizes of the parts and the prices at which they were to be sold, I was told they experimented and were trying to eliminate subsidies—except that the sunk capital costs of the national water carrier were not included in the charges.

There seemed to have been some reluctance to see us on the part of the Minister of Agriculture. Our only other governmental meeting in Israel was at the Foreign Ministry, where we met first with the Deputy Foreign Minister and a younger man who was apparently involved with water negotiations. They both then accompanied us to a meeting with Foreign Minister Shimon Peres. As it happened, Eliakim Rubenstein—the man in charge of the official negotiations before Oslo—had been meeting with Peres and was asked to stay.

The meeting with Peres went off about as well as one could expect. Everyone had done their homework and appeared to understand what we were proposing to do. They showed considerable interest, although Peres was more interested in what the Jordanians had to say than in nearly anything else. Peres, though, was skeptical about the proposal of the Gazans concerning wastewater. When we put it to him as one way of possibly overcoming mutual suspicions, he waved his hand and said, "Too complicated. We are advising [the Palestinians] that we will help them build a desalination plant in Gaza, which they should fuel with natural gas brought from an Egyptian field in the Sinai. Then they will be independent of us. They will never consent to having their system intertwined with ours."

Later, when I told Jad about this, he simply said, "He is wrong."

At the conclusion of our meeting, Peres said very deliberately, "Israel will remain officially skeptical of your project, because if we were not to remain officially skeptical, your project would die. However, if your other clients get on board with you, you will not find us behind."

That convoluted intention, a harbinger of the difficulties in bringing about cooperation, seemed to me about as much as anyone could expect.

The meeting with Peres was the end of the water project-related part of the trip. Ellen and I attended a board meeting of the New Israel Fund and returned home on January 26.

3
Preparing for the Cyprus Meeting

During the London trip, it occurred to me that it wasn't going to be as simple to use the model for negotiations as I had once thought. Looking back, I realize I had inserted a section in the basic paper about this but failed to emphasize its importance.

The problem was that when one country adopts a policy, it confers benefits and imposes costs on itself and other countries. The model measures these as changes in producer and consumer surplus, and they can be quite large (actually, there is a change in the area under the national policy demand curve of other countries as well—this is the social benefit surplus). It follows that negotiations must take place not only over who owns the water but also over what policies will be permitted in a jointly managed system; this might provide an incentive for countries not to participate in a jointly managed system as a way to avoid restrictions on their water policies. The property rights question and the question of where the water gets pumped and used might not, in fact, be analytically separable—and later in the project, we were able to show that cooperation would always benefit all the parties.

Much of February 1994 was spent on matters related to this in one way or another. From a substantive point of view, a lot of my time on the project was spent in repeated runs through the model and its computer output, helping spot bugs and locate problems. Some problems were less obvious than others. For example, the Palestinians had produced data and estimates that assumed East Jerusalem as their territory. Not surprisingly, the Israelis included East Jerusalem as part of Israeli Jerusalem. Since our project was not going to resolve the Jerusalem issue, it would be necessary to have two versions of the model.

While we were in Israel, Theo Panayotou had asked why a similar approach would not deal with land problems. I pointed out that thanks to desalination, we had an unlimited supply of water at a finite price. Not so with land. But were it technically possible to reproduce Jerusalem in infinite amounts for a relatively small sum of money, a similar approach might solve the Jerusalem problem.

Even two models, though, did not completely simplify the matter of Jerusalem and water. Palestinian population projections for East Jerusalem presumably depended on the assumption that it will be part of Palestine. We would need to get the Palestinians to rearrange their population projections in a scenario in which they would not have Jerusalem—something they would be very reluctant to do. And there were other problems as well: the treatment of the Israeli settlements and the fact that the Palestinian and Israeli parts of the model both assumed that they would get to pump from the Mountain Aquifer as much as they think they should. While truly efficient management would have pumping places independent of who owns the water, it was plain we would have to differentiate scenarios to satisfying quantity constraints, with the Israelis pumping less in the Palestinian scenario than in the Israeli one.

Later, when we wrote our book on the project's economic model,[1] we dealt with these problems by letting the Palestinians and Israelis both use their specific claims to Jerusalem, once the necessary changes had been made. We found that with respect to cooperation it made little difference which choice we took. Consideration of problems such as these made it clear that a principal purpose of our next meeting would have to be agreement on a list of different policies and scenarios, including pumping scenarios and different conveyance arrangements. In this connection, the Jordanians appeared to have given us a conveyance-cost matrix for a conveyance system that did not exist. This also needed to be discussed.

There were some important additions to the project during this period. Yuval Nachtom returned home to Israel (where he continued to work on the project) and was replaced by Aviv Nevo, an extremely bright and helpful graduate student at Harvard who had been a student of Zvi Eckstein's in Tel

Aviv.

February brought difficulties with the project's relations with Jordan, and particularly with Munther Haddadin. He thought it inappropriate for him to come to Cyprus and criticize the work of his own team. My view was that he was feeling insufficiently consulted by the Jordanian team, so I offered to go to Amman after the Cyprus meeting, scheduled for the beginning of March, to brief him. He was quite receptive to this, but no such trip took place. Still, we talked to Maher Abu-Taleb and emphasized the need for frequent communication with Munther.

During this same period, Uri Shamir came twice to the United States and spoke with me. He had passed the papers on to people at the World Bank who were showing considerable interest in the project.

Meanwhile, we still needed Palestinian-Jordanian support in the multilaterals. We could request an intersessionary program from the Americans, or simply seek general support from interested donor nations. I tried to get both Jad and Munther to commit to this approach; both were well disposed in principle but reluctant to do it before they had seen some model output. I had hoped that such model output would be available by the time of the Cyprus meeting—but it was not to be. The project was not far enough along, and then tragic events intervened.

Hebron and Its Aftermath

It was Purim and Ramadan when, on February 25, 1994, an Israeli settler dressed in his army reserve uniform entered the crowded Tomb of the Patriarchs in Hebron where Muslims were at prayer. To the horror of the world and the shame of most Jewish people around the world, Dr. Baruch Goldstein—a native of Brooklyn, New York—opened fire, killing twenty-nine people and wounding a well over a hundred others. Obviously, the consequences of this act were larger than its effect on the water project. Here, however, I can discuss only the latter.

Upon hearing the news, I first telephoned Bisharah Bahbah and then Jad, who was already in Cyprus. I told each of them I knew they did not hold me personally responsible for the events at Hebron, but also explained that on *Yom Kippur* (the Day of Atonement) Jews confess for the sins of the entire community. We bear a collective responsibility, and I wanted them to know I felt deeply ashamed. It was a deeply emotional conversation. Both Bisharah and Jad were appreciative; Jad even faxed me a note thanking me in terms I much appreciated.

The Cyprus conference, however, was in turmoil. Within a couple of days after Hebron, both the Palestinians and the Jordanians had pulled out of the peace talks. The following Monday morning, the news came that the Jordanians would not attend the Cyprus conference. It seemed important to me to go ahead nonetheless, even though I could not believe the conference would actually occur as scheduled.

There was another shoe to drop, which took a little longer than I expected—perhaps because Jad was a bit hard to reach. On Monday evening, February 28, just as Ellen and I were packing to go to the airport, the news came that the Palestinians were also pulling out.

"You can only go ahead with this conference if you are determined to burn your friends," Jad told Bisharah. Obviously, there could be no question of doing so. Only one person actually went to Cyprus: Atif Kubursi had purchased a non-refundable ticket before the conference was called off. He said he had a wonderful time.

What to do next? Faxes flew back and forth about a possible new date for the conference. Jad argued strongly against the first proposal to meet before the next meeting of the multilaterals in Amman at the end of April; I felt I had to take his advice. Sometime near the end of May, or in early June, was proposed next. Then there were ideas for partial meetings. When it appeared that the Jordanian team and Jad would all go to a water conference in Philadelphia in early April, there was a proposal to meet in Cambridge right after—with at least one Jordanian representative attending. But Jad had been instructed not to go to conferences with Israelis until further notice. So, while he did agree to meet with me in Philadelphia on the afternoon of April 10, such partial meetings could certainly not substitute for a full-dress project conference, which still had to take place. When that would happen obviously depended on the status of the peace negotiations.

There were other political difficulties as well—to put it mildly—caused, in a time of extreme sensitivity, by communications going out too quickly and without proper authorization. The first of these incidents occurred in early February. Gideon Fishelson received the Palestinian report and then sent some questions about it to Nevo and others. The questions were not posed in diplomatic form, and Hillel—believing Jad had seen them—sent a very long set of replies to Jad as well as to Gideon. He cleared this with no one.

Jad was quite upset, considering it ill treatment of him that Fishelson received his report when he had not received the Israeli report (although there was no Israeli report at the time).

Clearly, we lacked sufficient management control over our participants. Fishelson's comments were not intended for Jad, and Hillel's reply should not have been sent to Jad. There was no bad intent, though; it was easy to forget this was a highly political project with highly political participants (the Israelis, in particular, sometimes behaved as technocrats). We could have cleared all this up quite easily in a face-to-face meeting.

Then a much worse incident occurred. Because Jad pushed so hard to receive the Israeli report, Fishelson sent him a draft without first clearing it with anyone. Jad got it just after the massacre in Hebron, and he was outraged. Not only did the report use Israeli maps and tables that referred to the West Bank as "Judea and Samaria," but rather than clearly identify the status quo it projected the current situation as what *would* happen, not just Israeli claims. (Of course, a similar thing had been true of the Palestinian report.)

Perhaps naïvely, I replied to the outrage in Jad's first fax, which objected to an Israeli and a Palestinian version, by telling him I hadn't seen the report yet and by being quite reasonable on the subject of East Jerusalem. I also agreed that the term "Judea and Samaria" had to go, stating it was doubtless unintentional.

Jad's next fax was addressed "Dear Colleagues," not "Dear Frank." He threatened to resign from the project unless something was done. I spoke with Hillel and then phoned Jad. Reaching him in Bethlehem was not easy. At his home, I was given another number. At that number, they denied he was there. But I explained who I was, and they got him. I took that as a good sign.

While we had a productive conversation, it would have been easier had I actually seen the report. But we agreed something would be done, and I implored him to remain on the project. Soon thereafter, Hillel produced a large set of suggested changes to deal with the problems of the Israeli report. Fishelson agreed to them and produced a more sensitive version.

Again, I felt there was no substantive issue with the report, but I came to realize that the initial document was totally insensitive to the feelings of the

Palestinians and the result was nearly to blow the entire project to smithereens at a time when extreme political delicacy was needed. I hoped my meeting with Jad the next week would further calm things down.

Incidentally, it was during my phone call with Jad that I expressed my willingness to come see him on April 10 while he was in the United States—an offer I later repeated.

First Substantive Results and Some Substantive Issues

In March, while on vacation, and continuing into early April, I wrote a long project status report for internal use. I tried to remind everyone what we were doing and, in particular, that claims regarding where water should be pumped are not the same as property rights.

I also tried to set in perspective our results so far—and there were five particularly interesting results to report. The results for Jordan were ridiculous because of convoluted conveyances. The value of the water in the Mountain Aquifer was not, in fact, great. The location for optimal pumping of the Mountain Aquifer shifted to Palestine as time went on. By 2020, a desalination plant in Gaza would be essential. And finally, heavy Israeli subsidies to agriculture didn't seem to have much effect on the other countries.

On April 10, I met with Jad and Maher Abu-Taleb in the lobby of a hotel in Philadelphia. Sari Nusseibeh joined us. He was an intellectual leader who later became president of Al-Kuds University; I knew of him from having joined with others in 1991 to demand his release from an Israeli prison, where he had been jailed for his work with Peace Now and then adopted as a "prisoner of conscience" by Amnesty International. I recall mentioning to him that I had begun reading an English translation of the *Koran* (which I am ashamed to say I never completed), and when Sari asked me why, I told him that since I was working with Muslims I naïvely thought I would be able to understand them better.

It was a pretty good meeting that helped us straighten out a certain amount of misunderstanding in the project. Jad had thought the Israeli version deprived the Palestinians of a conveyance system—which was not true. We agreed on a number of ways to present things. Jad and Maher agreed to give us the information on wastewater treatment and the destination of wastewater I had asked for to handle recycling.

Our discussion made me realize that there were a number of ways in which the "single pipe" algorithm of the Eckstein-Fishelson model was causing us more and more difficulty. The original Eckstein-Fishelson model assumed all water flows ran north to south and that the solution, therefore, would be a price in the north plus prices in the south differing from that in the north by the known conveyance costs. This phenomenon made it easy to guess solutions: simply pick prices in the north and move them up and down until demand equals supply for the system as a whole. But as time went on, this approach proved more and more difficult, for several reasons.

First, the algorithm presumed water flows only in one direction and so, in effect, there is a "single pipe" that can branch as the water flows. But this isn't always true. In Jordan in particular, there were a number of possible conveyance connections through which different districts could ship water into other districts, and it was not clear which way the water has to flow. The result of this was a mess when we came to analyze the Jordanian part of the model. We tried to do so by assuming how much water would flow in each of the various directions, and it was very hard to get that to make any sense.

Second, while the results concerning where it is optimal to pump from the Mountain Aquifer were very interesting, they were essentially trial and error results without an adequate model of the hydrology of the aquifer—that is, a model of how pumping at one place affects costs at another place. It seemed such a model had to be built, and I assigned Fishelson and David Grey to work on it. We recognized, however, that with such a model the single pipe algorithm would not usefully optimize the pumping locations.

Third, brackish water and particularly recycled water needed to be treated in the model. It might have been possible to do this in the short run for recycled water by assuming one knew into which districts the recycled water would run and assuming that recycled water would flow to appropriate agriculture at cost. But this wasn't really an adequate solution; it would require a second network for recycled water, with other constraints on its use. The single pipe algorithm wasn't very suitable for this.

Finally, it was not clear the single pipe algorithm would really be suitable for the various constraints on the hydrology of the Jordan River and the Sea of Galilee. And further, it could produce very strange results. For 2020, if Israel had the water and cooperation broke down, the solution for Palestine would, of course, be a desalination plant in Gaza. Water would have to be piped back up to the West Bank, which would be very expensive. The single pipe algorithm correctly predicted the desalination plant in Gaza with enough water, but still ended up with water prices in the north cheaper than in the south. We could have fixed up the algorithm to get the right result, but we kept running into these anomalies in which the assumed form of the solution broke down.

All this made me very receptive to Bob Dorfman's urging of a point he had pushed long before with less success: that we should turn to an algorithm with explicit maximization, and maximize the sum of producer and consumer surplus (what is maximized in a competitive market subject to policy constraints). He urged the use of an algorithm called GAMS—the General Algebraic Modeling System that became known in the late 1980s—set up for explicit optimization. We agreed Bob would hire an assistant, N. Harshadeep (Harsh, as he is called, once explained to me that he has no first name in the sense Americans do), who was completing his Ph.D. on water problems and begin with a conveyance model for Jordan, expanding to take in demand curves and, finally, the entire model.

This looked extremely promising at the beginning of May. Conveyance model runs were sensible. However, there was a bug in using demand curves. We hoped, however, to have the entire model based on GAMS by the end of that month and to be getting production runs by early June. Harsh was a very important addition to the project.

The algorithm was not the only substantive development after the meeting in Philadelphia. Another had to do with the treatment of capital costs. While they have to be recovered, capital costs don't have to be recovered in the price of water (as opposed to fixed charges such as hook-up charges). The price of the water should reflect only the marginal cost—the cost that varies with use. Capital projects could be evaluated in the standard manner by seeing whether the discounted sum of buyer surplus additions, which come with the project, exceed the capital cost of the project.

This had two consequences. First, we could put in the operating costs of all potential conveyance links and ask which one would be worth building, which might be a useful thing to do given that the Jordanians were refusing to provide scenarios involving such capital projects. They preferred to do it themselves, saying they feared their scenarios might be "misrepresented" since they didn't fully understand the algorithm and the way the model works.

The communication problem mentioned earlier clearly created problems. We planned to give them a working computer program as soon as it was ready, and a fully documented version in Cyprus.

Second, this proposition applied to desalination plants as well as to other capital projects. Hence, the eighty cents per cubic meter we had been using for desalinating costs was too high, since it included capital costs. The clear implication was that with desalination plants on stream the water in dispute would be worth even less than our model had been predicting—something we learned to fix later.

During this period, two people began again to play a larger and important role in the project. Bob Dorfman's revitalization, a result of the transfer to GAMS, was a pleasant outcome. It was marvelous to see him contribute so much. Zvi Eckstein also returned to the project. There were some difficulties of reintegration because he was then in Israel, but since he was most interested in using the model for management purposes and since that was an aspect I had been largely neglecting, I asked him to write a paper on it for Cyprus, which had been rescheduled for early July. Lenny and I figured we would go to Damascus and Amman the preceding week. That was all good news. On the down side, Munther Haddadin's daughter did not get into Harvard.

Before leaving for Israel at the end of May, Hillel Shuval introduced me to Yehuda Bachmat, a mathematical hydrologist who had joined the project. He had been the head of the Israeli Hydrological Service and agreed to undertake the modeling of the hydrology of the Mountain Aquifer in conjunction with an as-yet-unnamed Palestinian. We expected this would enable us to have serious optimization runs on the pumping patterns required. I realized that there would be no need to explore different pumping patterns from the Sea of Galilee to see which would be best. We already had the facility with GAMS to provide possible conveyance links (for example, semi-unlimited capacity to those links) and ask what the optimal pattern of pumping for the three countries is.

The first three weeks of June were a period of particularly difficult, pressure-filled work. It seemed the transfer to GAMS was successful, and there was a considerable prospect we would have a lot of output in time for the Cyprus meeting—something we really needed. Both the Palestinians and Jordanians were pressing hard for copies of the computer program and for more results. It was difficult to explain to them that the lack of communication was not because the central team was going too quickly, but because it was going too slowly. Production runs for the model were set for two weeks before we would leave the United States, and I would have about ten days to write the entire thing up.

At the end of the first week, Atif Kubursi, the Lebanese economist working as a technical consultant for the Arabs, came down from Canada and spent two days just as we were supposed to be doing "production runs" of the model. But, as it turned out the program was not ready, and we kept discovering more and more difficulties, mostly associated with the introduction of recycled water. This continued right up until the time I had to leave for the Middle East.

Atif was very helpful. During a discussion with him, I realized one has to give consumers a discount for profits made on recycled water to induce the appropriate flows and that the introduction of profitable recycling lowers the price to consumers and raises the shadow value of fresh water, since water consumption does not then produce a valueless effluent. The latter point, which at first seems paradoxical, came to me overnight during Atif's brief visit.

Unfortunately, Bob Dorfman disappeared from the project for a while about this time, for health reasons. With the help of Aviv, Harsh, and me, he had written an excellent description of what we were maximizing and how the computer program worked.

Aviv, Harsh, and I now went through a period in which model runs showed results that couldn't be true, so we had to go back and figure out what we had done wrong. Aviv left for Israel around June 15, and Harsh and I worked night and day—me days and Harsh nights and some days—to get out a paper in time for the conference. I told him it was nice working with someone in a different time zone.

As the writing of the overview paper progressed each day, it became plain there were some results that couldn't be true and the program had to be revised. This continued right up to June 22, when we finished and printed the paper, and then hours before my departure when a really important addendum was written because I realized subsidies in one country (say, Jordan) could actually help other countries because of the effects on the balance of trade in water.*

The effects of speed showed. There were still minor mistakes. For example, I had grossly simplified Jordanian policy toward industry by assuming that industry received water for free, whereas the policy was actually that industries that drill their own wells are not charged scarcity rents but have to pay their own costs.

There were other areas as well. I discovered the Jordanian supply estimates included water from its claim on the Yarmouk River. We needed to redo everything parametric on the amount of that water. Further, the model called strongly for the installation of a pipeline from Jerusalem southward to Bethlehem and Hebron. When I pointed this out later at the conference, I was told that pipeline already existed.

Despite all this, the results were absolutely fascinating and the model and paper were powerful. Harsh's computer program was becoming quite easy to use and fascinating to the user—even more so when additional bells and whistles were added.

Back Corridors of Power

While all the model-related work was progressing, Ellen and I were invited to the house of Peter Gomes, Plummer Professor of Christian Morals at the Harvard Divinity School and Preacher to the University. He and his official residence were pretty impressive. The May 7 party honored Rabbi Julia Neuberger, who was to give a sermon the following day—Mother's Day—in Memorial Church. We had fond memories of meeting Julia and her husband previously, and were pleased to attend.

A ball of fire, Julia Neuberger is an English rabbi married to Anthony Neuberger, then an economist at the London School of Economics. She has been the Chancellor of the University of Ulster—chosen, she says, because the Protestants and Catholics could not agree on someone else.** Since that party, she has gone on to even more impressive things, including becoming

---

* This refers to harm as regards the price of water to or from other countries; it does not take account of competition in agriculture or other water-using products.

** Ellen says this is similar to the Nusseibah family in Jerusalem traditionally holding the keys to the Church of the Holy Sepulcher because the Christian sects cannot agree on who should hold them.

both a Dame and a Baroness and thus a member of the British House of Lords. She is also now the rabbi of a London congregation.

Julia and Anthony were at Harvard in 1991–1992; Julia had a Harkness Fellowship at Harvard Medical School and spent time at Beth Israel Hospital (BI) and the Harvard School of Public Health. Her specialty at the time concerned medical ethics and dealing with death. We met them at a dinner hosted by Dr. Benjamin Sachs, who was the head of obstetrics and gynecology at the BI (and also my third cousin), and his wife Vicki. We liked the Neubergers and saw them again once or twice. When they went away, we all said it would be nice if we saw each other again—and we still see Julia and Anthony fairly frequently in London or Cambridge, Massachusetts.

At the Harvard party, Julia inquired whether I was doing anything interesting and I mentioned the Water Economics Project. She thought she had heard it discussed while in Jordan the previous October. I pointed out that was unlikely, since the conference that really pushed the project forward had occurred at the end of that month.

Julia reminded me about the Jewish-Christian-Muslim cooperation project, to which she was the Jewish representative serving with Prince Philip and Crown Prince Hassan bin Talal of Jordan, the brother of then King Hussein. I remembered Julia joking two years before about there being no Jewish royalty; apparently, a Rothschild joined that project in the meantime. She asked whether I would like to become involved; I said yes, since bridging countries and cultures was the sort of thing I seemed to be doing. But nothing ever came of the initiative. More importantly, though, she asked whether I would like her to put in a word for me with the Crown Prince, who has become her close personal friend. I promised to call her after consulting Lenny Hausman.

Lenny said he and the Institute needed no particular entrée to the Crown Prince, but that it might be useful if Julia could tell the Crown Prince that I was someone he ought to take seriously. This coincided with some other project-related news. In conversation with members of the American delegation to the multilaterals, Hillel learned that the German delegation had proposed a study of supply and demand for water. We wrote to the delegates calling their attention to our project. More momentous, though, was that Hillel also learned that Munther had told members of the American delegation he was opposed to our project, which was not what he was telling us.

Because of this, Lenny and I thought it quite important that the Crown Prince see me personally and, if possible, privately. We did not believe he was getting objective views from his advisers. (At roughly the same time, Lenny sent a fax to Hanni Mulki, the Head of the Royal Scientific Society in Jordan, stating there had been some difficulty coordinating with Jordan, among other matters.)

I got back to Julia that same weekend and faxed her a good deal of information. Shortly thereafter, things began to happen.

Julia left a message on my answering machine from Germany. She said she had spoken about me at length with the Crown Prince, and asked for my schedule in Amman. There would only be one afternoon Lenny and I would be there together, but I told her I'd be as flexible as necessary.

Two faxes came one day at the end of May. The first was from Hanni Mulki, who asked me for some materials and also told us Munther would be glad to attend the Cyprus meeting, which we had not expected since he had refused our invitations. I had even offered to brief him in Amman before or after the Cyprus meeting, believing he would regard that as appropriate and be gratified by the recognition of his importance. Plainly, something had happened in Jordan.

The second fax was from Julia, who had spoken again with the Crown Prince when she visited for drinks *en famille*. She sent along a copy of a fax to her from his secretary in London that said the Crown Prince would be happy to see Lenny and me.

Lenny and I called Julia and spoke to her for some time about this. She made several observations. One was that the Crown Prince, who expected to become King sooner rather than later given his brother's illness, was very forward looking and would be easy to talk to if we spoke with him as an equal; I replied that I do not make a habit of talking down to royalty. Julia also related that he was skeptical of our project, and that Julia and Anthony had encouraged him to share his doubts directly; I observed I didn't think the Prince really knew much about the project. Then I discovered Julia had not received the basic papers I had initially faxed so she could speak more knowledgeably with the Prince or pass the papers on.

I could have been wrong, but I suspected Munther had briefed the Crown Prince. Maher Abu-Taleb, the only person in Jordan I was absolutely sure understood what we were doing, was certainly too junior to have done so. In any case, our visit, scheduled for June 27, was looking to be a very important one.

Meanwhile there were other political developments. The Syrians were quite interested in seeing Lenny and me at the end of June or beginning of July, which suggested they might be serious about peace negotiations with Israel or, at least, interested in having a model of their own. Also, from June 20–23 in Paris, the United Nations sponsored a "seminar on Palestinian trade and investment needs" held by the "Committee on the Exercise of the Inalienable Rights of the Palestinian People," which had been established in 1975 by UN Resolution 3376.

Various experts had been invited to Paris for a roundtable discussion, including Stanley Fischer, a friend, former student, and colleague who had just been named deputy managing director of the International Monetary Fund. He couldn't attend, and asked me if I would like to go instead. I called Lenny, who had also been invited and asked to help in recruiting Israelis. His view was that the conference was getting off the ground very late, as the Palestinians discovered they have to talk to other people besides just Palestinians, and that it might be very difficult to get people to attend on such short notice.

The formal invitation came a day or two later. I decided against attending; I could not in good conscience give a Water Project results paper before the Cyprus meeting, and I didn't even want to give the previous autumn's prospectus papers to politicians without the full consent of my colleagues. Furthermore, I would certainly lose a good deal of time that could be spent on the more important task of writing the Cyprus report.

My decision was strengthened by something that happened the next day. Lenny called to say there was now a summit on the Middle East economy—in connection with the World Economic Conference held each year for business and political leaders in Davos, Switzerland—scheduled for November 6–8 that President Bill Clinton would be attending, presumably along with political bigwigs from the region. Shimon Peres had apparently proposed during the previous Davos meeting that there should be a similar conference on the Middle East. Fischer had been asked to help prepare the program, and Lenny, whose advice had been sought, suggested putting me on the program. The time certainly seemed right to present the project results.

My parents and grandparents would have *kvelled* at the thought of me hobnobbing with such people.

Second Middle East Trip

Ellen and I left for the second project trip to the Middle East on June 23, stopping in Vienna to celebrate our 36th anniversary and then heading to Jordan, where we arrived the evening of June 25. Munther had requested I phone, and we made an appointment for early the next morning. I was quite annoyed to learn he had not yet received the paper I had worked so hard to get out and that I had specifically requested be sent to him first to prepare for our meeting. It turned out that courier delivery to the Middle East took a while; papers sent on June 22 from the United States began to arrive during the afternoon of June 26. Some Cyprus participants didn't even get their copies until after the conference began.

I met with Munther and Ali Ghezawi on the morning of June 26. Ali had attended the original, disappointing project conference in 1992. The next year, he was working in the Ministry of Irrigation and felt he could no longer participate directly. I had met him that January. Then he ended up coming to Cyprus (as did Munther) and, during the conference, was moved to the Foreign Ministry in a fairly high position in water negotiations.

We had a very good meeting, during which I briefed them on the state of the project and our results so far. They were very interested. Munther, in particular, picked up on the point that Jordan subsidies would help, not hurt, the other countries—although he insisted that these were not true "subsidies" at all, but were justified by other problems farmers have, a difference I attributed to be largely a matter of language. In fact, when I got to this point the next day with the Crown Prince's advisor, Munther told him, "You have to listen to this."

I went from my meeting with Munther and Ali to Maher Abu-Taleb's firm, where I met with Elias Salameh and Iyud Abu-Moghli, who was working on the project. Maher was in the United States attending a seminar at the Harvard Institute for International Development for several weeks (he did not come to the Cyprus conference.) Our discussion was similar to the one in the morning, and I spent part of the afternoon helping Abu-Moghli get his copy of the program up and running.

That evening, Ellen and I were invited for a family dinner at the home of

Najeb and Jacqueline Fakhoury, friends of the royal family. Jacqueline had represented Princess Basmat at a conference in Krakow on women's rights, where she met our friend Mariam Maari. She had gotten along well with Mariam and most of the Israeli delegates, especially our two old friends Galia Golan and Naomi Chazan. Indeed, the only one she didn't like was Yael Dayan, Moshe Dayan's daughter. She told us that later she had taken a lot of flak from the newspapers for being friendly with the Israelis, and had decided not to dignify the attacks with an answer.

The Fakhourys (and most other people we talked to in Jordan) were very excited about the likely prospects for a serious breakthrough in peace negotiations between Israel and Jordan. King Hussein was in the United States during our visit and had called for moving forward separately. Major talks were scheduled to begin in mid-July.

Lenny Hausman arrived in Amman early on June 27th and then a car came for us from the palace, where after a short wait we were taken in to meet with Ahmed Mungo, the Crown Prince's chief policy advisor, and a woman named Rhonda (whose last name I have forgotten). Munther joined us a bit later. We were told that with the King out of the country, the Crown Price was doing double duty and would not be able to meet with us.

It was our only disappointment on the trip. I was quite annoyed about this, as was Najeb Fakhoury, whose reputation for influence was important to him. The appointment had been made long in advance, and as late as the previous afternoon we had understood the meeting would take place. Moreover, the Crown Price met with the delegation of Project N'shama, and however laudable their visit, it was not nearly as important as ours. Apparently, Project N'shama's pull was higher than that of Julia Neuberger. It would have been nice if the Crown Price had at least said hello.

Apart from this disappointment, though, we could not have had a better reception in Jordan. Either Julia Neuberger or someone or something else had come through for our project in a serious way. The meeting with the Crown Prince's advisor went very well, and it became obvious that Munther was now a supporter (unofficially, as I'm sure he would want me to say). At the end of the meeting, he asked me why he should go to Cyprus to hear our performance a second time. I told him he wanted to see what the other people would say. He said he was particularly interested in Jad's proposal for an environmental profile and, at his request, I rescheduled that presentation so that he could have a more convenient travel schedule.

Incidentally, by this time, it was obvious that interest in the Cyprus meeting had exploded, with a very large number of people coming and semi-official interest from all three countries. In addition, before leaving the United States I had spoken with Linda Heller Kamm, a member of ISEPME's water committee and a co-chair of Americans for Peace Now. She wanted to know whether she should attend. I told her there was about a 70 percent chance it would be a historic occasion, and a 30 percent chance it would be a complete fiasco. She came.

What accounted for the sudden, rather favorable change of attitude in Jordan and, especially, on Munther's part? I could think of several possible reasons. One was that by June, there had been a breakthrough in the prospects for Israeli-Jordanian negotiations and agreement, so the atmosphere was just much more favorable. On top of that, Julia Neuberger's intervention had caused the Crown Prince to take a favorable interest in the project. The results of the project thus far had also made it clear that the model could be a very useful tool; more important, the results were favorable to Jordan, and in particular would support Jordan pursuing its own water policies—especially low prices in the Jordan Valley.

Munther had become an ally of sorts. He is a complicated man who, apparently, comes from a very important family. He was once shot in the hand while doing engineering work that involved water. I am told he is also an accomplished poet in Arabic. It was decided that he would attend the Cyprus conference, along with Ali Ghezawi.

Visit to Syria

Late on June 27, Ellen and I flew from Amman to Damascus. Lenny stayed behind in Amman and joined us very early on June 29.

We were slightly nervous about going to Syria, partly because of our unfamiliarity and partly because we expected it to be a more hostile climate than other places in the region. We were not used to traveling in totalitarian societies.

The Syrians, though, were very nice to us. Kamal Hamdad, who had taken care of Lenny on a previous visit, met us at the airport; he was an official in the Ministry of Economy, which was hosting us. At the airport, he took us to the VIP room where we waited for our luggage and he began to discuss the touring program that had been planned, stressing in particular taking us to Kuneitra, the capital of the Golan Heights.

Prior to our visit, Ellen and I had discussed the issues likely to arise. We should assume our hotel room would be bugged, I had said, and she should suppress her natural inclination to shout "F**k Hafez Al-Assad" into the light fixtures. Our long-time friend Robert Rifkind said it would have been more effective if Ellen had said, "Darling, isn't it wonderful what President Al-Assad did on his last secret visit to Tel Aviv?" We also realized that Hamdad, who was very polite and helpful, would probably report any overheard conversations, so we should refrain from political discussions. Our primary concern, though, was not to emphasize to the Syrians our frequent visits to Israel. We assumed they knew we were Jewish.

Without thinking, though, I mentioned we had been to Kuneitra before—a visit that could have occurred only by entering from the Israeli-occupied area before the city had been returned to Syria. Realizing what I had just done, I hit myself (mentally) on the forehead (and have claimed ever since that I had acquired a large flat spot in that area). Fortunately, no one commented on my gaffe. Indeed, it developed that Hamdad was originally from Banias in the Golan, having left there at the time of the Six-Day War in 1967 when he was ten years old. We chatted about our memories of Banias.

I suppose we also left no doubt about our religion since we asked especially to see the second-century Dura-Europos Synagogue in the National Museum of Damascus. We did get to see it—a fantastic sight with major frescos depicting biblical scenes, unusual in that the scenes are of human figures.

Hamdad escorted us everywhere for the next four days. But first we were driven to the Sheraton Hotel in Damascus.

The next two days were spent in tourism and, generally speaking, without propaganda. On Tuesday, we asked to go to Palmyra, a staggeringly impressive archaeological site. The following day, we visited the National Museum as well as a folk museum, and saw other sites around the city. English-speaking guides were provided in Palmyra and the museums, and so we got rather more out of our visits than we could possibly have done on our own.

On the second day, Hamdad observed that we were on a street with Jewish shops. There was no political discussion.

After our long day to Palmyra and back on June 28, we were cleaning up at the hotel around eight o'clock or so when the phone rang. It was the Minister of Economy, Mohammed Al-Imadi, who was downstairs in the VIP Room along with an old friend of his from Kuwait, who I think was an official of the World Bank representing the Gulf States. An invitation to drinks turned into a major dinner outside at the hotel's Ishbilia Oriental Restaurant. This was extremely pleasant and almost entirely social. Minister Al-Imadi was the only one who managed to refrain from the enormous meal. I did note that he appeared to be accompanied by a large coterie, some of whom I assumed were bodyguards.

We found the city to be much less westernized than Amman. There were fewer signs with the Roman alphabet. And while there were pictures in Amman of King Hussein all over, and occasionally of Crown Price Hassan, they were not nearly as omnipresent as the pictures of Hafez Al-Assad in Syria, especially in Damascus. Moreover, there were many pictures of Assad's eldest son who had been killed the previous winter in a car accident. There appeared to be a definite cult of personality in Syria.

When Lenny arrived, we began a series of meetings. The first meeting was with Al-Imadi and some of his assistants at the Ministry of Economy. The major conference paper had arrived a day or so before, but Al-Imadi had not read it, instead (wisely) sending it on to the Ministry of Irrigation. He listened with some interest to my presentation, but his principle reaction came later at lunch.

We then went to the Ministry of Irrigation, where we met with Minister Abdul Rachman Al-Modani and an assistant, Daoud Majed, who headed the Ministry's International Program and also translated (although the Minister speaks some English). Majed had read the paper.

At one point in the meeting, Modani observed that it was difficult for Muslims to think about water as a commodity. I suggested that in that case they should take advantage of the fact that the Israelis probably could do so. The Minister said the reason was that water is mentioned more than ninety times in the Koran but not at all in the Torah. I instantly pointed out that was absolutely wrong. But this was all pretty good-humored.

This meeting went on at a high level of interest from all participants. More than that, Majed followed us out afterwards to tell us how revolutionary he thought the paper was. Apparently, he had read it quite intensely.

We next went to the Foreign Ministry, where we met with the Secretary of State for Foreign Affairs (a chief civil servant). He was suffering from a terrible cold but had come in just to see us. This meeting was quite different from the previous ones. The Secretary began with a twenty-minute harangue about how Syria was not participating in the multilaterals and would not negotiate over subjects such as water "until the land is given back."

Eventually, I was permitted to make my presentation. At the end, Lenny suggested the Syrians might want to build their piece of the model to get ready for when negotiations would be appropriate; Syria could also gain from having such a model just for domestic use. Lenny also pointed out we were not the multilaterals. I'm not sure whether the Secretary understood what was being suggested. In any event, he ended the meeting with a repeat of his opening harangue.

We were somewhat disappointed, but Hamdad—to our surprise—told Lenny the next day it had been a very good meeting. He pointed out that the Secretary was well known for cutting meetings short when he thinks they are pointless and that we had had a particularly long meeting with him. This was even more significant in view of the state of his health.

Or perhaps Hamdad was just being polite.

Running very late, we were next taken to the U.S. Embassy for a short meeting with Ambassador Christopher Ross and an assistant. We discussed what we had been doing, and the ambassador suggested we shouldn't read too much into the fact that the Syrians had not been willing to see us in January but now were willing to see us in June. He said that might just be a matter of bureaucracy.

Ambassador Ross was both charming and a bit unusual. He spoke excellent Arabic, having spent considerable time in Lebanon. (The Syrians remarked to us that he sounds Lebanese.) He told us that when hostages were being released through Damascus and it came his turn to speak at the press conference, he noted that just about everyone there was from the Arabic press so he spoke in Arabic, after which CNN identified him as "another Syrian official."

Our conversation continued over a big lunch with all the people we had met that morning, except for the Secretary from the Ministry of Foreign Affairs. Ellen joined us. I had mentioned to Ambassador Ross that the Syrians were surprised that Ellen had not come to our meetings. With a smile, he said we had set back the cause of women considerably, since some time had been spent educating the Syrians that women should be full participants.

The conversation at lunch ran somewhat more freely than at the formal meetings. For instance, the ambassador revealed to Ellen that he was never consulted when U.S. Secretary of State Warren Christopher came to Syria for talks.

By lunchtime, Minister of Economy Al-Imadi had had time to think about what we had said in our earlier meeting, and was now quite struck with the proposition that the value of the water in dispute was very small.

"It won't buy an airplane," he remarked.

He also saw the possibility of applying the same method to other water disputes, some of which he identified.

Much of the talk at the table, however, was from Daoud Majed. He was embarrassingly enthusiastic about the paper as a work of genius. It was obvious that the Ministry of Irrigation, and perhaps even the Ministry of Economy, was dying to get their hands on the model, if only for use internally. I promised Majed the computer program if he could get permission to ask for it.

I had the general impression that we had generated a good deal of excitement in the domestic Ministries and that they might very well put some pressure on the Foreign Ministry to be allowed to participate—but we would have to see. Later, when I mentioned these events to Munther Haddadin at the Cyprus meeting, he suggested several times that we might involve the Syrians by inviting them to participate in connection with their joint water issues with Jordan. This would keep them from negotiating with Israel. We expected to follow it up.

That night, Lenny and I were interviewed on Syrian television in a wide-ranging discussion of peace issues and economics. I was quite shocked when the interviewer first asked Lenny why Israel needed any security guarantees since "it is well known that the Arabs have never started any war against Israel."

In his reply, Lenny said words to the effect that it was true Israel started the 1982 war and also the 1956 Suez War. One could argue about the 1967 war, although I would take the position that, in fact, the Arabs started it. But from my perspective, that the Arabs began the wars in 1948 and especially 1973 cannot be in question.

The interviewer wasn't all that old, but presumably he could remember 1973. I think this was a case of propaganda really succeeding (or being necessary to repeat).

By the time the interviewer got to me, I told him I wasn't willing to talk about anything but water, and that's what we talked about.

On the morning of July 1, we were taken to Kuneitra in the Golan Heights for what was, I'm sure, the standard propaganda treatment. We met the Governor of the Golan, who explained to us in great detail the sufferings of the people and how the Israelis systematically destroyed Kuneitra when they withdrew in 1974—some of which, I'm sure, was true, but a lot of which was also a considerable exaggeration.

Lenny pointed out that there were two tragedies, the other having to do with shells lobbed into Galilee from the Golan Heights before the 1967 war. The Governor replied quite irrelevantly that one ought not to think just of the Golan Heights, because the Galilee Heights on the other side of the valley were even higher. That has about as much force as observing that the Alps are also high mountains. In any case, neither Lenny nor I sat entirely still for this, although we were polite and emphasized our belief that the Golan must be returned in connection with a secure and lasting peace.

We then toured the Golan a bit. The most interesting part was when we stood at the top of a large hill with the Israeli-occupied section across a deep valley on the other side. As we approached the top of the hill, we could hear voices in the air. These turned out to be a number of Druze families with electronic amplification equipment shouting across the valley to their families on the other side who were replying through similar equipment. We were told this goes on every day because the families are only occasionally allowed to meet in person. It was quite a visual and auditory experience. One could see verdant fields beginning at the UN-patrolled ceasefire line. The Syrian city of Kuneitra has been left a bombed-out wasteland since 1976.

Hamdad and his family were very sentimental about his boyhood home and tended to come up every Friday for a picnic. He asked Ellen to take pictures in the direction of Banias, but although she sent the pictures to him, she never heard from him again.

From the Golan, we went directly to the Damascus airport and flew on to Cyprus for the conference.

4

The Cyprus Conference

The Cyprus Water Conference took place, at last, over the first few days of July 1994. It began with an opening dinner on July 1st and continued for three very full days and evenings. Discussions even continued on July 5, beyond the official close of the conference.

Interest in the project and the conference was very high by the time of its opening. Avraham Katz-Oz, the chief Israeli water negotiator, could not attend and sent a deputy, Moshe Yizraely, who also brought with him an expert adviser from Rehovoth, Amotz Amiad. Other Israelis included Dan Bitan of the Truman Institute and Shaul Arlosoroff, a well-known water expert who had been at the World Bank. Of course, Zvi Eckstein also came. Gideon Fishelson had an operation for stomach cancer the week before and could not attend.

The Jordanians sent Munther Haddadin and Ali Ghezawi. The Palestinians sent Marwan Haddad, their second or third water negotiator, although he originally came because of his interest in joint management. In the week or so before the conference opened, Jad Isaac added a number of other Palestinians.

The American delegation to the multilaterals thought about sending someone, but eventually decided not to do so and instead asked me to report to them upon my return to the United States.

The conference was held at a terrific site: the Four Seasons Hotel Limassol. It had both a swimming pool and a beach. After arrival, but before dinner, Ellen and I went for a swim in the pool, which was tepid. The next day, we decided to try the beach. The swim in the sea was more refreshing, but it was somewhat off-putting to have two renowned water experts independently point out to us that the presence in the water of a certain algae that grows only in sewage suggested a high level of pollution. We returned to the pool.

The opening dinner was fairly informal, with welcoming speeches from Lenny Hausman and Anni Karasik describing the work of ISEPME and a small welcome from me. I assured those with official positions that they would not be asked to commit themselves to anything but also stressed the importance of what was to come. Most of the attendees had not yet received the paper, and so it was handed out that night.

I was seated at dinner between Jad and Marwan Haddad, whom I was meeting for the first time. At the time, Marwan was director of the Water & Environmental Studies Center (WESC) at An-Najah University in Nablus, where he was also an associate professor. Jad seemed pretty upbeat, but told me that he had been hospitalized with a heart condition the week before and needed to stop killing himself with work. He also commented that he was supposed to be in Jericho in a couple of days to be sworn into the Palestine National Authority, but had decided to ignore it. Coincidently, Yasser Arafat arrived in Gaza and then went on to Jericho during the time of the conference.

Zvi Eckstein expressed to me some concern about the acknowledgment of his earlier role in the project. This was not unexpected. We had had a similar phone conversation six weeks earlier, and Lenny and I had discussed the matter in Syria. I wanted to be fair and, if necessary, generous. Zvi (and probably Bob Dorfman) deserved a lot of credit and appropriate joint authorship. Zvi's model was a major contribution to the project in its early stages—even though, as I expressed earlier in these memoirs, microeconomists generally speaking just know how to build such a model. Still, some credit was due given that Eckstein and Fishelson actually went ahead and did it before anyone else, and did it independently.

Conference Day 1

The first full day of the conference, Saturday, July 2, ran from 9:00 a.m. to 5:00 p.m. It is no exaggeration to say that I either spoke or answered questions for the entire eight-hour period, except during the breaks for coffee and for lunch. I used every expository device I know. I mentioned my sister Joanne as an example of opportunity costs, as I've done in lectures to students.* I sang a temperance song about water. Part of this was not only to inform with substance but to try to be personally impressive and appealing. It must have been quite a show.

Still, even people who had heard part of this before had a difficult time adjusting to the proposition that water, despite being necessary for life, can have a low economic value.** The fact that value is formed at the margin and not on the average is something that seems very natural to economists but not to other people.

Running through the conference was the question of whether to adapt the model in two ways: to have many different qualities of water and, especially, to be inter-temporal, meaning a model where what one does to the aquifer at one time is reflected in what happens later. Our model at the time was a steady-state model. The question reflected high interest in the use of a possibly expanded model as a decision tool for water management and sharpened my impression that the high level of intellectual interest in the model and the proceedings of the conference were fairly remarkable.

Substantively and with few exceptions, the first day went pretty well. The exceptions, though, were quite embarrassing. They took place in the afternoon, when I began by presenting specific results. The problems had to do with the input data.

The model runs called decisively for a pipeline connecting Jerusalem to the southern cities of the West Bank, Bethlehem, and Hebron. Without it, the prices generated by the model for those cities tended to be much higher

---

* At the time, my sister Joanne ran a firm that placed pictures on soap that would not disappear as it was used. She continually reported to me that the firm was taking in a lot of money, but I could never tell whether the firm was profitable because she did not pay herself a wage. When my daughter Abigail became friendly with some MIT economics graduate students, the first thing they wanted to know from her was whether she really had an Aunt Joanne.

** Adam Smith had discussed this sort of proposition two centuries earlier when he contrasted value in use with value in exchange. Smith pointed out that air was free and diamonds expensive because diamonds are scarce and air—while necessary for life—is not.

than in Jerusalem and northward. Unfortunately, it turned out that there already was such a pipeline; neither the Israeli nor the Palestinian team had noted its absence in the model. It was, of course, some comfort that, with such a logical pipeline missing, the model called strongly for its construction. Still, better not to make such errors.

There were also a number of apparent problems with the Jordanian input data. Jeremy Berkoff, a development economist from the World Bank, was particularly insistent on this, and before leaving the conference on the next day he prepared a paper describing all the problems he saw—and three in particular. These included the controversy over a source in the Dead Sea district. Berkoff also noted my misinterpretation of Jordanian policy toward industry; the Jordanians did not supply water to industry free, but instead permitted industries that dig their own wells to obtain the water by paying only the drilling and pumping costs (and had a similar policy for agriculture in the northern highlands). Another problem concerned the Yarmouk River. The Jordanian team's report included as a source in the northern highlands some part of its waters, and we had misinterpreted this by taking two alternative versions as though they were different sources. We had also failed to note that the availability of that water depends on agreement with Syria and (Israel). The model had to be run parametrically on the amount of water available to Jordan, Israel, and Palestine.

The Yarmouk River problem illustrates a recurring feature of large projects such as this one: if there is something in the back of your mind you think you don't understand, it will come back and bite you. From time to time, it had occurred to me that I didn't really understand what the situation was about the Yarmouk or what we were doing about it. It was not hard to fix, but I should have paid more serious attention to it.

In any event, we agreed these things would all be fixed up. Munther deputized Berkoff to help us, and it looked as though we could work everything out when Berkoff returned to the United States on August 1.

The day did not end at 5 o'clock, however. Lenny and Anni had set up a dinner with all the Israelis to discuss how the model might be used in negotiations and related subjects. This was part of a general program Lenny and I had agreed to in Syria about exploring how to take the model forward. The Palestinians were also invited to attend, but said that it was premature to meet on such subjects with the Israelis and they would rather meet with us alone—which we did the following night.

The discussion with the Israelis started off in a fairly depressing way. I had assured those with official positions that they would not be asked to commit themselves, but Lenny asked straight out how they felt about using the model in negotiations. I thought this put Moshe Yizraely, whose first exposure to the model had been that very day, pretty much on the spot. I didn't think matters were helped much by a relatively long explanation by Anni about how ISEPME had been so successful with its transition book.[1]

Yizraely's reply was not encouraging. His view was that the multilateral negotiations did not matter at all and that he didn't see that the model would be at all helpful in the bilaterals. Later, he relaxed this, when he considered the probability that negotiations on water would not take place for a couple of years.

I found his attitude quite disappointing, although he and his advisor Amiad seemed somewhat more forthcoming in private conversation. Their attitude appeared to be one of "I don't see how this will aid my negotiating position," rather than seeing how this might be a way to assist the peace negotiations in general to the benefit of all parties, including Israel.

Unfortunately, this matched what Jad told me two days later about the negotiations in general. The Israelis seemed focused on how little they could give up, with little regard for *being sure a peace agreement would be actually reached and remain workable*. It was a predictable attitude on the part of middle- and low-level bureaucrats, but not one that boded well. If, as I feared, it would characterize the water negotiations, then any influence the project was to have would have to be acquired by going to high-level people who would be willing to take a more general view of what could be accomplished.

Shaul Arlosoroff was somewhat more forthcoming in the conversation that night with the Israelis, and even more so as he began to think about what the model would do. He emphasized, as he did in the plenary sessions, the need for an intertemporal model that would be useful as a management tool. He said there had been some experience with model building previously, in which they had ended up with something no better than the accounting system of Mekorot (the technical firm of the Israeli water authority). I didn't think he really yet understood how our model worked or what it could do.

I was relatively unhappy at the end of the evening. There had been the embarrassing set of input errors in the results that afternoon followed by the somewhat depressing conversation at dinner with the Israelis. Being told what a terrific expositor I was didn't make-up for having the substance partly wrong. But Zvi Eckstein cheered me up by pointing out that we could do things like deal with drought years simply by running scenarios with supplies considerably reduced. He convinced me that the model in its present form would be worth a stage-one book after which we could go on. He also took a much less pessimistic view than I of the dinner conversation.

Conference Day 2

The next morning had been reserved for hands-on use of the model by the different parties. I had quite naively supposed they would find this so fascinating that they might even perceive how they would negotiate using it. Such was not the case. Only the Palestinians spent a long period exercising the model and running different scenarios. The Israelis said they would run it when they got home. The Jordanians were few in number and, at first, did not participate.

Then a somewhat curious incident ensued. About midway through the morning, Munther said he would like to have the training experience and asked who would train him. I told him he could have Aviv Nevo, if he wanted an Israeli. He politely refused. I then suggested Harshadeep, but Munther said Harsh had been around "too many Jews." I next offered myself, but Munther chose Harsh.

Munther made reference to "too many Jews" several times. I was not quite sure how to read it. He may very well have been suspicious that a number of the Americans involved were Jewish. During a coffee break later, when the subject of Jad's proposed environmental profile of the Jordan Valley came up, someone asked me what the model would bring to that part of the project and why I would be personally involved. I replied truthfully that I didn't think there was a large analytic content to it, nor did I have the appropriate intellectual capital. However, Jad had asked that I do it because he trusted me, and I had accepted because I thought that was important.

At that point, Munther, who was listening, said, "You know, we trust you too." He was quite serious. He could also be quite charming. At one point during our time in Cyprus he was sitting in the Jacuzzi at the hotel and, not knowing I was observing, he began to sing himself an American lullaby, "All the Pretty Little Horses."

The afternoon was spent, in part, on Jad's proposal on which there was general agreement. Apparently, it was important and not merely for substance. I was later informed by Charles Lawson from the U.S. Department of State and others in the American delegation to the multilaterals that the same proposal has been floated in other fora and aroused a great deal of controversy.* Not so at the Cyprus conference.

That evening, we had dinner with the Palestinians and asked them the same general question we had asked the Israelis the night before. They were much more reassuring. The Palestinians, whatever the tensions among them, all appeared to support the model with enthusiasm. This was particularly true of Marwan, with whom I developed a fairly friendly relationship during the conference. It appears that plain speaking and no nonsense, combined with weird humor, goes down pretty well.

At the dinner, Jad proposed we have someone—perhaps David Grey—examine the different aspects of international law on which the different claims to water were based and try to set forth how much water would actually go to each party under the different principles. To my surprise, no one had ever done that; even more surprising, when it was proposed the following afternoon, Yizraely opposed it as probably undoable. It really did appear that the Israelis couldn't even agree (or perhaps didn't want to know) what the amount of water in dispute really was.

Conference Day 3

The morning of July 4 began with a presentation from Yehuda Bachmat on modeling (with a steady-state model) the Mountain Aquifer. It was beautiful and appreciated. The presentation was theoretical and addressed how to calculate the effects of pumping in one place on the costs in all others. It would require estimating the parameters involved, which required data.

---

* I visited the American delegation to the multilaterals in mid-July for an unsuccessful meeting. The State Department officials, led by Lawson, listened politely but were not going to go further. About a decade later, when former Secretary of State George Shultz approached the State Department about the Project, he was told they already knew all about it. That was downright false; Lawson and others had been exposed to the Project directly only before it really got underway and later only by word of mouth. But the message from then Secretary of State Colin Powell to Shultz was a clear "don't call us; we'll call you."

Bachmat not only knew what he was doing as a hydrologist, but had managed a structure that would fit very nicely into our model exactly as it was supposed to. There was general appreciation for this.

The former head of the Israeli hydrological service, Bachmat had the needed data in his office. But there was a complication: the Palestinians wanted to appoint someone to work with Bachmat. That was fine with him, but would the Palestinians be permitted access to the data? Bachmat, Marwan, and I discussed this; we agreed that we would have to try, and if there was a problem we would go in and bang the table pretty hard. I thought we probably had enough clout to clear that one up.

Zvi Eckstein followed Bachmat with a discussion on the use of the model for management. It was a perfectly good paper but ran into problems that I did not fully understand. Apparently, for three days before the conference and continuing during the conference, there were negotiations between the Truman Institute, the Palestine Consultancy Group (PCG), and ISEPME on the creation of a study for joint management. Apparently, the negotiations were quite delicate, one issue being why the two regional groups needed ISEPME at all. There were occasional references to this in the meetings, with people observing that they couldn't participate in this or that unless those negotiations reached a successful conclusion.

In fact, the negotiations were so delicate that Anni felt compelled to make a public statement that Zvi was speaking only for himself and had no knowledge of those external negotiations. I didn't fully understand what was involved.

For the last formal part of the conference, I summed up what was to be done. We thought of the model in two stages. Stage I was the existing model as a steady-state model. It was to be adjusted by correcting the input errors and incorporating the Bachmat model. We hoped to have a book-like manuscript by November or December (in reality, while we produced plenty of articles over the years, the book did not appear until 2005!). Stage II of the model would involve expanding it to a dynamic model and including lots of different quality parameters for water. I undertook to write up a description of the mathematics of how this would operate. Arlosoroff, who was the most insistent on the need for a dynamic model, kept warning of the possibility that there might not be enough information on the dynamics of the system to enable this to be done.

In general, different people had different horses to ride all through the conference. There were sensible concerns such as the need for a dynamic model and the need to incorporate different quality parameters. Other concerns made sense, but I thought they might not be easily accommodated in our framework; environmental preservation was an example. Still other concerns were like those raised by Musa Nimah from the American University of Beirut, who kept insisting that the first amounts of water must be terrifically valuable because they are necessary for human life. It is no simple matter to educate people about value in exchange versus value in use, but I think we ended the conference having been pretty successful on that issue.

People were very tired at the full final session (including me), and tempers began to fray a little. Jad, in particular, became irritable—but he was somewhat provoked by Moshe Yizraely.

I also made a political mistake. In listing the things to be done, I included providing information concerning the pipeline from Jerusalem to Bethlehem and Hebron. I happened to mention it after writing down the names of the Israeli team as the group that would do this. Jad bridled and asked why this should be an Israeli task when the pipeline lay entirely within Palestine territory. Mistakenly thinking this had concerned Jerusalem instead of the route of the pipeline, I said rather sharply that I didn't care who provided the information so long as someone gave it to me. But Jad was quite right, as I told him afterwards. More surprising was that we got through all this time without more problems.

Arrangements were made for certain work. The teams working on the model in the region felt the work had been too centralized at Harvard and wanted to have more of a say in what went on. They were quite right. It was very important that they feel ownership of the Project. So, I committed to make periodic trips to the region, perhaps every six weeks, for smaller meetings. We hoped we all could do this in the same place and we thought that it might be possible, given the thawing of Jordanian-Israeli relations. If nothing else, we could probably all work in Jericho.

We did not schedule the first of those meetings, but it would have been sensible to have one around the end of October at the time of the First Middle East Conference. Originally slated for Marrakesh, it had been rescheduled for Casablanca.

We also committed to provide copies of the program to a very large number of people. But at the last minute, I realized the $800-a-shot GAMS license meant a huge expense. So, I had to renege, temporarily, and say that we had begun negotiations through the World Bank to get a reduced rate—which was successful.

Munther had to leave the conference halfway through the last day, but before doing so he took an action I need to report. Munther was quite insistent that he (and official Jordan) would not correct the errors of the Jordanian team. He appointed Jeremy Berkoff to work with us and said they would tell us if we hadn't gotten it right. But outside the main sessions of the conference, Munther had a conversation with Lenny that concerned not only the repeated suggestion that we try to get Syria to work in connection with Jordan, but also something else.

It turned out Munther had his own consulting firm that often competed with Abu-Taleb's firm. Munther made it quite clear to Lenny that after the first phase of the project is over, Abu-Taleb was to be out and he was to be in. This was a fairly serious use of muscle in a way that would certainly be considered inappropriate in the West. However, given Munther's political clout, we had no choice but to accede. We needed to maintain very good relations with him. Fortunately, Abu-Taleb's record was not so sterling as to make us feel that his firm was the only one that could do the work.

Still, there was something quite unfair going on. In conversations in late July with Abu-Taleb back in Cambridge, Lenny and I suggested that if any action were to be taken, it would have to be taken in Jordan, perhaps through Abu-Taleb's father—who was the former chief of the Jordanian General staff. But ISEPME couldn't get involved.

In any event, while the Jordanian part of the model in substance had become my chief concern for the time being, Jordanian political reaction had become the least of my worries.

The Final Evening

After the conference ended, Ellen and I had drinks with Jad and Atif Kubursi. A man named Gaby Kiddy, the local representative of the Christian Council of Churches, had invited them out to a restaurant; we were invited to continue the conversation at dinner. We were all very tired. Jad, who was quite depressed, spoke to us at length about his feelings regarding the peace negotiations and what was happening in the Palestinian camp.

Jad viewed the Palestinians as woefully unprepared for negotiations, partly for cultural reasons. In the Arab world, when people make peace, they hug each other and then fix everything up afterwards. That, of course, is not what happened in the peace negotiations. Instead, the Israelis arrived with a team of lawyers and wanted to put everything in writing. The Palestinians, by contrast, did not use their own experts.

While we were all in Cyprus at the conference, Arafat had arrived in Gaza and posed in front of a citrus canning factory. Jad told us that Arafat had been told (presumably by Jad and others) as far back as 1985 that the water in Gaza would not produce enough fruit to keep that plant in operation for more than fifteen days a year. Yet, the plant had been built because the consultant had received a bribe of some sort.

Jad also cited the Cairo Agreement signed in early May. While the Paris Economic Agreement between Israel and the PLO had been quite favorable for the PLO, Jad's view was that the Cairo Agreement superseded it and took back a number of the favorable points, severely limiting what could be done in Gaza. In general, he made it sound as though the Israeli attitude was, as I stated above, to see how little Israel could give away, without considering whether the process as a whole could succeed.

I told Jad I didn't know when he asked me how one managed to maintain economic development while keeping social stability. After some discussion, he asked who might be able to assist on such a question. I suggested Dick Eckaus, my colleague in the MIT Economics Department (who later agreed). Atif thought him an excellent choice.

During dinner, Jad also declared, "No one will run this model."

I was quite surprised. "But you will," I replied.

"Why will I?"

I think this had a lot to do with Jad's internal problems in the Palestinian structure.

The following day, I had a conversation with David Brooks, a natural resource economist working on Mideast water, concerning these matters. He came to my room, since he said he preferred to have it quite private. In part, the conversation concerned what I already knew but with more detail. It turned out Jad would have liked to be in charge of water but wasn't going to be. Instead, he had been put in charge of the environment. The chief water negotiator was to be Riyad El-Khoudary, president of Gaza, in part because the PNA felt it had to include someone from Gaza.

Jad was described as someone who didn't know anything about water. Marwan Haddad, who did, was to be number two (or perhaps number three, according to the American multilateral delegation). But Marwan and Jad did not really get on. Jad was discouraged and was contemplating retirement to a more academic life.

Some of this was confirmed by an event soon afterwards. Stanley Fischer asked me who should be invited for the water program in Casablanca. I said he had to have Munther and suggested Shaul Arlosoroff and Jad. Stan said Jad was very controversial and suggested El-Khoudary. I suggested Marwan. It concerned me that the view of Jad as controversial could be harmful to our project.

5

Improving the Model and Correcting Defects

In the wake of the Cyprus conference, we spent a long time in the fall of 1994 trying to correct input errors and discovering other errors in the model, improving the write-up, and improving our understanding. There was a host of issues to address. The first concerned the Jordanian report.

We met with Jeremy Berkoff, the World Bank economist, in August, and he was tremendously helpful. It appeared the Jordanian report claimed water from projects that were not yet built and assigned water to districts without really understanding how the model results were going to be used—in part because the central modeling team had failed to communicate well, for which I took responsibility. For example, because the dam of the Zarqa River catches much recycled water from Amman, the Jordanian report put it in the Amman district, despite the fact that it is down in the valley below and it makes much more sense to assign it to the Jordan Valley for conveyance reasons. There were also issues as to the existence or inclusion of pipelines and of conveyance costs. Perhaps most important, though, we simply had not treated the Yarmouk properly. The Jordanian report counted different amounts of water from the Yarmouk, depending on negotiations, as supplied to Jordan. We couldn't treat it that way. Instead, we had to take as an input the amount of water available after Syria took some and after accounting for what was needed to maintain the level of the Dead Sea. The question of how much Jordan gets—an interesting result of the model—was to be an output, not an input. It took a long time to get all these issues resolved.

It also took a while to determine where, in fact, the pipeline system south of Jerusalem was and to account for the use of the Yarmouk in Israel. We eventually had to subdivide the Kinneret district to account for that use.

Then, in mid-autumn, we considered how to deal with capital charges. I asked Bob Dorfman to think about this; he came back with a very tidy analysis based on a paper by Harold Hotelling, the mathematical statistician whose work had become quite influential among economists. It essentially says that capital charges should be recovered in the capacity shadow prices. That is, if capacity is optimally designed, the present value of all capacity shadow prices will be the marginal capital cost of the operation. That's fine, except when there are increasing returns to scale. Since, in fact, the model calculated capacity charges, we didn't have to resolve the issue right away. In the longer run, it would make sense to put optimal capacity design into the model.*

The calculation of consumer surplus was another issue and one of those maddening examples involving something that drives people crazy but turns turns out to be fairly trivial. We were probably fortunate, however, that the mistake was made. When calculating the changes in welfare in one country due to water subsidies in another country, I had noted before the Cyprus conference that the increase in producer surplus appeared to outweigh the decrease in consumer surplus. Puzzled, I thought long and hard and discovered that the international trade effect applies—for instance, that when Israel subsidizes agriculture and takes more water from the Jordan River, that increases money flows from Jordan and Palestine, assuming Israel owns the water. That is a very important proposition. In writing the report in December, however, I became aware of the fact that the model, even after accounting for this effect, was still showing increases in producer surpluses outweighing decreases in consumer surplus. A couple of days of dogged endeavor followed. At least twice, I thought I had proved—incorrectly—that because of recycled water such an effect was possible. In the end, we found a mistake in the calculation of consumer surplus.* Had the bug not been

---

* Doing so would require a multi-year model, which is somewhat more complicated. In the late 2000s, in fact, we did develop a quite powerful multi-year model called MYWAS that deals directly with capital expenditures as they are made (or forecast).
* It was not a sophisticated mistake in the calculation of the integral under the demand curve, but had to do with how we treated the fact that water demand is sufficiently insensitive to price, so that the integral is infinite unless one restricts the price range

61

there, I probably would never have discovered that the international trade effect applied.

Other errors were discovered. For example, at some point we had managed to block the Israeli National Water Carrier in the model and were sending water round via Jerusalem to reach southern points. We discovered some other minor defects as well, especially when Jad came to visit to give a seminar in October and we looked at the model results with him.

In general, central modeling would begin with Harshadeep doing some runs. Then we would sit down together and go through them to see whether they made sense. In part because Harsh was working on his doctoral thesis and in part for other reasons, this process did not really converge until the main report was drafted at the end of December.

All of this took a lot of time and communication with the people in the region, so reaching a mutual understanding of what was wanted happened slowly. Communications elsewhere weren't a lot better. We still had no answer from the World Bank to our request for a GAMS license discount.

Casablanca

Amidst all the work on the model, the long-awaited First Middle East Conference took place in Casablanca at the end of October 1994. I duly attended. Its move from Marrakesh prompted some inevitable thoughts about the movie *Casablanca*. I considered paraphrasing Captain Renault when I spoke, saying "I am shocked, shocked to find that there is a conference going on here."

My son Abraham, though, came up with a better line, and suggested I begin by quoting Humphrey Bogart's character Rick: "I came to Casablanca for the waters."

I suspected most of the audience would not know what on earth that was about. But still, when I learned that I could only speak for a few minutes in a small session devoted to water, I first thought that I would be using up one-third of my lifetime allotment of fame. Then I thought of paraphrasing the Rick character even further: "I see that Casablanca is not the place for water. I was misinformed."

From my point of view the conference was largely a waste of time. The

---

involved.

event was largely formal, with people beforehand saying how wonderful it was going to be and afterwards saying how wonderful it had been. It was immensely crowded. It's possible much went on in back rooms, but that wasn't very visible. Among others, the Israelis considered it to be major that it took place at all. Again, it was the phenomenon of self-congratulation that everyone sat together in the same room.

In terms of water, it was important that I went, even if not much happened. Our water panel was charged with discussing the joint projects that would be undertaken for water cooperation. Each of us on the panel got to talk for what turned out to be seven minutes, which was ridiculous. In addition, the room allotted was small and not all the people who wanted to come could attend.

The presentations varied. Riyad El-Khoudary, the head Palestinian water negotiation and panel member, did nothing but list the wrongs done to the Palestinians by the Israelis. It was not at all productive. Jad Isaac was present, and was very annoyed by this. He had come a long way. A year earlier, it would have been his own principal message.

El-Khoudary's behavior on the water panel matched, to a certain extent, the general Palestinian approach to public meetings at the conference, which was to complain much more than to offer constructive suggestions about all the things that needed to be done. For instance, at the lunch given by the Palestinians to discuss business, Yasser Arafat said that he knew businessmen wanted guarantees but that the only guarantee he could offer was that they could call his office. It was one manifestation of a general feature of the Palestinians: they were centralized and used to referring everything to the head office for approval, and not really prepared to do business on a day-to-day basis. It was quite disappointing.

The water panel's Israeli speaker, Avishai Braverman, was an economist and president of Ben-Gurion University. (Some years later, he became a Labor member of the Knesset.) He had headed a water study, about which I had known little, that looked at water quantities and balances and concluded that recycled water would be very important. I thought that was correct, but not that desalination would be required in about twenty years.

Munther Haddadin was, of course, Jordan's panel member. He had just finished negotiating the water and lands parts of the Israel-Jordan Peace Treaty. Predictably, he was not remarkably helpful, trotting out again an old chestnut: "If we were lost in the desert, what then would be the value of a drop of water?" I decided to put a refutation of that in the draft report.

The discussion was not terrific. Some people could do back-of-the-envelope calculations on desalination costs and saw the point about the value of water right away. One journalist claimed water would be the next cause of war in the Middle East and that three countries were currently preparing for that war, although he wouldn't say which. I was interviewed by a couple of other journalists, but nothing seemed to come of it.

A few other things about the Casablanca conference are worth mentioning. For instance, I spent a brief time discussing with Munther his role in negotiating the important Israel-Jordan Peace Treaty that had been signed a week before the conference. It included a number of complicated water provisions (said to be the hardest part of the treaty to negotiate). The Yarmouk water was to be shared. Israel was to give Jordan water or find additional water for Jordan. There were to be dams built to use the Yarmouk water more effectively. Finally, Israel was to desalinate the saline springs that made the Jordan River bed water unusable.

This was all very sensible. Indeed, our model suggested that the single most important project for the region involved more efficient use of Yarmouk water together with expanding pipeline capacity to carry water to Amman.

Casablanca raised an interesting and complicated question for me: What, if any, was the role of our project? One possible answer came when I was informed later, in August, that it appeared that there would be a treaty, and Yossi Beilin—a well-known Israeli statesman who over the years worked tirelessly to bring about a peaceful two-state solution and who is one of my heroes—spoke to Americans for Peace Now.* Asked about water, he replied, "Water? That's just money. The water in dispute isn't worth $150 million dollars per year, and we're going to negotiate our way out of it."

---

* I helped found Americans for Peace Now in 1981. Much later, I became the board chair, serving from 2006–2010; I am still on the board today. I moved away from explicit involvement in part because of the pressure of the project, partly because of my increased activity with the New Israel Fund, and because with the election of Binyamin Netanyahu as Israel's prime minister I foresaw that being publicly involved with these two organizations was not going to help me negotiate with the very right-wing Israeli government.

I knew for a fact that he got that figure from a paper prepared for him by Zvi Eckstein. It was, at the time, the number from our project (which became much smaller over time as desalination costs dropped). Further, after the treaty was signed, Dan Zyslowsky, a former water commissioner, was quoted in the Israeli press as saying, essentially, that water is just money and that the 50 million cubic meters per year Israel promised to Jordan just meant the Treasury had to find a way to make good on it in the budget.

These two remarks suggested we had influenced the Israeli way of thinking. But it was not simple to find out. Even when I had occasion to introduce Beilin at ISEPME after the treaty was signed, and he said he had not seen the treaty, and he may not have known then that I was the "water person."

Even if our project was having some influence on Israel, however, Munther Haddadin was giving no sign that our project had anything whatsoever to do with Jordan's attitude towards the treaty. Nor would I have expected him to do so, regardless of whether it actually did have some influence. There is certainly no internal evidence in the treaty that we did have an effect. In the treaty, water was not traded for other things at some fixed price. It was obvious that the water issues had been decided in general in connection with concessions made by the parties to each other in other areas.

Still, when Julia Neuberger visited Cambridge in early November, she told me she had been told by the Crown Prince and two of his aides that my project and I had been very influential in permitting Jordan to sign the water parts of the treaty. "They sang your praises," she said.

I suspected this was just another case of what Jordanians are known for: telling people what they think people want to hear. The Crown Prince must have known Julia would speak to me.

In closing on the trip to Casablanca, I should note that it was not all work. Before, and especially after the conference, Ellen and I went sightseeing in Morocco. We had a very enjoyable time. Our old friend Rhoda Fischer, Stan's wife, joined us, and because I was traveling with two women with the same last name (if one ignores the spellings), the suggestion arose that I was traveling with my harem.

Ellen and I had discovered that we could not return home by air in one day, so we decided on a layover in Paris for a first-rate dinner. When we checked in the next morning for the flight to Boston, the security official asked me the purpose of my coming to Paris. He was much amused by my truthful reply: "To have dinner."

The December Report

Much of December 1994 was spent writing a draft report, which was an important document in the life of the project. It seemed to have engendered a good deal of excitement among those who read it, both project participants and others. Its main conclusions are worth noting; the numerical results provided here are more accurate in our 2005 book about the project,[1] but the conclusions are qualitatively similar.

First, as expected, we concluded that the disputed water was not worth a great deal. For example, it would take until 2020 before 400 million cubic meters of Jordan River water would be worth $200 million—an amount that was small potatoes relative to the economies involved. I noted, however, that the present value of the amounts involved would sound like a pretty healthy sum, since one should use real (rather than nominal) interest rates for discounting. Still, I explicitly refrained from doing that in the report, because the sum would look so large. In fact, an appropriate comparison would also require taking the present value of the gross domestic products of the economies involved, which would be far larger.

The report also noted there would not be a regional crisis of water for human consumption. Even with the Palestinian assessment of greatly increased population and per-capita consumption—gross overestimates— the shadow price of water in most places in the region would not rise above roughly sixty to seventy cents in the next thirty years. Further, consumption figures in the model were at or above the actual numbers, while the shadow values were below what households were currently charged in many places and comparable to or below what they were charged in Boston.

The interesting exception concerned the shadow prices in Amman and the Northern Highlands region of Jordan, which would go through the roof by 2010 ($8 to $9 dollars per cubic meter in Amman; more than $17 in the Northern Highlands). Further examination revealed that this had nothing to do with the value of the water in dispute, but was an infrastructure problem—namely, the lack of facilities to convey water to those regions. Effectively, it would do the Jordanians no good to have more water in the valley below if the existing pipe to Amman was already at capacity. In runs of the model that relaxed the capacity constraints on the pipelines, the shadow prices in Amman and the northern highlands fell to about seventy cents per cubic meter, comparable to the values in the rest of the region.

The December report did conclude, however, that there would be a crisis if agriculture had to run on unsubsidized fresh water. That crisis could certainly be solved or at least greatly alleviated by using recycled wastewater—a conclusion that would stand up despite the fact that the report erroneously emphasized the recycling of water in Gaza.

Subsidizing water for agriculture was an alternative. The report investigated the effects of 50 percent subsidies by each country on the water economy of each of the other countries (neglecting the effect on competition in agricultural output), and it turned out subsidies would actually benefit the other countries. This was largely due to the international trade effect mentioned earlier, but it also required some ownership of the disputed water by the non-subsidizing countries, so that they got the benefit of the increase in the producer surplus on those waters. The report calculated how much of the disputed water each country would have to own, which appeared to be quite minimal.

With respect to desalination, the report concluded that it did not appear it would be profitable on the Mediterranean coast until at least 2020 and then only if there were to be a technological breakthrough. Running the model with desalination operating costs at a $1.50 per cubic meter gave shadow values on the Mediterranean coast that got up to roughly ninety-five cents per cubic meter. That became our prediction of the cost desalination would need to have to be a competitive technology in 2020.*

It was obvious that a very, very important use of the model would be the evaluation of different capital projects. At the time, the model was being run without any capital costs whatsoever, except for desalination plants. This was useful because it helped reveal whether certain projects, if available free of

---

* Technological advances have since brought the cost of desalination down, but our results (as of this writing in 2015) still imply that Mediterranean desalination will be efficient by 2020 only in strong droughts. This produces a problem for planners: since much of the costs are incurred when building the plant, once a desalination plant has been built it will be used even in more normal times. Israel has chosen to accept such a situation and has built very large desalination plants. Some of them require the builders, by contract, to achieve large outputs every year.

charge, would be used. If the answer was no, it was obvious it was a project not worth building—as in the case of desalination. Further, it indicated pretty strongly where one should look for useful projects such as the pipeline system to Amman.

Clearly, we needed to do some actual project evaluations, and that prospect excited people in the different countries.

II. 1995

6
The Next Trip to the Middle East

The week before Ellen and I set out again for the Middle East on January 19, 1995, Lenny Hausman and I traveled to Washington to meet with Dennis Ross at the State Department. A very pleasant and intelligent man, he was President Clinton's Special Envoy to the Middle East and headed a team of several people in the Office of the Special Middle East Coordinator. We spoke with him largely about how the project might assist negotiations with Syria—and later, in March, I wrote him a short memorandum on that subject.

I also met with Munther Haddadin while in Washington. He was, as usual, pleasant but totally noncommittal. He had not had time to read the entire report, but said, "You reached essentially the same conclusions."

On January 19, on our way to the Middle East, we traveled by way of Montreal, where I gave a talk to the board of the CRB Foundation about the water project. I had forgotten that Janet Aviad, a leader of Peace Now in Israel, worked for the foundation. There was a lot of interest in the project, but the time was limited, and I don't think I did a terrific job.

Meetings with Palestinians

We arrived in Israel on January 20 and began a very intensive week the next day, when I had an all-day meeting with an expanded team of Palestinians at the Palestine Consultancy Group in East Jerusalem. Marwan Haddad chaired the meeting. I was happy to discover Jad Isaac was there and was again part of the team. Indeed, it appeared he and Marwan were to be co-directors. There were also a lot of new faces at the meeting, including a hydrologist, a young economist named Nasser, and an official involved with water, Karen Assaf, the American or Canadian wife of a Palestinian who was to become very important in the project. It was an extremely pleasant meeting.

Many of those present had not had time to read much of the paper, so I made a fairly full presentation. There were lots of questions and excitement. I had come intending to get them to sign on to a report a couple of months down the road, but it was apparent that they very much wanted to exercise the model, to deal with it and with the Palestinian country report in a number of ways to satisfy themselves that this was what they wanted to do.

I could have no objection; indeed, I welcomed it—if only they would do it, and soon. I was encouraged by the fact that they were anxious to have Atif Kubursi and Harshadeep visit immediately and provide a week of intensive training on use of the model and on the economics involved. It would not happen before April, though, in part because of the difficulty putting together the schedules of all the Palestinians involved.

There was some concern over the novelty of what was being proposed and a wish to compare what we were doing with the possible use of other models. When Marwan asked how it was that this procedure was being tried out on them for the first time, I said it was because they were lucky. My answer amused the group, but then they asked why I didn't work on the water difficulties between California and Arizona rather than those in the Middle East. I got serious and explained that constructing models like ours requires much effort and time and a fairly large sum of money (at least before we learned how to create such models). Besides, I said, I didn't care as desperately about what happens between California and Arizona.

They got my obvious and sincerely made point and appreciated it.[*]

The Palestinians who had arranged my schedule had not wanted to hold a joint meeting with the Israelis, saying they wanted a chance to meet with me and have me meet with members of their team I had not met—which I understood after our all-day meeting. I suggested a meeting with the Israelis later in the week, to which they agreed.

Later that day, I met privately with Jad. He suggested we phrase the

---

[*] I told this story many years later when a group at Arizona State University became interested in our work and I visited them with Annette Huber-Lee. "I'm here now," I said, as part of my talk on our Middle East project.

notion of trade and water in terms other than one country selling water to another. He proposed that property rights in water should never be sold, but rather that owners could give leases allowing their water to be used for some period of time, with the model determining the prices. This seemed quite a good idea, one that would be politically more palatable than the notion that the inalienable rights to water were to be sold. It was reminiscent of the creativity in the Israel-Jordan Peace Treaty: Jordan does not sell land but, in effect, leases it to Israel or Israeli farmers. I promised to suggest Jad's idea the next day to Mohammad Nashishibi, then the Palestinian National Authority's Finance Minister.

That night, the Shuvals had Jad and his wife and Ellen and me over for a very pleasant dinner. When discussion turned to politics, however, Jad was again pessimistic, saying he thought the present version of the peace process would not get much further. Arafat had achieved the right to appoint the mayor of Gaza, he noted, but that it didn't look as though much more would happen.

I told Jad he was always pessimistic.

The next day, January 22, Issa Khatar and I were driven from Jerusalem to Gaza for a meeting with Minister Nashishibi. Issa was the director of the Palestinian Consultancy Group. Our driver, Rafi, was an extremely helpful cab driver that ISEPME used. Nashishibi's driver was to meet us at the checkpoint at the north end of the Gaza Strip.

I thought the check point would be quite a small affair, but it was enormous and chaotic, with a number of lines of traffic going in and out. There were Israeli installations all over the place, and then smaller Palestinian installations on the other side of the borderline. We first drove through and looked around and then attempted to park in front of an Israeli pillbox, which made the officer extremely angry. He chased us away to the side.

No one checked our passports or papers. Eventually, Issa asked the Palestinian police and we were able to find Nashishibi's driver, and we were driven into Gaza. The outskirts were as bad as I had been led to believe: buildings in ruins; garbage everywhere; the smell of sewage. The principal street in Gaza along which we drove, however, did not seem bad at all—certainly no worse than places in Morocco or Damascus. When I asked Issa, he told me it was by far the best of Gaza's streets.

We arrived at the PLO government compound—which was the right word to describe it: a bunch of low, relatively impermanent buildings that looked a bit more like a rundown army base than anything else.

The meeting with the Minister had been arranged the previous November, when I spoke with Nashishibi at the ISEPME board meeting he attended and he invited me to come.

I spent more than two interesting hours with Nashishibi. He was often interrupted (once by the arrival of the Mufti of Jerusalem), but plainly wanted to give us his undivided attention. He had read the report and was very interested in the project. He had gotten the point about water not being worth war, and he wanted to discuss project evaluation.

Issa and I left Nashishibi's office floating on air, and I thought the meeting could not have gone better. It was apparent that he saw the model as very useful for the Palestinian government, at least in its own water policy. We had no definite commitment from him, and we had asked for none. Still, Issa—normally very cautious—told me we should ask for an hour with Arafat.

Political events brought us down to earth.

Back on the street, Issa commented that there seemed to be very few people about. He wondered why. The street seemed very crowded to me, I noted, but I guess it wasn't the usual crowd.

While we had been meeting with Nashishibi, there had been a bombing. Two Palestinian suicide bombers had attacked an important junction in Beit Lid, a strategic crossroads between Tel Aviv and Haifa, killing 20 Israeli soldiers and one civilian. It may have explained the "emptiness" of the street, but we still did not know about it.

We returned early to the checkpoint and waited for quite a long time at the spot beside the Israeli pillbox, where we were to meet Rafi to take me to a New Israel Fund meeting in Tiberias, and we needed another taxi for Issa. There was an immense traffic jam coming south into Gaza that stretched as far as the eye could see. Evidently, the Israelis were checking papers very carefully.

I asked Issa why that should be, particularly in the middle of the day. He had no idea. We walked across the border to a very rundown Israeli roadside restaurant and phoned Rafi several times, but never reached him. Eventually, we walked back to the pillbox, again crossing the border. At no time were we asked for papers. We even saw Nashishibi's car go by, taking him to Amman, and waved at each other.

It was only when Rafi finally got to us that we learned the reason—the bombing.

Rafi drove us to a much better restaurant and gas station on the Israeli side, where the other taxi met us. Issa told me that he thought he should get away as quickly as possible, before he "got into trouble." I realized later that what he meant was the possibility—one that quickly became a reality—that the Israelis would close the borders of the occupied territories and he would have problems getting home.

Rafi drove me to Tiberias with no further incident. Ellen met me there.

I was very, very keyed up over the water project. As we came over the hill from which you can see the Sea of Galilee,* my first thought was not how beautiful it was but that it represented 580 million cubic meters of water taken for the national water carrier somewhat to the north.

That night, after the finance committee meeting of the New Israel Fund, I gave a talk about the water project. The next afternoon—Monday, January 23rd—Ellen and I returned by taxi to Jerusalem. Despite our slight concern that the way though the West Bank would be closed, we traveled without incident.

Meetings with Israelis

The next day, I went to Tel Aviv University to meet with Zvi Eckstein, Gideon Fishelson, and Yuval Nachtom. Anni Karasik came, as did Hillel Shuval, along with Dan Bitan from the Truman Institute, who was joined by an assistant.

The first part of our meeting was taken up with personal conversation and criticism of the report, some of which had to do with details. For example, a pipeline for recycled water we had said would be available in 2010 had already gone into operation. There were, though, three more important topics that arose. The first was that Hillel, especially, emphasized the need

---

* People are very sensitive about nomenclature. At the Saturday meeting, a Palestinian had objected to a *New York Times* map I was using for illustrative purposes that had Jerusalem marked as part of Israel. But that did not lead to a serious problem. Another example: I settled on "Sea of Galilee"—the Arabs want to call it "Lake Tiberias" and my own tendency is to use the Israeli term "Kinneret"—despite that no one in the region uses that name.

for a clearer and more extended discussion in the report. He wanted a glossary, material on the role of shadow values, and some clarification of what was meant by "value." The latter point had already come up at the meeting with the Palestinians. It is so easy for economists to fall into a trap of assuming that everyone thinks about things as they do, uses the same vocabulary, and has done so forever.

The second important topic concerned something that had come up during my meeting with Nashishibi. We had discussed the Red Sea-Dead Sea Canal, and I had made the point that desalinated water from the canal would be delivered near the Dead Sea, from where it would have to be pumped (at some expense), because it was needed elsewhere. Nashishibi had said that that there would be a large demand for desalinated water south of the Dead Sea in the northern part of the Arava, because of the considerable expansion of farming expected there.

I suddenly realized the importance of what had seemed only a detail in the model output. Every model run had predicted desalination (even at $1.50 per cubic meter) in the Wadi Arava district of Jordan. Since that district, as drawn by the Jordanians, had only one spot on a seacoast, I had interpreted that in the report as desalination at Aqaba, and thought it was about more water for the town and its immediately surrounding area. I did wonder why the same result was not true at Eilat, and what would happen if the two were interconnected across the border, but simply put that aside—ignoring that one ought never to put aside results one does not fully understand, especially in a complicated model. In fact, the Jordanian team had drawn the Wadi Arava district as very large, extending from the Dead Sea in the north all the way down to Aqaba. Since the model assumed zero conveyance costs within districts, the desalination at Aqaba would be used to supply water for agricultural demand in the northern Arava, very far away. The misleading finding that this was economical then suggested it would pay to desalinate water and bring it to the northern Arava if it could be done at less than $1.50 per cubic meter, since this was then the highest cost used for desalination. That, in turn, suggested that desalinating water with the Red Sea-Dead Sea Canal might, in fact, be a good idea. It did not turn out that way.

I suggested to the Israelis that since it would be easiest if the Jordanians would divide the Wadi Arava region north and south, they should similarly not include the section near Eilat, either in demand or supply, but only the northern Negev. They had done this already because the south Negev was entirely cut off from the rest of the water system. It was an example of miscommunication that can arise when work gets quite decentralized in a big project, even among people who fully understand the issues.

The Israelis agreed to divide the Negev and model the south.

The third important point was that Gideon seemed to have found something wrong with the amount of recycled water in Gaza. The model predicted that by 2020, more than 100 million cubic meters of water would be recycled in Gaza and sold to the Negev—far too much. There were two things wrong with that prediction: one had to do with the lack of a sewer system; the other was that the predictions of urban water consumption per capita in Palestine in 2010 and 2020 were unreasonably large. A calculation taking predicted prices and predicted water consumption suggested Gazans would be spending 5 percent of their income on water—far more than the generally expected 1 percent.

Of course, that problem arose from the Palestinian country report and Jad's prediction regarding per capita consumption coming up to Israeli levels. (Jordanian predictions were more moderate.)

The Israelis and I discussed this at some length. We agreed the Palestinians would have to correct this, but it would be very delicate. The Palestinians might hold as a matter of principle that their water consumption would be equal to that of the Israelis, make it a part of their negotiating position, and insist that, hence, their per capita entitlement should be the same as that of Israelis.

We agreed this would have to come up and be handled delicately at the joint Israeli-Palestinian meeting scheduled for the coming Friday. It certainly could not be presented as an Israeli criticism. I agreed to bring it up as one of several things I had discovered that week in the model that needed attention.

The meeting then turned to raising funds and related topics, and Zvi Eckstein left. He had already said he wouldn't be able to come to the Friday joint meeting, and I learned later that he he told others he was quite annoyed and was "finished" with the project. It was an unfortunate turn of events, but Zvi was no longer central to the project. He had been very important in getting it started in the autumn of 1993, but after he left the Boston area in December of that year, he had inevitably become considerably more peripheral—partly by his own making.

Perhaps more important, Zvi's original intellectual contribution had disappeared from the project. He had a really good idea of solving the model by using the fact that, at an optimum, price must be such as to make supply equal to demand in every district, and, when water flows from district A to district B, the price in district B must be equal to that in district A plus transport costs. That was more or less the case for the Israeli system—and one could solve the model by trial and error, choosing a single price in the north and letting all the other prices adjust by adding conveyance costs. But it was a very inflexible method for dealing with different conveyance distances and uncertainty about the direction of water flow. It was also cumbersome when dealing with recycling systems. Introducing GAMS proved a godsend in this regard, but also meant Zvi's particular insight was no longer playing a role.

This produced a continuing problem for me. The ultimate products of this research would surely bear names other than my own, including Zvi. But I was now responsible for the main effort, and I was doing all the writing. To make matters worse, it was politically impossible for me to put Zvi's name on central reports without including a Palestinian. I presumed this issue would go away in the fullness of time, when everyone was ready to sign off, but meanwhile I kept producing reports (with suitable disclaimers) over my name—which did not help relations with Zvi (who, of course, would not mention it to me directly.) Zvi also felt he was being shut out of our communication with the Israeli team, which tended to go to Gideon, who was the expert on water and agriculture.

Zvi was also miffed because we had prevented him from speaking directly with Jad after the incident in the spring of 1994, following the Hebron massacre, when Jad discovered the Israelis had commented on his report and a draft of their report proved quite offensive to him. There also appeared to have been an administrative foul-up in which Zvi was not kept in the loop regarding the joint management study that had been proposed between the Truman Institute, the PCG, and ISEPME.

Late that afternoon, after the much-annoyed Zvi had left the meeting, I sought him out in his office for a frank discussion. I believe I succeeded in calming things down, but could not be sure what was going to happen in the future.

After Zvi's departure, the meeting turned to a discussion of proposals for the joint management section of the project. Brooks had promised as early as the Cyprus conference to provide money for Palestinian and possibly Jordanian participation. A proposal had been drafted for him. Unfortunately, he now appeared not to have funds.

A potentially more influential funder had appeared, however. The previous spring, the German delegation to the multilaterals had announced a desire to fund supply-and-demand studies of water in the region. We attempted to communicate with them to point out that we had already done that work and they ought to join with us. We got back some polite letters during the summer of 1994 saying they weren't yet ready to talk to us but were quite interested. That seemed to be the end of it, but now it appeared the Germans were getting ready to fund. Indeed, they were currently in the region receiving proposals and talking to people.

So much for our system of international communication!

The Germans had appointed a team from a consulting company, GTZ, to help them with the decision. Dan Bitan had spoken with them, as had Issa Khatar, but the Germans were very noncommittal. Anni had attempted to get a meeting with them to no avail, with no response to her calls or faxes.

When our meeting broke up, we had decided more needed to be done to reach the Germans.

The next morning, the Germans finally got in touch with Anni and agreed to a meeting that afternoon. The Israelis had closed the West Bank borders; unable to get Palestinians to Jerusalem, Anni, Issa, Jad, and I all went to Bethlehem's town hall. Two Germans came, both resident advisors for matters relating to water, one for the Bethlehem district and the other for Ramallah.

It was a wonderful meeting. I made a full presentation and joined in a dialogue with Jad, who knew the players. We conveyed the importance of what we were doing, right down to the possibility, as Jad noted, of using these techniques to improve the city system of Bethlehem. The Germans were obviously very impressed, but they were also anxious for political cover. They were told they could get a letter from Arafat designating the Palestine Consulting Group the official people to work with.

The Germans were very happy to hear this. If it could be done, they said, they would be delighted to work with us. When the German team more directly responsible for such decisions came to the region in early February, they suggested, matters would be decided.

We left feeling something really important had been accomplished. German support would not only mean considerable financial relief, but would also provide us with a kind of official *entrée* into the multilaterals. The four of us discussed the possibility of getting additional money in the form of joint funding both from the Truman Institute and through the Jordanians. Issa and Jad again asserted that it would be an easy matter to include Riyad El-Khoudary, president of Gaza University and chief Palestinian water negotiator, and get a letter from Arafat.

Meetings with Jordanians

Early on Thursday, January 26, I went to Amman for the day and returned in the early afternoon. I still find it quite remarkable that one can do that. Still, it was not a journey without obstacles—something I had already heard it could be. While I was still in the United States, I learned that the Allenby Bridge is the only entrance into Jordan for which one needed a visa in advance. This caused quite a flurry, but the Jordanian embassy was very cooperative and produced a visa for me in about five minutes when I showed up there before seeing Dennis Ross. But there were other nuisances to come.

Rafi picked me up at 7:00 a.m. in Jerusalem, and we drove to the Allenby Bridge, scheduled to open at 8:00 a.m. and close at 3:00 p.m. The people who work at the bridge didn't show up on time, though, and we weren't allowed to pass through until about 8:10 a.m. I was one of the first people into the processing center. The processing was easy enough—although there was an immense Israeli exit tax—but then it developed that I had to wait for the shuttle bus to come through from Jordan.

The year before, Lenny, Ellen, Theo, and I had crossed the Allenby Bridge from Jordan into Israel in a much simpler way: we were driven by the Jordanians to the bridge, presented our papers, walked across, and waited for a taxi to take us to the processing center. It seemed the arrival of peace has made things more complicated. Not only did I have to wait for the shuttle bus, but once it arrived, I had to wait for enough passengers to fill it up. Finally, at roughly 9:00 a.m., the bus left for the bridge. We crossed, but still had to have our passports checked by the Jordanians at the border. Even that, though, was not the end. The Jordanians, emulating the Israelis, had set up a major processing center some considerable distance from the bridge. There I arrived, and was quite relieved to see Iyud Abu-Moghli waiting for me with a car. Not finding me, he had been about to seek special permission to drive to the bridge. In any event, we were soon on our way to Amman, arriving around 10:00 a.m.

My meeting with Abu Moghli, Maher Abu-Taleb, and Elias Salameh lasted for two or three hours. I think it is fair to say that the first thing that came across was their sense of excitement on reading the draft report. Maher said he now really understood what the project was about. He and Salameh talked about taking it to their ministers so they could see the usefulness of the project for planning purposes.

I found the same sense of excitement everywhere on the trip.

On substantive matters, the Jordanians confirmed that there were actual plans to build pipelines to bring water to Amman, and promised to provide a list of projects being either contemplated or designed. They agreed with Hillel, who had suggested I had been much too sharp in the report when pointing out their necessity—something that must have been well known to the Jordanian government.

When I brought up the question of division of the Wadi Arava district, the Jordanians found it hard to see a problem. They agreed, however, to divide the district as requested. I had the feeling they didn't really understand what was involved, so I was not altogether surprised a few weeks later, after I had returned home, to receive a fax from Abu-Taleb saying that he had reviewed the matter, and they saw no reason to divide the district since they had accounted for all the water sources. I received no reply to my return fax pointing out that the reason for subdivision was a matter of demand, not supply.

The meeting proceeded. I inquired about the confusion I thought existed regarding the waters of the Yarmouk. In the runs described in the draft report, we had assumed 250 MCMs of Yarmouk water flowing down from Syria. We had Jordan taking 100 MCMs per year at the King Abdullah Canal, the capacity limit the Jordanian team had given us; the Israelis taking a small amount more at Bechan; and the rest effectively being turned down the Jordan Valley, becoming available again to the Jordanians in the Jordan Valley at a higher cost.

The cost matter here was of some importance. We had been using about

8 cents per cubic meter as the cost of conveying water in the Jordan River bed, because Hillel had told us something would have to be done to clean up the river bed or to provide a pipeline through which the water could run. Of course, there would be no pumping costs, and since we were not running the model with capital costs, the capital cost of the cleanup or the pipeline should not have been included. Hence, we ought to have run the model with zero transport costs. The Palestinians and, later, Shaul Arlosoroff, were quite definite about this.

In any event, it was not clear to me where the Yarmouk water actually went. In Elias's view, the Jordanians took up to 100 MCMs per year and might be able to take additional amounts later depending on whether various projects were built. The Israelis took a limited quantity and transported it to the Sea of Galilee. Elias claimed, however, that the banks of the Yarmouk have large ditches and boulders, and the Israelis were taking the rest of the water, hiding it in the ditches, and transporting it to the Sea of Galilee.

This was, to put it mildly, not the Israeli position. Arlosoroff said the Israelis did no such thing, and I was inclined to believe him—despite that statistics for disputed areas often had a high level of uncertainty. The remainder of the water came down in the winter rainy season and simply flowed into the Dead Sea, unusable because there were no storage facilities.

I asked Elias whether the annual capacity limit of 100 MCMs on the King Abdullah Canal represented the amount the Jordanians *actually* took or the genuine capacity *limit*. It was the amount they actually took, he said. When I inquired how much the canal could, in fact, handle, he gave an annual figure far in excess of the Yarmouk's flow.

Again, we seemed to have misunderstood each other. The seasonal nature of the flow made the calculation useless.

The remainder of the meeting was devoted to a discussion of the German funding issue and the on-going problems with Munther Haddadin. Our Jordanian team was quite anxious to continue working. They had contacts with Germans in Jordan and planned to speak to them about the project. They had no objection to including Munther, but didn't want to be squeezed out in the cold. They pointed out that Munther—who was, of course, Jordan's official water person in the multilateral negotiations—had already made a proposal from his consulting company to the Germans.

Actually, there was considerable danger that the Germans would, for political reasons, end up contracting with Haddadin, El-Khoudary, and and the Tahal firm (in Israel), and receive from them merely rehashes of our existing reports. If that happened, we were sure that the German effort would come to nothing.

The meeting ended amicably, as these meetings always seemed to do. Communications difficulties with the Germans seemed to be a function of using faxes and letters, and never surfaced in person.

I was driven back to the bridge and arrived in Jerusalem about 2:30 p.m.

A Successful Week

Late that afternoon, I had a long discussion with Yehuda Bachmat about the model. He had made a serious and successful effort to understand what was going on in the project and had a number of quite sensible questions. Despite having had only one phone call with the hydrologist appointed by the Palestinian team to work with him, he told me he had essentially finished his work. He said that while he had had access to raw data on wells, he was unable to pass those data on even to me except in processed form. This was likely to make it difficult for the Palestinians to participate fully. Still, he promised to provide his model of the Mountain Aquifer on a diskette within a week, along with a write-up.

We had foreseen this problem as far back as the Cyprus meeting. I told Marwan Haddad then that if access to data became a problem, I could not promise to solve it, but I promised to bang on the table and make strong representations to Israeli officials. In mid-February, it became apparent that this would be a problem, with the Palestinians suggesting that there was no point in their working with Bachmat unless it was solved. So, I wrote to Gidon Tzur, the Israeli water commissioner, reminding him of his promise the year before to help—and requesting assistance. I received no answer.

Many weeks later, we had still not received the promised diskette and write-up from Yehuda Bachmat.

On Friday, January 27, we held a joint meeting of Israelis and Palestinians. For a time, it looked as though the meeting might not take place at all, since an Israeli closing of the borders made it difficult for Palestinians to travel. We had repeatedly invited the Jordanians, but they insisted it was still very difficult for them to leave the country and so could not attend. Certainly, we were unable to hold it at the PCG offices in East Jerusalem as planned, and in the end, we had to meet at Jad's institute in Bethlehem. Certain Palestinians, notably Marwan Haddad, were unable to get there.

Despite these difficult circumstances, it was a very upbeat meeting. The Palestinians had already attempted to run the model. Perhaps in part because of electricity failure, they had failed to get it to converge with one of the standard scenarios. During a break, they had me watch while they tried to run it again—and, and after some considerable time, the model did in fact converge (they were using a 486 rather than a Pentium processor). I was glad of this, knowing it would always be best for the project if the Palestinians were secure in their knowledge and felt they were being treated fairly and equally.

The issue of technical support for the model was a recurring one, and people in the region were anxious for Harshadeep to come and assist in making the model more user friendly—something we consistently tried to oblige.

The meeting's substantive matters concerned the problems I had encountered during the week. I began with the Wadi Arava-Elat issues and then tackled the issue of per-capita water consumption in later years, especially in Gaza. I tried to show considerable sensitivity to what I thought might be the Palestinians negotiating position, which—as it turned out—seemed not to have been necessary. All the Palestinians agreed there was a serious problem and that the per-capita estimates were out of line. Jad promised to review them, and everyone agreed that the model had to be realistic to be taken seriously. It was the very point the Israelis had made to me the previous Tuesday.

The meeting then turned to how to proceed with the Germans. Jad, Issa, and I reported on our meeting; Issa and I also reported on our very successful meeting with Nashishibi. There was a lot of discussion about what to do about Jordan. We agreed the best thing would be to try to get the Jordanians (our team plus Munther) to form a consortium, and that we would ask Lenny Hausman to try doing something about that. There was even hope that we might get support for all three countries. Certainly, that was what the Germans ought to have welcomed.

The meeting broke up a little before lunch time, but not before I had a moment alone with Jad. We agreed that given the circumstances of the Beit Lid killings earlier in the week, his pessimism that I had noted at the dinner with the Shuvals had not been overstated.

In terms of the water project, it had been a tremendously successful week—despite the Beit Lid bombing and what was happening to the peace process. Nevertheless, I had the feeling that we were on our way to something of historic importance. Others shared my optimism.

It was Friday afternoon. As always when in Jerusalem, Ellen and I walked to the Western Wall. Over the past 15 months, I had taken to praying for strength in assisting the peace process through the project. This time, when I touched the Wall, my hands and arms tingled.

7

Moving Ahead

I wish I could say that the feeling of optimism I brought home from the January 1995 trip continued, but with one big exception, I cannot.

Jad wrote a proposal to the Germans, but because of the closing of the territories, no one from the Palestinian Consulting Group felt able to go to Gaza to talk to El-Khoudary—and for some reason they chose not to use the phone, so El-Khoudary was not, in fact, really included. Further, the letter from Arafat that Jad and Issa had seemed so confident about obtaining was not forthcoming. Sari Nusseibeh, the head of PCG, was asked to speak to Arafat about this, but felt he could not use his political capital when the amount of money was so small, not more than $200,000.00.

So, as of mid-March, it still seemed possible the Palestinians would succeed in arranging a consortium, and the German proposal was still alive, but it was by no means a certainty.

On the Jordanian front, a similar effort appeared to have gone nowhere. Lenny spoke to Jawad Anani, the former Jordanian Minister of Information, and asked him to try to form a consortium. Apparently Anani had enough clout to do so, but as of the end of February he would not discuss any results on the phone, promising (at least as I gathered) to discuss it when he attended ISEPME's business conference at the beginning of April.

Then there was the issue of the Germans and the Israelis. I saw Shaul Arlosoroff on February 13 and learned that Dan Bitan had met again with the Germans, but that he had not been particularly happy. Apparently, Israeli officials were telling the Germans that they must deal with Tahal Consulting Company and Avishai Braverman. The Germans were aware that if that was all they did, they might simply again get the study already done for the World Bank. We reminded them that it might make sense to cooperate with the Truman Institute, but as of mid-March, nothing had happened on that front.

Indeed, except for my talk with Shaul, nothing happened in early 1995. In part, that was due to the inevitable slowdown when communications are long-range. In part, it may have been due to the fact that I had expended so much energy in writing the draft report, and then on the January trip that I was unable to bring myself back to pushing full-out on the project.

A Postscript to the January Trip

My talk with Shaul, though, did produce some good news. It was, in effect, a postscript to the trip.

Shaul had not been available in Israel when I was there because he was recovering from surgery in Philadelphia. When I saw him on February 13, he seemed fine. He was then spending a long recovery period at the home of some Philadelphia friends—more accurately, a mansion that had, among other notable things, an early Monet painting. We spent the day together.

It was an extremely good meeting that demonstrated, as I saw it, a change in attitude on Shaul's part. At the Cyprus meeting, he had been friendly but rather skeptical. At that point in time, he had thought that the project was just an academic exercise. After reading the draft report, however, he decided there was something important going on. This mattered not only because Shaul was by far one of the most knowledgeable Israelis about water, but also because he had become the head of an eight-person committee (that included Gideon Fishelson) charged with reporting to the Knesset and various ministers on Israeli water policy. He wanted me to submit a revised version of the draft report to his committee the following summer as an official document and appear before the Knesset committee to testify.

Shaul was a veritable goldmine of information and constructive (usually) criticism. Most of our discussion was about truly substantive matters, but he was also concerned about phraseology, since he wanted the document to be taken seriously in Israel. He was speaking on these questions from the point of view of someone who would want to find reasons to trash the document, he told me, and he was adamant that I had to use the words "Palestinian entity" or "Palestinian autonomy" rather than "Palestine." We also discussed the use of the phrase "water in dispute," which he seemed to think meant water that was up for grabs as opposed to water over which there were conflicting claims. From an Israeli point of view, of course, relatively little of the water is "in dispute" in the first sense, and I had to be clear that I was using the term in the second sense.

We discussed other, more substantive topics. Shaul offered a quite convincing view that the report's emphasis on recycling water in Gaza and shipping it to the Negev was unrealistic. He pointed out that the model was charging only 10 cents per cubic meter for recycled water based on a wastewater treatment plant already in place. But there was no sewer system in Gaza, and the capital costs involved in building a sewer system and such a plant would be quite substantial. Those facilities, he insisted, would not be built in time to have an add-on recycled water plant help very much.

Moreover, the Negev was already supplied by an existing recycling system (in the model) from Tel Aviv. Plans were well underway for an "eastern conveyor" to bring recycled water down from the north. There would be plenty of recycled water available in the Negev, and, said Shaul, it would make more sense for the Gazan aquifer to be recharged with such recycled water than to have the Gazans recycle and sell water to Israel.

As for pipelines to Amman, Shaul said there were pipelines and pumping stations already under design and being let out for contract. They would bring water either from the Golan or presumably from the Yarmouk or Jordan River bed in cooperation with Israel, as in the peace treaty. According to Shaul, the water not taken from the Yarmouk by Jordan, or taken in relatively small amounts by Israel, was winter runoff going down to the Dead Sea. Storage facilities (dams) would make that winter runoff usable.

Further, Shaul said that Elias Salameh's calculation of the capacity of the King Abdullah Canal was quite wrong. In fact, the canal would be capable of taking the entire daily *average* flow of the Yarmouk, but not its heavy winter flow. Moreover, Shaul could not understand why the Jordanians were saying it costs 12 cents per cubic meter to take water in the canal when it has only gravity flow. It was something that required investigation, and illustrated Shaul's strong belief that the model had to be made seasonal, so it could adequately handle winter and summer flow problems. Later, I discussed that question with Bob Dorfman, and we concluded that it would be fairly

90

straightforward—although quite a bit of modeling—to make that adaptation.

Shaul also explained to me something he said I had gotten completely wrong. In meeting with Dennis Ross in January, I had suggested that a simple calculation showed water from Turkey would not be economical. That calculation assumed 8 cents per cubic meter per kilometer pumping costs, observed the distance involved, and established that the Turks would have to charge essentially zero for the water and zero capital costs for such a project to be worth doing, given the shadow values in the north of Israel that the model was producing.

Shaul believed there would be no pumping costs—it would all work by gravity—and that extensive new pipelines would not be necessary. The water would be brought in existing riverbeds basically into the Litani, with the Litani then connected to the Jordan at the point where they are only 10 kilometers apart. Of course, the question of bringing water from the Litani alone was also very important, not to mention that it all assumed close cooperation with Lebanon, which seemed very unlikely.

When I explained to Shaul what I had discovered about the model's continual call for desalination at Aqaba and the possibility that this meant the Red-Dead Canal would be useful, he was quite skeptical. In his view, there were brackish water sources at the north end of Wadi Arava that could be used to supply agricultural demand at a much lower cost. More generally, he thought we needed to investigate desalinating brackish water. Brackish sources could simply be included as sweet water sources at an appropriate desalination cost.

The brackish water problem, though, was actually more complicated. It was linked to the issue of modeling effects over years, a subject Shaul had emphasized at the Cyprus conference and that would not be easy resolve. He had in mind things such as deliberately overpumping the Mountain Aquifer in some years, making some of the water brackish, desalinating the brackish water, and replenishing the aquifer later. At one level, the existing model could already handle such policies. To the extent that a dynamic model of what happens year to year was required, it was not clear that enough was known about the hydrology of the aquifer (or the other water sources) to enable this to be done.

Of course, a really important use for a model of such inter-year effects would be the inclusion of decision under uncertainty, given the stochastic nature of rainfall.

By the time our discussion ended, Shaul and I had talked through about half the draft report. He left with me his copy, annotated with many, many useful and detailed comments.

Six Quieter Months

Very little happened with the project in the weeks right after my conversation with Shaul. In March, though, I heard again from Julia Neuberger, who told me I hadn't quite gotten her earlier communication about the Crown Prince and the treaty right. It wasn't that we were important in the Jordan-Israel treaty itself, but that we were important in the discussions surrounding the treaty. That made perfect sense, and matched exactly what Shaul had suggested when I told him about the Crown Prince's statement: that we allowed those higher up than the water negotiators to take the view that water is not all that valuable and that water issues could be settled in the context of a larger agreement. I do not know whether we affected the Israelis as well as the Jordanians on this, but I finally understood why the Crown Prince would have said that the water project had been very influential in permitting Jordan to sign the water parts of the treaty. This did not exactly make me unhappy!

When I called Lenny to tell him about this, he told me that he had attended a meeting of the Harvard Seminar on Negotiation that was addressed by a Norwegian diplomat who had been fairly high up at the time of Oslo discussions and risen higher since. The diplomat began by saying the next war in the Middle East would be about water. He was quite surprised when seminar participants told him about our project and he became even more interested when he learned I would be speaking in Stockholm. We sent him a copy of the paper. Apparently, the Norwegians had a big project on water in the Middle East about which we had never heard.

During this same period, my economist colleague Tom Schelling suggested my name to the National Research Council for a committee on new sources of water in the Middle East.

Things were largely quiet for most of the next six months, but things still happened—mostly toward the end of that period. Much of my project time was concerned in one way or another with cooperation on related projects or with funding, which had become a major concern. ISEPME was in deficit, and money for the project was very tight. While we had been turned down for a grant from the CRB Foundation, there were several other funding possibilities on the horizon.

When I mentioned to several friends the need for funding, two of them—both lawyers—offered to try. One sent my papers to foundations to which her firm was connected, but I heard nothing in reply. The other, a close friend of more than twenty years Robert Rifkind, turned out to be a trustee of the Revson Foundation. While he said he had no influence whatsoever there, he turned out to be very helpful.

When Lenny Hausman spoke with him later, Rifkind said that the eminent Harvard economist Henry Rosovsky had been kind enough to suggest that this work was Nobel Peace Prize material. Rifkind then sent materials to Eli Evans, the Revson Foundation's executive director, and in the late spring Lenny met with Evans, who seemed quite interested in supporting us, particularly our efforts to have a policy impact. But we never saw a penny from the Revson Foundation; the foundation was involved in something else.

It turned out Revson was sponsoring the National Resource Council's joint project on new water sources in the Middle East—that was the committee Tom Schelling had suggested me for—and Evans asked me and Lenny to meet with Alex Keynan, a biologist who was highly placed in the Israeli Academy of Sciences. That meeting happened in early July in Woods Hole, Massachusetts, where Keynan spends each summer visiting the Oceanographic Institution.

Keynan was an older man. We had many friends in common. He had been involved for some time in trying to organize scientific cooperation between Arabs and Israelis, which he had found worked best under an outside cloak—in this case, the American National Academy. The academies of the United States, Israel, Jordan (the Royal Scientific Society), and a Palestinian institution had agreed to have a joint project on the optimization of water resources in the region. Nothing had yet been done. Instead, there was to be a meeting in Amman in September to decide who would be on the panel.

Optimization of water resources clearly should have meant us. The problem was whether the people involved would realize that it is what economists do and what the project was about. Keynan was obviously well disposed to that, but word came back that the academies were interested only in "scientific" activity and not at all in conflict resolution. Still, that also still should have meant us.

I saw the list of people nominated to serve on panel: "primary" and "secondary" candidates. I was in the former group, and was the only one in "environmental economics." I thought that perhaps I stood a good chance of being named. Elias Salameh was listed for Jordan. Uri Shamir was reported to be one of the people who would help decide the panel's membership, and he was quite amenable toward us (while saying that the Americans would surely choose their own delegates). It remained to be seen what would happen (ultimately, I was not selected for the group).

Shamir was also involved in another group that had shown interest in the project. The Rothschilds had established Yad HaNadiv, a foundation in Jerusalem that gave money, generally anonymously, to support a good many things, some of which I benefited from, such as the Jerusalem Guest House in Yemin Moshe where Ellen and I often stayed during academic visits to Jerusalem. They also had a Music Center in Yemin Moshe. Yad HaNadiv had also set up a water research institute at the Technion, headed by Shamir. The Rothschilds had heard about ISEPME in general and the water project in particular, and later that July, Ariel Weiss, the foundation's executive director, came to Cambridge and spent the morning finding out more about us. We were to submit a proposal.

Another potential funding opportunity came in the spring when, while traveling, Hillel Shuval met some French diplomats (I think) who told him the French were quite eager to sponsor a major international project on water in the region. Hillel told them about us, and they were very interested. We took out the joint proposal of the Truman Institute, Palestine Consultancy Group, and ISEPME that had been prepared for David Brooks's Canadian operation (since out of money), rewrote it fairly thoroughly, and sent it to the French. On Hillel's advice, we asked for $2.5 million. The focus was joint management with the help of the model.

As of September 1995, the proposal was still very much alive. Indeed, we knew it had been passed on favorably by the committee and kicked upstairs to the decision makers. We believed there was a recommendation that we be given an interim grant of $200,000 to explore matters further. Even that would have been a big help.

In August, while traveling in Switzerland, Hillel met somebody connected with the Euro Bank who was also interested in such matters, and said the bank might be interested in a piece of it. We sent him essentially the same proposal we had sent the French.

Meanwhile, nothing ever became of the German money. The Germans did give the Palestinians a grant, but it all went to Marwan Haddad, and was not part of our project.

We also pursued funding from the Japanese. I had become friendly with Asit Biswas, who then chaired the Middle East Water Commission (and later founded the Third World Centre for Water Management, based in Mexico). We thought Biswas might help, because it was the Sasakawa Peace Foundation that was supporting his own research. He informed me, however, that the foundation would have no money for new projects in 1995 or 1996. He did offer to use his good relations with other Japanese foundations to smooth the way for us—but with a caveat. He had experienced such trouble getting ISEPME to answer his mail and faxes that he wanted any grant applications to come out of my office at MIT. Ellen suggested I tell him that while any application had to come from Harvard, I would have my office handle all the correspondence so he would not be embarrassed by a failure to answer. I also apologized to Biswas personally.

That April, we began to get word that the U.S. Agency for International Development (USAID) was going to issue requests for proposals (RFP) on resolving conflicts over water in the Middle East. We thought at first that we would bid on this jointly with the Harvard Institute for International Development (HIID). But then we began to get calls from other people asking for copies of our report and whether we might be interested in working with them.

I discussed this with Howard Raiffa, who would have liked to work with us, but was involved with a firm called Conflict Management Group (CMG) that was itself tied up with a Washington consulting firm called DAI. Howard took me over to the CMG offices. A call was placed to DAI, and a meeting was organized for later in the month at ISEPME.

It turned out that the person from DAI who was to head the team, Peter Reese, had written the paper on which the RFP turned out to be based and was absolutely insistent that he would be the team head. While he had heard about our project, Reese claimed he had been unable to get in touch with me before April—something about how MIT couldn't find me or confused me with Stanley Fischer. Nevertheless, we made an agreement that we would do the economics and have a substantial independent piece of the action. The only problem was that the preliminary description of the forthcoming RFP suggested serious interests in intra-boundary disputes, and it was agreed that our place might have to be rethought if, when the RFP officially came out, it concentrated so much on such disputes as to change our views as to what our role could be.

When the RFP was issued in June, with a return date only a few weeks later, we found ourselves first stalled and then absolutely dumped by Reese and DAI. I think Reese did this to keep us from forming an alliance with other possible contractors, leaving him and his group with a clear field.

This caused us a lot of difficulty and annoyance. We talked to at least two other groups, and eventually settled on an alliance with a third one with Management System International (MSI), led by Bert Spector. The team also included a group called Search for Common Ground. Howard also said he would work with us if we won the contract.

Our proposal, which featured the project's model (and me), was submitted in June. I thought we had little chance of winning. Then, just before Labor Day, we were notified that we were in the "best and final round," and we were asked a number of questions we had to answer quite quickly. These concerned what would happen if the parties to a dispute did not accept the model, how the model would be presented and used, and so on. I emphasized in reply that the model was not something that would be given from outside but would be built by the parties themselves. What we had to offer was a method.

As September unfolded, we didn't know whether we would get the project—either because we might lose, or because it was not at all clear that USAID would even continue to exist.* We weren't even sure we wanted to get a project with an RFP that concentrated on intra-boundary disputes and wouldn't even begin to address international disputes until about three years

---

* The previous year, U.S. Senator Jesse Helms (Rep.-N.C.), who chaired the Senate Foreign Relations Committee, had introduced legislation to abolish USAID and replace it with a grant-making foundation. While it did not become law, President Clinton was at the time making noises that he would make some changes in how USAID's work was handled, as part of his efforts to win greater cooperation from Congress for his foreign affairs agenda. So, things were a bit up in the air.

out. By then, we thought, the negotiations involving Israel would be over (which turned out not to be a particularly astute forecast). We weren't certain we wanted to spend large amounts of time on relatively small dispute areas in the region.

Other Developments

Meanwhile, the possibility that our methods might be applied to other water conflicts and other water systems was beginning to spread. Indeed, Lenny Hausman was doing his best to spread it. That May, he traveled to the Persian Gulf and other areas in the region. He pushed the Water Project in Saudi Arabia, Kuwait, Oman, and Turkey.

People Lenny met with in those countries expressed considerable interest. In July, I was invited to visit Saudi Arabia to discuss their water problems; but Saudi Arabia closed down for vacation in August, and no date was ever fixed. When the Kuwaitis visited ISEPME in August, I was away; still, there appeared to be considerable interest in having me train Kuwaitis to do our sort of thing themselves. When the Omanis expressed interest in a big conference on water problems and our project and methods the following spring, I hoped we would be ready if it happened.

During the same period, March to September 1995, very little happened in terms of model development. We were still waiting for the completion of the Palestinian report, which was supposed to contain new scenarios and new data. The Palestinians had been at work since Harshadeep and Atif Kubursi finally went there at the end of May for a productive and quite intense visit. I looked forward to reading about what they had done.

There were, though, some substantive developments on other fronts. Yehuda Bachmat delivered his model of the Mountain Aquifer early in the summer. We tried to incorporate it into the full model. There was no technical reason why we could not do this, but, inevitably, there turned out to be problems when the full model was run with the Bachmat model included. At first, the model had us pumping amounts out of the Mountain Aquifer far greater than those our other experts told us could be pumped as renewable amounts, and at costs far below what everyone said were the actual pumping costs. When we communicated with Bachmat about this, we determined we might have overlooked constraints on the total amount that can be pumped. But the constraint he imposed (330 million cubic meters per year) was well below what we believed to be the amount of water in the aquifer. We had not, however, included a separate part of the aquifer called Beer Sheva-Hebron; perhaps that was where the rest of the water could be found. We tried to clear this up.

When we did receive the Palestinian report, it looked as though the pumping amounts were cleared up. The 330 million cubic meters referred to the Western Aquifer, which is one of the three sections of the Mountain Aquifer. The low cost, though, continued to be a problem. Two parameters needed to be given from outside the Bachmat model. One was the cost of energy; the other was the average efficiency of pumps. We had taken the cost of energy the Palestinians used, which was comparable to that used by the Israelis. Bachmat suggested using $0.75-0.8 per cubic meter for the average pump efficiency. But this yielded wildly low estimates. Zvi Eckstein suggested that our estimated pumping costs might be loaded with overhead costs. I hoped not. In that case, all our estimates would involve much too high costs.

We also began to investigate some other infrastructure matters. The previous January, Shaul Arlosoroff had suggested that we might investigate a pipeline from the Jordan River to Nablus. We did so, extending connections from Nablus to other Palestinian cities as well as to the Israeli water carrier. Preliminary findings suggested that such a pipeline system would be highly desirable. Indeed, they suggested it would be an efficient way to feed the South with water coming *into* the national carrier from Nablus, not flowing out of it.

If one thinks about it, this makes sense. Suppose you wanted to design a national water carrier without regard for political boundaries. In that case, you would probably not want to pump water out of the Sea of Galilee, bring it over the hills, and then all the way down, having to pump some of it back up. Rather, you would want to let the water flow naturally down the Jordan and pump it up further south—for example, to Nablus.

This problem was complicated by the fact that pipeline pumping costs depend on the diameter of the pipeline, and hence on the flow to be handled. We tried to find out enough about this to be able to deal with optimizing the size of the pipeline. But most of the runs to do things like that would have to wait for the new Palestinian report, which looked like it would be a serious job.

As all this transpired, I was also busy preparing the paper to give at the

Stockholm Water Conference. This required first an abstract, then a relatively short paper to be included in the proceedings, and finally a proper paper for publication. I was pleased that Asit Biswas was putting together a special issue of the *International Journal of Water Resources* in which my paper, together with the others at his workshop at the Stockholm Conference, would be published.[1]

Putting together these drafts produced some revealing features. When I circulated the short proceedings paper, nearly everyone signed off on it—but not Marwan Haddad, who was one of many named in the acknowledgment footnote as having participated in the project. He asked that his name be removed. His issue was with the statement that the owner of water who uses it buys it from itself —essentially, that whatever water one uses from one's own well, rather than selling, is a kind of lost profit and thus represents an opportunity cost. He insisted that did not apply to a poor country such as Palestine.[*]

When I looked through the paper and compared it with what I had written in the report of the previous December, I found that, in the interest of brevity, I had taken out all the persuasive passages about why my statement was the right way to think about things. I promptly put those passages back in for the longer version and sent the text back to Marwan. But I received no reply, possibly because Marwan was busy in the negotiations themselves.[**]

In contrast to Marwan, the Palestinian Consultancy Group pressed strongly for me to list all the people who were working on the new Palestinian report and to list the PCG itself, as well the names of Issa Khatar, its director,

---

[*] This difficulty with Marwan reminded of an encounter much earlier when I attended a conference at the Vatican in 1963. P.C. Mahalanobis, a rather famous Indian scientist and applied statistician, gave a talk on advising the Indian government whether to import wheat, import fertilizer to make the wheat, or import machinery to make the fertilizer to make the wheat. He proudly announced that he had opted for the most roundabout method. Bob Dorfman asked him politely whether he had taken into account that the hungry people would have to be fed while the roundabout method was coming on stream. Mahalanobis, a rather difficult man, gave an angry reply that amounted essentially to this: "These people are starving! A poor country can't afford to understand opportunity costs!"

[**] The reason the rich country would not buy the water of the poor one is that the poor one would not agree to sell unless it was in its own interest to do so.

and Sari Nusseibeh, its chairman. I was very glad to do so, and glad that they continued to insist that they be mentioned in this way in press releases. We had come a long, long way.

8

The Trip to Scandinavia

It turned out the Norwegian diplomat who had spoken to the Harvard Seminar on Negotiation was Jan Egeland, number two in Norway's Foreign Ministry. He had sent me a fax at ISEPME that was mislaid for a month—something that happened too often—before Shula Gilad got it to me.

Shula had taken over from Anni Karasik as ISEPME's deputy director for policy programs. We worked together a lot. She was very energetic and very helpful, but sometimes she dithered over things that were not important and failed to act on things that were.

Fortunately, no permanent harm was done by the delay. I wrote to Egeland and apologized. He had offered to see me on my way to Sweden, which was arranged. I had sent my paper for the Stockholm Water Conference ahead to Norway.

My first meeting in Norway, in mid-August, was with Jan Trolldalen, a professor involved as a technical advisor to the multilaterals. We had a pleasant luncheon and a friendly talk. He seemed interested in our water project, and we promised to stay in touch. Unfortunately, he had not looked at my paper until the previous midnight and then read it hurriedly—which turned out to be a pattern. When I met the following morning with Egeland and an advisor, Ralph Hansen, only Hansen had read the paper. Egeland had read only the abstract. They were very polite and very friendly, but had not really done their homework. I found this a discouraging visit.

From Oslo, Ellen and I traveled to Sweden for the Stockholm Water

Symposium. Asit Biswas had arranged for me to speak on the afternoon of the first plenary session. I believed I was giving a paper of the sort no one had ever heard before and that, given the impasse in the water talks between Israel and the Palestinians, it would generate a lot of excitement. I was completely wrong. It was as though no one paid any attention whatsoever. In the question period, only Biswas asked a question addressed to me. Granted, the facts that the room was very hot, and that by the time I spoke, only 170 of the 500 participants remained in the audience, may have been a factor.

Still, it was quite dispiriting.

At least I got a bit of feedback. Professor Malin Falkenmark, who headed the symposium's scientific committee, mentioned to me that my work was very new, and (my idiom, not hers) it was like trying to drink from a fire hose. At a social event on the Tuesday evening of the symposium, hardly anyone said a word to me about my paper.

We did, nevertheless, have a perfectly wonderful time in Stockholm. The Symposium was well organized, and the Swedish hosts were very hospitable. The entertainment culminated in a banquet in the city hall at which the Stockholm Water Prize was awarded to Water Aid, a British organization assisting in underdeveloped countries. The king and queen attended, along with about a thousand of their closest friends.

Stockholm was, like all of Europe at the time, in the midst of a serious and long-lasting heat wave. No one in the 1990s expected hot Scandinavian summers, so buildings were not air-conditioned. It was a problem both in Oslo and Stockholm, but the formal banquet was a special case. Most of those present were dressed in finery, and everyone was sweating to death.

At the end of the banquet, the master of ceremonies got up and asked us all to "rise and remain standing while their majesties pass out." He added "of the room," but not before the wave of laughter overwhelmed him. He said he would speak in Swedish for the rest of the evening.

By the Wednesday of the symposium, the near silence about my paper began to change. I discovered that the more than 50 or 60 copies I had left for participants had disappeared. That same day, I was the middle of three speakers at a press conference, and when I said what I had to say about water not being worth war, at least one or two of the reporters sat bolt upright. I ended up being interviewed for Radio Sweden and featured in the symposium's press release.

Meanwhile, ISEPME had issued its own press release the previous weekend, which had been picked up by the Israeli press. and then on August 17 by the *Financial Times* of London, which seemed to have taken it directly from the Israelis. This caused a great amount of excitement.

Apparently, more people read the *Financial Times* than I had thought. One of my cousins, Ruth Sachs, was corresponding with my daughter Naomi about genealogy, and sent her a copy from London. Shula called the hotel from Cambridge and asked the front desk to see to it that I was given a copy. When I returned home the following Saturday, my secretary, Theresa Benevento, had already faxed me a copy of the article.*

The *Financial Times* led to interviews on the BBC, Radio Vienna, and Radio Australia. The Palestinian and Egyptian newspapers *Al-Quds* and *Al-Ahram* ran articles. Most important, Peter Passell interviewed me for the *New York Times*—my hometown newspaper—and wrote a column about the project, which appeared on Thursday, August 24. Requests for the paper also began to pour in—including, quite remarkably, three or four from various departments of the World Bank. The Bank's representatives at the symposium hadn't seemed to care about it at all.

The day of the *Times* appearance also brought an event that I felt tested my moral character. I had returned from Sweden and was already on the road again, in Los Angeles to testify in a damages action involving Texaco. I knew the *Times* article was being published, so when I awoke at about 5:00 a.m., still on Eastern time and even Stockholm time to a degree, I looked outside my hotel room door for it, but had to wait a bit longer. A half-hour after I read the *Times* article, CNN reported that the Israelis and Palestinians had had a breakthrough in the water negotiations. I knew should be happy about that, but there was a sense of letdown. I guess I wanted our work to be part of such an agreement and felt as if we were making no impression.

An hour later, CNN clarified that the agreement was only a stopgap to

---

* Indeed, Theresa had sent the fax on the day that the article appeared in the *Financial Times*, about the time I was reading it in Sweden. This was quite amazing. Among her other good qualities, Theresa's efficiency won her the accolade "Theresa, the Wonder Secretary" in my family—but I had not previously thought she read the *Financial Times* with her morning coffee. I was disillusioned to find out that Rudi Dornbusch, my next-door professor at MIT for whom Theresa also worked, had in fact turned the article over to Theresa.

permit the signing of the redeployment agreement (Oslo Two). Strangely, I felt a little better, although, of course I wanted a serious and effective agreement to be reached.

The Escrow Plan

Just before leaving for Oslo, Lenny had asked me a pointed question about what I would recommend be done if the parties signed on to the project's plans. In particular, what should they do in the short run? He asked this in connection with his upcoming meeting with Aaron Miller, Dennis Ross's assistant, scheduled with the State Department on Tuesday, August 15.

In Oslo, Ellen and I had discussed this and had come up with a very good idea that we then elaborated on in an exchange of faxes with Lenny and Shula. After my return, this became a paper of several pages' length.

Our proposal was to establish an escrow fund. The parties would agree that, pending a resolution of the property rights issue, they would purchase water from the fund in quantities and at prices given by the model. Negotiations over property rights would continue, but these would then plainly be seen to be negotiations over ownership of the escrow fund. Ownership and usage questions would thus become explicitly divorced.

The plan had several merits. First, it made it possible to solve an impasse in the water negotiations and get on with the peace process. For another, it made it clear that no one was giving up sovereignty or claims over property rights. Further, it could be presented in terms of selling short-term permits to *use* water, not selling water itself. Finally, debts to the fund owed by a party using more water than its eventual property rights settlement could be described as going for the construction of infrastructure for the benefit of the other parties. (This matched the Israeli position that what they really wanted to do is find more water for the Palestinians.*)

I did not mention this plan when giving my paper in Stockholm, but I did tell Peter Passell about it, and he included a few statements about it in his *New York Times* article.

Meanwhile, Lenny met with Aaron Miller. First, though, he went to see

---

* The Oslo 2 water agreement was intended to last for a temporary time, but instead lasted for years on the issue of how much water the Palestinians would be allowed to pump. That amount is pitifully small, as described below.

104

Charles Lawson of the U.S. delegation to the multilaterals. Initially, Lawson was at best noncommittal about our project, but when Lenny brought up the escrow plan, he became seriously interested and asked whether he could accompany Lenny to Miller's office. Miller himself was very excited over the escrow plan, possibly because he saw it as a way out of the demands for U.S. funding of infrastructure.

This was all very heartening, and we continued to push the escrow plan. However, no real reply came from the State Department. So, we began to urge the plan on the negotiating parties themselves—which was not easy.

There were other developments that were quite interesting.

Partly in preparation for an upcoming trip to the region in mid-November, and partly because I felt the time had come again that we had to write a more substantive piece, I decided in mid-September that we should revise the report from the previous December, using the new Palestinian data and not waiting for the incorporation of the Bachmat report.* Ours was the sort of project in which waiting for everything to fall into place would mean there would never be any output. It was a hell of a job, but I completed a 150-page revision and sent it off to team members for review.

The results were striking. Principal among them was the proposition that the Palestinians needed to interconnect the West Bank districts, connect to the Jordan River, and connect again to the Israeli national carrier. Were that done, everyone would benefit. Most striking of all, Israel would benefit to the tune of perhaps $40 million of buyer surplus every year beginning in 2010. It was a major opportunity for international cooperation. The Israelis lie upstream on the Jordan River, so the Palestinians would pass water through their territory back into Israel and finally to Gaza, where some of it would be recycled and sold back to Israel for use in the Negev.

I needed to make people understand what kind of future this held, particularly since the model now showed water prices remaining only around 60 cents per cubic meter (shadow values in 1990 dollars) up through 2020. Desalination facilities on the Mediterranean coast would not be needed (with anything like current technology), but would be needed in Aqaba, where the

---

* We had, incidentally, come to understand why the Bachmat costs were so low. They included only energy costs, whereas the costs we were given for pumping from other sources included all costs, including capital costs.

105

Jordanians would need to plan for importing water from southern Israel.* The model predictions were for a very bright future as regards water, and for enormous gains from trade—but it's not easy to make people see that there are enormous gains from trade as opposed to fixed solutions. In part, that was what my forthcoming trip was about.

The occasion for the trip was my testimony before the Arlosoroff committee, but it had gotten much wider. In September, I gave a dinner talk to a "dinner and dialogue" session of ISEPME attended by large numbers of Israeli and Arab students, including relatively senior people (e.g., Wexner Fellows**). This has had some important consequences. First, Eliezer Yaari, who knows a great deal about Israeli media, became very interested and offered to help prepare me for an interview on Israeli television as well as arranging the interview itself. He also turned out to be the cousin of an old friend of mine, Menachem Yaari.***

More important, a man named Shmulik Merhav was at the dinner. He became really fired up about the project, as did his partner Tal Ronan, who visited from Israel a couple of weeks later. They ran a consulting firm that specialized in "paradigm breaking" and had the ear of Shimon Peres, who was really interested in new ways of thinking about things. Apparently, I opened Shmulik's eyes to a wholly new way of dealing with water.

Tal went to see Peres and set off what might have been an important chain of effects. Apparently, Peres remembered speaking with me in January 1994, but said, "We have an agreement. Is this still relevant?" He was referring to the interim Oslo II agreement for withdrawal reached with the

---

* Jordan superseded that view in later years by the exploitation of the Disi aquifer, a fossil aquifer in the south. As of this writing, the Jordanians had built a major pipeline to take the water to Amman and also entered into an arrangement to sell water to Israel in the south and buy it from Israel in the north. Some of the southern water is planned to come from desalination.

** The Wexner Foundation sponsors graduate student fellowships for leadership development at a wide variety of institutions.

*** Menachem Yaari is a distinguished economist, winner of the Israel Prize (among other honors), past president of the Open University, and has served as a member of government committees. His cousin Eliezer Yaari was later the executive director in Israel of the New Israel Fund at a time when I was the Fund's president—and I came to know him well.

Palestinians in late September, which made available an additional 28.6 MCM per year and suggested the Palestinians could develop up to a total of 70 to 80 MCM, including the 28.6 MCM from the eastern aquifer.

The agreement was fine for the near term, but a glance at the Palestinian estimates for household consumption suggested that this amount of water would be woefully inadequate 10 to 15 years later. Conversation with Uri Shamir suggested that the Palestinians might have inserted this number as the amount they *could* develop from the eastern aquifer, thus confusing supply with demand.* If this was true, it was a dangerously unstable agreement and would undermine the peace. If not true, water still remained as an issue for "final negotiations" slated to begin the following spring. In any case, there was not a permanent agreement in any safe sense—something borne out over time.

We sent along material to Peres on the escrow fund as well as my Stockholm paper, and he apparently became interested. We tried to set up a meeting with him during my upcoming trip to Israel.

Peres did do something else that led in a different direction. The Amman conference (the sequel to the one in Casablanca) was held at the end of October. Before that conference, Peres said he thought he might wish to discuss water and this escrow fund with Crown Prince Hassan of Jordan.

I should no longer have been surprised that Julia Neuberger reappeared in my life at about this time. She was in Boston, and we went to a party for her. I mentioned I would be in Jordan in mid-November, and she commented that the fact that the prince had not seen me struck her as "bizarre."

"He owes me one," she said, and told me she would fax him to suggest again that we meet.

When news came of Peres's remark a few days later, I sent a fax at four in the afternoon to Julia, asking that she fax the Crown Prince. By the time I returned home at 5:45 p.m., she had attempted to call him—he was traveling—but, failing to reach him, had spoken with one of his aides in London. They agreed we should fax the escrow material to the Crown Prince,

---

* Incidentally, Uri said the Palestinians sent Marwan Haddad home from the negotiations a week before they were finished because he kept protesting that the numbers were too low.

and Julia sent a fax to him directly.

Meanwhile, Shula had been discussing these questions with Jawad Anani, who was extremely helpful, even though he and I had yet to talk in person. He had also faxed the Crown Prince (and, I think, spoken with others, including Hanni Mulki, head of the Royal Scientific Society). He urged the Prince to see me and also to suggest to the French that we have some relatively official sponsorship. He suggested that Peres might wish to talk to the Prince about this. Meanwhile, Tal had apparently suggested to Peres that the Prince might want to talk about it. I suppose this how marriage brokers operate.

At the time, it looked as if would finally have an appointment with the Crown Prince when I made my trip. The details had yet to be arranged.*

The other big event was the visit of and later discussion with Uri Shamir in late September. Uri was now committed to using the model as at least one valid approach to thinking about water. We tried to get his Water Research Institute at the Technion to be a principal and probably *the* principal Israeli partner. Uri was receptive, although his own role in the negotiations made him personally have to go a bit carefully. He and I and were scheduled to go to Amman on November 15 and speak with Hanni Mulki about joint management.

Meanwhile, interest sparked by the newspaper articles continued to grow, Moshe Syrquin, who edited a journal for the World Bank, phoned in early September and asked whether we would write a paper—which Bob Dorfman and I did, Bob thoroughly rewriting pieces of my Stockholm paper to make it less propagandistic—in time for a (slightly extended) deadline of the second week in October. We hadn't yet had time to circulate it to the team to see who would sign.

Requests for papers came into Shula every day. I had an email from a Norwegian student (who, at least in part, was working with Trolldalen) who wanted to study water banking and couldn't find people in Norway with whom to speak.

There were a few personnel changes as well. First, Peter Rogers of the

---

* I would not have imagined at the time that it would actually be more than a decade later before I finally met with Prince Hassan, who was then no longer the Crown Prince.

Applied Sciences Department at Harvard joined our (more or less) weekly meetings. He hadn't done very much yet, but there was great potential. Second, I took on two MIT "UROPs"—students working in the Undergraduate Research Opportunities Program for credit as a part of their MIT program. We tried to assign them projects out of which they would get something and not merely be exploited. One surveyed water complexities around the world, and the other researched what was known about desalination costs.

Money, meanwhile, continued to be a major problem. There appeared to be a big snag with the French. Hillel's sources said it was being held up in a debate between the French Foreign and Environmental ministries. The latter were supporting us heavily; the Foreign Ministry appeared to have problems, perhaps because they thought money should not be given to nongovernmental organizations but only officially to governments. If we could get some support from the governments, it would obviously help.

I have commented on the possibilities of interest from Israel and Jordan. In Palestine, I was not sure what was happening. Jad Isaac appeared to have declined in political influence. Moreover, Issa Khatar told Shula that he had encountered opposition on the grounds of "how can we sell our water." I needed to tell the Palestinians that it might very well be that as their population grows, they would be the ones who would be buying, and also that they would not have to sell unless it benefited them. I was scheduled to meet with Nabil Sha'ath, then the Palestinian Minister of Planning and generally an important figure in the Palestinian government, just before I went to the Middle East. I would also meet, during my trip, with Nabil Sharif, then the Palestinians' head person dealing with water. We tried also to schedule a meeting with Abu Allah, a member of the Fatah Central Committee, who a decade later became the prime minister of the Palestinian National Authority.

Returning to money, there was obviously the prospect of support from Yad HaNadiv. Moreover, we received a letter from David Brooks saying he was almost ready to fund the Palestinian participation in the Project—a prospect we had thought was gone. Further, someone in the Cairo office of his organization wrote asking whether we had thought of approaching the Syrians to do some separate modeling of Syria and Lebanon that might, in the fullness of time, join on to the main model.

Finally, our USAID proposal was apparently very much alive. We were in the "best and final" round. I had not supposed that that would happen. Given what else appeared to be going on, I was not sure I wanted it to.

9

The Beginning of the Dutch Initiative

On November 4, 1995, Yitzhak Rabin was assassinated. The world, especially Israel, went into shock, and the Israeli government was in a state of flux. I had dropped trying to meet with any Israeli government official on this trip, but because of an extremely successful meeting with Nabil Sha'ath in Boston, I decided to try to meet with Yossi Beilin.

Sha'ath was then the Palestinian Minister of Planning. Over his career, he was also the Foreign Minister, a negotiations leader, and even a temporary Prime Minister. It did not hurt that he had training as an economist. At one time, he even came close to marrying Mariam Maari.

Nabil Sha'ath and I met at dinner in Boston on Sunday, November 12th. I had made occasional attempts to get to see Sha'ath from time to time—as he knew—but they had been unsuccessful. This time, Lenny and I, our wives, and two young assistants from the Institute had dinner with Sha'ath and his wife. We discussed the water project, and I gave him a packet of materials to read. The following morning, Sha'ath and I had breakfast together and continued our meeting, just before I left for the region.

A very intelligent man, Sha'ath had read the materials overnight, understood them, and became convinced of the importance of our work. I gave him the full report he had requested. He wanted to take the ball and run with it, very hard indeed. It was the highest we had ever gotten with this kind of clear enthusiasm, and it was extremely important.

Sha'ath wanted to know why he had never heard of this before. He told us that the people we had been talking to, including the people on our team, were all very well, but they were neither the decision makers nor did they have the ears of the decision makers. He wanted to spread this news widely enough that it would become the common parlance of people around Arafat when they talked about water.

He had a number of suggestions, beginning with a list of people I should see on my trip— principally his distant cousin, Ali Sha'ath, who was a Deputy Minister in his Ministry of Planning. He also suggested bringing in Karen Assaf, who had been at the team meeting the preceding January, and Samih Al-Abed, another Deputy Planning Minister. Those meetings were to be arranged for Monday, November 2, in Gaza. I delayed my return to the United States because of this.

But that was not the most important news. I learned from Sha'ath that the government of the Netherlands had announced at another Middle East Economic summit in Amman its willingness to "facilitate" inter-country projects that would lead to cooperation. This became known as the "Dutch Initiative."

Sha'ath thought our Project should be a prime one for the Dutch to approve. He had been in meetings with several other people from the region, taking place under the auspices of Jan Pronk, which were to be kept confidential. Pronk was the Netherlands Minister of Development Co-operation, a post within the Dutch Foreign Ministry, and was in charge of attempting to resolve conflicts around the world. The people involved in the meetings were all some sort of minister of planning: Yossi Beilin, Rima Khalaf from Jordan, and (*mirabile dictu!*) Yussuf Boutros-Ghali, an Egyptian official.

Yussuf, the nephew of former Secretary-General of the United Nations Boutros Boutros-Ghali, is a former student and an MIT Ph.D. in economics whom I had taught in class, although he had not been my thesis student. We were somewhat friendly. I had seen him briefly at the Casablanca conference and, with considerable delay, had already provided him with some materials about the project. But I had done so as a matter of general interest and because I thought Egypt might be interested. I had no idea he was involved in such a committee. Hearing his name again gave me the feeling that some things are just meant to happen.

The committee was to meet in The Hague in February. Sha'ath wanted me to come there and spend a couple of hours with them answering questions. It was big news indeed.

My first appointment was to be with Rima Khalaf, who I expected to see in Amman. I tried to get an appointment with Yossi Beilin, too, and I wrote to Boutros-Ghali. In each case, I said Nabil Sha'ath had suggested we should meet. Even with the disordered state of the Israeli government, I thought that might have some serious effect.

Sha'ath also said he might be willing to approach one of the donor countries for money for the project. That, too, would be excellent.

After meeting with Sha'ath, I flew to Tel Aviv, where I spoke to the Arlosoroff committee, a group of smart people obviously very interested in the model. Their questions went to its use, the data that went into it, whether the outputs were sensible, and so forth. The entire atmosphere was one of friendly cooperation. I hoped they would make use of the tool. I liked meeting with them, and I'm quite sure they liked meeting with me.

Unfortunately, one member of the committee who could not be present was Gideon Fishelson. He had again been in the hospital, had just returned home, and was heavily sedated. I tried to see him on Friday but could only talk by telephone, as he was in too much pain to see visitors. He had had a recurrence of his cancer; a heart condition kept them from operating; and from all reports, he was slowly dying. It would be a great loss. In terms of the project, not only did he provide invaluable help and wise counsel, but it should also be remembered that he was the first to remark that "water is a scarce resource. Scarce resources have value, and desalination puts a cap on the possible value of the water in dispute." That remark started me thinking on the subject and, effectively, began the project.

It did not appear Gideon would live long enough to see its completion.

Meetings in Jordan

On Wednesday, November 15, Shula Gilad and I went to Jordan with Uri Shamir and Ariel Weiss of Yad HaNadiv. It appeared very likely that Yad HaNadiv would fund the Israeli effort in the ongoing project, centered at Technion's Water Research Institute (WRI). There was some prospect that they would fund the Raiffa Seminars as well. We discussed this on the way to Jordan. Uri had a set of people he would like to involve, including Gadi Rosenthal, an economist on the Arlosoroff commission who had impressed me very much the day before. Also, Uri was entirely willing to continue to involve most of the people who had worked on the project so far, principally Zvi Eckstein and Shaul Arlosoroff. (Gideon Fishelson, as mentioned, would not be available.)

Not to my surprise, however, Uri felt it important that Hillel Shuval not be involved—which was a problem. According to Uri (and Shaul had also indicated this), Hillel was known for particular positions, and any project with which he was closely associated in Israel would be viewed with considerable suspicion. Further, he could not keep secrets well. This meant I was going to have to let Hillel down as the project progressed—something I was not happy about. We decided to wait to do it until the Yad HaNadiv–WRI connection was firmed up.

We arrived at the Royal Scientific Society (RSS) after considerable mix-up about when we were due. Munther Haddadin and Ali Ghezawi were there and had been waiting for us for a considerable time. We talked to Hanni Mulki, who could not attend, by phone.

The meeting did not begin well, but turned out satisfactorily. Munther started by stating that the RSS couldn't possibly sign off on something in which it hadn't really been involved. He reiterated his position that we had used the wrong people—at one point referring to them as having been "picked up off the street." But with considerable help from Uri Shamir, Munther was persuaded to agree that the RSS would endorse the project as a useful tool, that he would work with us on revising the model to meet various objections, and that he would be involved and would urge the use of the tool by the various authorities.

Hanni Mulki had told me Munther would represent the RSS and that his (and I think Ali's) participation would be financed by the RSS. At this stage, I didn't think we could ask for anything more.

There was some residual doubt as to whether Munther agreed with Uri regarding the level of effort involved. Uri wanted the Israeli side to go forward not only correcting errors, but also improving the model in terms of things such as inter-year effects, articulation of the agriculture sector, and so forth. Munther, I think, simply had some difficulties he wanted ironed out in the model version at that time. I offered to do that either by returning to Jordan the following Sunday or in a meeting of a couple of days when I would be in the region in January for the New Israel Fund.

Munther's objections seemed to be of two types. One was plainly conceptual: he insisted that there could be social value to water even if that value was not expressed in subsidies. I believed this to be merely a matter of nomenclature and could be settled through some conversation. Second, Munther believed the recommendation not to build the pipeline from Ma'an-Disi* to Amman was wrong; he blamed it on what he thought was our failure to treat the quality of water in the Amman area adequately.

I first said that might be true, but on further reflection, I thought it was not. The model was showing that in the presence of a large pipeline to bring Jordan River water to Amman (and that must be fairly high-quality water), the shadow value in Amman simply stays well below what it is in the south of Jordan. I didn't think Munther understood that or, just as likely, he had already made up his mind—for whatever reason—about the Ma'an-Disi pipeline, which I knew had been under serious discussion.

Of the model's result that one needed to bring Jordan River water to Amman, Munther said, "Every child of 16 knows that." In retrospect, I thought Munther had the problem I often observe in people faced with fairly complex models: if a model says something an individual believes to be true, the model is obvious and unnecessary; if the model says something that individual believes to be false, then the assumption is that the model must be mistaken. But I believed we would get on.

It was fairly evident that the meeting was satisfactory from the Jordanian point of view. When Hanni Mulki phoned in the middle, Munther spoke to him in Arabic for a while and then put me on the phone. Mulki reiterated that Munther would represent the RSS and told me to hang tight. He thought he would get me an audience with the Crown Prince on Sunday. (We had received a fax from the Prince's office saying he would be out of the country, and he was as Mulki and I spoke.) I suspected that arranging things with the Prince was contingent on my having reached a meeting of minds with Munther.

Another track proved to be very difficult. Nabil Sha'ath had strongly suggested that we speak with Rima Khalaf, the Jordanian Minister of Planning. But he had asked to keep the existence of the meetings involving

---

* The aquifer in question is the Disi fossil (i.e., non-replenishing) aquifer. At this stage of the project, the Jordanian team referred to the district involved as "Ma'an-Disi."

the Ministers of Planning confidential. We tried to arrange the meeting by having Najeed Fakhoury call Rima Khalaf and tell her Sha'ath had asked us to meet her. She agreed to a meeting on the day of our meeting at the RSS, but Fakhoury had insisted that Munther was *really* the man to see regarding water and reported to us that Khalaf had said the same thing. Of course, we couldn't tell him explicitly why Sha'ath had asked us to see her.

In the end, we did not see Khalaf that day and decided to enlist Jawad Anani to call her at home, explain what was going on, and see whether we could make arrangements to meet on Sunday when, I presumed, I would be returning to Amman to see the Crown Prince.

It was evident that Lenny's recruitment of Jawad Anani had been very important. Not only was Jawad enthusiastic about what we are doing, but he also obviously had enormous influence. Just before we left the RSS, Munther said to me, "You have a very important friend, and I will tell you his first initial: Jawad Anani."

Meetings in Palestine and Israel

Thursday, November 16, began quite auspiciously. We were surprised to hear from Shula that Yossi Beilin's office had called and that we had an appointment later that afternoon. The government was in considerable disarray following the Rabin assassination, but evidently the recommendation of Nabil Sha'ath was enough to get a meeting.

First, though, Shula and I went to Gaza to see Nabil Sharif, the Palestinian Water Commissioner. It was a comedy of errors. We arrived approximately on time, but the guards at the Israeli checkpoint would not let Shula across the border. She had come into Israel on her Israeli passport and was leaving on her American passport, and therefore lacked the entry form all foreigners have and that is needed when leaving the country. She had had the same problem at the Allenby Bridge the day before, but had some higher-up support dealing with the situation.

Shula phoned Nabil Sharif, who laughed, said he had similar problems a great deal of the time, and offered to drive out to the checkpoint to see us, rejecting the notion that I should come to his office myself. But he never found us.

After speaking with him once more we had to leave, but rescheduled the appointment for Monday, when we would be coming to Gaza again to see various people in Nabil Sha'ath's ministry.

The event at the checkpoint turned out to be lucky for us in one way. When we arrived at Beilin's office, we were told everyone was swamped. The only thing that kept them from canceling was that we were coming from Gaza.

In fact, we did not meet with Beilin* but with the Alon Liel, director-general of Beilin's Ministry of Planning and Economics (who was to become a close friend and tower of strength for the project), his assistant Amir Tadmor, and another person. There was a somewhat comic scene in which I indirectly and, I thought, subtly, attempted to find out whether they knew about the meeting of the planning ministers. They did indeed, and after they informed me about it, I let on that I already knew, which made things a lot easier. Our meeting was very short. We gave them materials for the first time, so there was no chance for reading in advance. They were, however, quite interested, and I was hopeful something would come of it.

Apparently, the directors-general of the Ministries were to meet on November 30th to organize matters for a later meeting of the ministers, using professionals and academic help. That's where they saw us coming in; they didn't think academics would participate in the February 1st meeting of the ministers themselves (the meeting to which Sha'ath had referred). But we would see.

Friday, November 17, was a stand-down day, which was just as well. I was exhausted. That evening, though, I went for dinner at the home of Hillel and Judy Shuval. Yehuda Bachmat and his wife were there, along with another couple unconnected to the water project. There was some discussion of the project and of its future. I informed Hillel and Yehuda about the planned involvement of Uri Shamir and mentioned that anyone who participated would have to be acceptable to the authorities.

Dealing with Hillel was going to be difficult. I needed to find a place for him, if possible, on the central team, since it seemed plain that he could not

---

* As I came to learn, Beilin and Sha'ath had communicated concerning the project and the Dutch Initiative (although this may have been later). Further, I believe that the fact that I did not see Beilin, and that even at the full meeting I ultimately attended in The Hague, he barely nodded when I said hello was because he did not want the others to think I was supported only by Israel.

be closely associated with the Israeli part of the project.

The next day, we held our project meeting in East Jerusalem. Inevitably, I suppose, most of the Palestinians had not read the report. It had arrived late, but also their travel into PCG headquarters was somewhat restricted. So, I spent a good deal of the time in exposition.

The tone of the meeting was extremely positive, and there was a general sense of good will. Even the issue of Bachmat's data not being available to the Palestinians seemed to have disappeared. They might be going to get it, and anyway, they seem not to care very much any more. Jad, however, was absent; he was in Sweden.

We took decisions on several substantive issues. A committee of Arlosoroff, Haddad, and Abu-Moghli agreed to examine the question of conveyance costs and try to reach a resolution. In practice, this was likely to mean Arlosoroff doing the work and the others then commenting.

Hillel, Shaul, and others pointed out that Jordan River water, even if released down the riverbed from the Kineret in large quantities, would need to be treated. It followed that our big conveyance system for the West Bank would have to involve a pipeline, not the riverbed. Discussion of the benefits from the conveyance system, estimated at roughly $70 million per year, made it clear that building such a pipeline was affordable. We observed that discounted at 10 percent, this meant one could afford a $700-million-dollar pipeline.*

There was general agreement that the report overstated the amount of water in dispute and, therefore, its value. We agreed Marwan and Hillel would start off from the Johnston Plan—an old proposed division of water ownership that had never been generally accepted—and try to produce estimates of the actual amounts of water in dispute and use these to show the benefits from trade. It was a good idea—but was never acted upon.

Zvi Eckstein suggested that we try certain alternate scenarios that did include trade. In fact, the discussion in the report as it then stood mixed up two issues. One of these was the low value of water in dispute and the other was gains from trade. We needed to illustrate the latter explicitly, and this

---

* In fact, we were wrong. Because the model runs in real rather than nominal terms, the appropriate discount rate is not 10 percent but something considerably lower. This made the discounted value of the benefits much greater than $700 million.

would do it.

This also raised the question of treating Jordan River water that had come up when I met with Munther Haddadin in Jordan. A glance at the figures involved appeared to show that he was wrong when he said our conclusion about not building the pipeline from Ma'an-Disi to Amman was mistaken. In our runs for 2010, without treatment costs, the shadow value in Ma'an-Disi exceeded the shadow value in Amman by a considerable margin. Treatment costs could not be large enough to overcome that plus the cost of pumping the water a very large distance. It did not surprise me at all to learn that Munther had authored an almost-finished cost-benefit study recommending the Ma'an-Disi pipeline.

On Sunday, November 19, Shula and I had two meetings. The first was with Nabil Sharif, the Palestinian Water Commissioner and chief negotiator, who we had failed to see the previous Thursday in Gaza. He had been involved in tossing Marwan Haddad out of the Oslo II negotiations a week before they were over.[*] We met at a Kibbutz hotel near Abu Ghosh, an Arab village near Jerusalem. The meeting turned out to be with the joint Israel-Palestinian water committee. Uri Shamir was there.

Nabil Sharif was a nice, friendly man. He had obviously never heard about our project in any detail, and it wasn't at all clear to me that he understood what we were offering him in terms of a tool for planning. Still, he took the materials and thanked us.

We then went on to Ramat Gan, where we saw Marwan al-Muasher, Jordan's first ambassador to Israel, along with Imad Fakhoury, one of the sons of Najeeb and Jacqueline Fakhoury. Imad was the economics attaché to the embassy. They were quite interested and seemed to understand what was going on.

We were running out of copies of the report and needed to get more. While we were in the Jordanian Embassy, Shula asked Rafi, the cab driver, to find a place to get the copying done. Rafi located a convenience store with a Xerox machine that would do one page at a time. The proprietor, an older man, was attempting to turn out 10 two-sided copies of a 150-page document.

---

[*] Incidentally, there was universal agreement that Oslo II and its water provisions would, if made permanent, be a disaster waiting to happen—and, unfortunately, those provisions lasted far too long.

Rafi told us it was clearly a mistake on his part to do the job there, but the old man's pride was now involved. It seemed impossible to stop him. Since they seemed well into it, Shula and I went to wait at a nearby coffee shop where Shalev Gilad, Shula's husband, later joined us.

Three hours later, we emerged with an uncollated set of papers from which it was barely possible for Shula to make a couple of decent copies for the next day.

We all have to learn how to be a little tough about these things, but I was pretty amused.

Back to Jordan and then to Gaza and Back to Israel

Monday, November 20, was a record day in terms of driving and border crossing. Quite early in the morning, we went to Amman, where we had secured a meeting with Rima Khalaf thanks to phone calls from Jawad Anani in Washington, whom we had to let in on the secret about the Dutch Initiative and the meeting of the planning ministers. Dr. Khalaf took a while to realize that we were talking about the Dutch Initiative; in any event, she kept pointing out that she would certainly consult with Munther Haddadin. I tried to make clear to her that that was perfectly all right and that the only reason we were seeing her privately was that we did not feel it was our business to tell Haddadin about the Dutch Initiative.

For someone who had never seen our project, she seemed reasonably interested.

We then went again to the Royal Scientific Society, where we met with Ali Ghezawi. Hanni Mulki was at a meeting. We had received a message saying it was impossible to arrange a meeting with the Crown Prince, but that one was promised for my return in January. Good luck with that, I thought.

We discussed possible financing for the Jordanian part of the project. Jan Trolldalen phoned Ali while I was there, and Ali and I discussed briefly the possibility of Norwegian funding. The RSS had given us a letter approving the project, but also laying out the various reservations Munther had that we needed to discuss in January. I was quite confident these could be overcome.

We had secured similar letters of approval from the Palestinians (Nabil Sha'ath) and the Israelis (Yossi Sarid, the Minister of Environment, and Gideon Tzur, the Israeli Water Commissioner, who apparently approved our project without fully knowing what it was he was agreeing to). We used these letters to show the French that we were legitimate.

It was hard to tell how things stood with the Israelis. The people in Yossi Beilin's office, whom Shula called again, told us they would be really interested, but they didn't know where their office stood since, the day or so after I left, the government had been reformed, with Beilin becoming Deputy Prime Minister. His former ministry might have been disbanded. On the other hand, we were informed that Peres, before the assassination, had expressed a desire to meet with me. We would try to reinstate that for January.

Jordan's Crown Prince also, at least formally, indicated strong interest.

From Amman, we returned to the Allenby Bridge and crossed the border. Rafi met us with sandwiches for lunch and drove us to Gaza. This time, we got over the border and, although there was confusion about which car of two cars were there to meeting us (we probably took the wrong one), we were driven to the Ministry of Planning to meet Ali Sha'ath, Nabil Sha'ath's deputy, and Said Abu Jalallah, who was in charge of sewage and infrastructure planning. They were both engineers. Nabil Sha'ath had suggested we get some others to attend from the West Bank, but they could not obtain permission to come to Gaza.

The people we did meet with knew two things about our work: ours was a serious water model, and Nabil Sha'ath had obviously recommended it. Nevertheless, it was an extraordinarily positive meeting. They seized upon this as a method for their planning, and we tried to organize cooperation as quickly as possible. They wanted a hard copy of the model and a lot of technical information. They also asked us to develop an arrangement in which they could send people to Harvard for training. We suggested that they write us a letter requesting it, and then we could provide funds for this purpose from a grant for technical assistance to ISEPME administered by Bisharah Bahbah.

Said emphasized that because recycling and sewage facilities require very large sums of money, they would be impractical. But his interest picked up considerably when I pointed out that the Israelis would benefit quite a lot from some of these recycling facilities and that they might, therefore, provide some financing aid.

There was an interesting personal moment when Ali Sha'at asked me what Harvard would get out of the project. I responded with what I *wasn't* getting out of it. I pointed out that I wasn't getting paid. I admitted that being the author of the various documents would typically be a big thing, but said with some embarrassment that I was already a famous economist and didn't need publications or publicity for my career. I was doing it, I said, because I cared a lot about what happened in the region. I added that I had been treasurer of American Friends of Peace Now for many years, which might help him understand my personal interests in the Middle East.

Ali certainly understood that.

Surprisingly, it was the first time anyone had asked me that sort of question directly, although others had hinted at it. Shula told me others had asked her about me. I was happy to see that my reputation as an economist carried some weight in this regard.

After the meeting, we drove to Jerusalem and had dinner out. Avishai Braverman, whom Sha'ath had asked us to consult as well, joined us for dessert. We gave him the report and had a friendly chat. We did learn that at Ben Gurion University, of which Avishai was the president, there was another water institute in competition with that of Uri Shamir.

Home and Fundraising

I returned to the United States on Tuesday, November 21, exhausted but feeling very good about what had been accomplished—although the situation in Israel remained very much in flux.

Back home, much of my attention was on raising money for the project. We were flat broke, and Lenny was desperate. The project was putting ISEPME deep into deficit.

We had a number of possibilities and promises, but no actual payments. It seemed quite likely that Yad HaNadiv would finance the Israeli part of the project, at least if we could find financing for other parts. It still seemed possible that David Brooks, who was away until December 18, would finance the Palestinians. And the Jordanian Royal Scientific Society would finance meetings with Haddadin and Ghezawi, but not major participation. We needed to find funding for the Jordanians.

The Revson Foundation had made a lot of promises the previous spring, but we had yet to see a nickel. I spoke with Bob Rifkind and asked him to phone Eli Evans, Revson's executive director, to find out what was happening. Similarly, the French proposal seemed to be bogged down in the Foreign Ministry. We hoped the letters we sent would have some effect.

A fax from Asit Biswas said that the Nippon Foundation might be interested. I suggested that they could take over the entire project for about $2.5 million for about three years or take over support of one country or take over support of the central program. Meanwhile, the USAID proposal was still alive.

On November 27, Rudolf Dolzer, former national security advisor to the chancellor of Germany, visited me. He had been sent by Nazli Choucri of MIT's Political Science Department. While I was totally overcome by jet lag that day and did not give my usual exciting performance, Dolzer still seemed quite interested. He asked for a complete package of materials and returned to Germany, promising to stay in touch. Choucri and Myron Weiner, also in MIT's Political Science Department, scheduled a meeting on December 13th to discuss possible joint activities and funding applications.

On a different front, we experienced a slight setback—or perhaps not. The meeting of the directors-general of the various planning ministries took place on November 30 in The Hague. Alon Liel, who we had met in Yossi Beilin's office, did not attend because of the shakeup in the Israeli government. Instead, it was Amir Tadmor, who had come in late to that Beilin meeting and so did not realize we were offering something connected with the Dutch Initiative.

Ali Sha'ath was also there, but in our meeting with him in Gaza he hadn't seemed to know about the Dutch Initiative and we hadn't brought it up.

The result of all this was that Shula learned from Tadmor that our project had not been raised at the meeting in The Hague. However, Tadmor said it wasn't too late and that he proposed to put us on the agenda. Shula was to call him several times a week to see what was happening.

Meanwhile, I tried to get in touch with Nabil Sha'ath about this, but failed to reach him. So, we sent him a fax.

III. 1996 to Early 1997

10
Continuing On, Despite a Loss

While back in Cambridge we learned of the death of Gideon Fishelson on December 13. Ellen and I arrived in Tel Aviv late in the afternoon on January 14, 1996, and went to dinner at the Dan Hotel. It was part of the memorial events (or *shloshim**) for Fishelson. Munther Haddadin, the guest of honor, was there with his wife, Lexi. Also present were Uri and Yona Shamir, Zvi and Hadassah Eckstein, Shula Gilad, Estee Landau, and Miram and Eliakim Rubenstein, the latter then a judge in Jerusalem who had been the principal negotiator for the Israelis on the Jordanian Treaty.

It was an interesting evening. Munther had once said he would come to Israel only after the visit of King Hussein, which had happened, but still I told Munther that coming for Fishelson's memorial was a big gesture. Jordanian professionals were definitely not encouraged by their peers to interact with their Israeli counterparts.

The Haddadins had entered the room laughing. Munther explained that he had said to his wife, Lexi, as they walked to the hotel, "Darling, I never expected to be walking along the beach at Tel Aviv holding your hand."

"Oh?" she replied. "Whose hand did you expect to be holding?"

A good time was had by all.

Much of the conversation was taken up with reminiscences of the negotiating days, with stories passed back and forth among Uri, Munther, and

---

* Traditionally, the thirty-day point after a death is considered a change in mourning, and the gathering then is called the *sloshim*, or "thirty."

127

Eliakim Rubenstein. A lot of the stories were about jokes, particularly from the Israelis, that had been told to lessen the tension. There appeared to be genuine satisfaction with the state of the Jordanian-Israeli agreement on which all of them had worked—but less satisfaction about the unfinished business between Israel and the Palestinians.

Rubenstein shared a story about some Israeli telling him that it was a mistake to hold some of the water negotiations between Israel and Jordan at Beit Gavriel, a villa on the shore of the Sea of Galilee. Seeing all that water, this Israeli argued, would harden the Jordanian position. Rubenstein responded that the Jordanians probably already knew the Sea of Galilee was there, and later, when the negotiations began, he took Munther to the balcony, told him what the Israeli had said, pointed to the Sea of Galilee, and said, "I want you to know that is not water."

Another of Rubenstein's stories concerned the negotiations that took place at some royal establishment in Jordan. Noah Kinarti, Israel's chief water negotiator, fell asleep when the negotiations were over. Rabin, in another room, asked Rubenstein where Kinarti was, and Rubenstein replied that he was doing something he had never done before in his life and probably would not get a chance to do again: snore on a royal couch.

Munther pointed out that when he and Rubenstein first met, Munther hadn't been particularly friendly. But Rubenstein had told him he knew who Munther was, and proceeded to reel off a good deal of Munther's biography and family background. If he knew so much, Munther asked, why was he there leading the Israeli position? Rubenstein replied that it was because he (Rubenstein) was a bad man. Munther asked how he got his wife to marry such a bad man, and Rubenstein replied that he had fooled her.

We all looked at Miram Rubenstein, who said he was still fooling her.

Rubenstein also commented on Munther's quite wide knowledge of Arabic history and culture and of Islamic tradition and poetry, commenting that it was the more remarkable because Haddadin is a Christian. Munther's reply was that quoting Arabic poetry to his wife, who is from South Dakota, is how he got her to marry him.

When the talk turned to Syria, it was clear that all of us considered the regime there to be tyrannical. Munther commented that the world had gone from a situation in which people in Amman would go and buy a trousseau for their brides in Damascus to a situation in which women in Damascus came to Amman for the same purpose.

Rubenstein shared a memory of our meeting two years ago and asked to see a current report of the project. Uri, in talking about water, emphasized the necessity for a scarcity chart for water, something he thought the project could provide.

The next day, Ellen and I joined the Haddadins and Shula Gilad for a tour of Jaffa. It was as dull as I thought it would be, but the day was somewhat enlivened by the interchanges between our archetypical Israeli guide and Munther. Inside a mosque, when the guide said that Muslims pray four times a day, Munther—who is a Christian—said, "They pray five times." When the guide kept referring to the Turks having done this or the Jews having done that, Haddadin wanted to know whether he thought there had ever been Arabs present. And so it went.

The Fishelson Memorial

The afternoon and evening were devoted to the Fishelson Memorial events, held thirty days (plus a little bit more) after his death on December 13.

We went first to lunch at the faculty club at Tel Aviv University, and then to the Fishelson household, where we were received by Dahlia Fishelson and her children as well as by the sisters of Jedda (Gideon's nickname, which means I have been told means "strong guy").

Before heading to the memorial assembly, Shula and I went to see Habash (Haim Ben-Shahar), an old friend who was the former president of Tel Aviv University. We discussed the project with him—he had received some materials the previous year—and when we finished, he said he would recommend to Yossi Beilin that he meet me personally. He was glad we were staying around for a couple of weeks.

Next was the memorial meeting—a great occasion. The small auditorium at Tel Aviv University was packed with about 200 people, including Jedda's family and his colleagues, as well as just about everyone involved with water in Israel. That Munther Haddadin had come from Jordan and Marwan Haddad had come from Nablus to speak in memory of an Israeli was a significant indication of a new age.

Among those present associated with the project were Hillel Shuval, Yehuda Bachmat, Uri Shamir, Zvi Eckstein, Shaul Arlosoroff, Marwan, Munther, and I. It was important that Alon Liel, Beilin's director-general with whom we had met in November, also attended. He came to this in part because he was unable to attend our meeting in Beilin's office the next morning.

First came the eulogies of Gideon, in Hebrew. Then Zvi Eckstein spoke about his work with Gideon on the prototype model out of which our project's model had grown. It was then my turn for the principal address—given in English. I described how the genesis of the project could, in large part, be traced to Gideon's remark at the London Conference in 1990, "Water is a scare resource. Scarce resources have value. But water can be replaced by desalination, so there is an upper limit to the value it can possibly have, and we can estimate it." I then reviewed the project as it had grown and summarized our results thus far.

I closed with a vision of trade and joint infrastructure essentially solving the region's water problem and taking water out of the peace negotiations. If that happened, I said, we would be able to say to all the peoples of the region what is written in the Torah, "*Kakatuv 'U'shaftem mayim b'sasson mimanay ha-y'shuah*'" ("As it is written: with joy shall you draw water from the wells of salvation").* When I said *Kakatuv*, slipping into Hebrew, the audience giggled and rustled. Then, the crowd recited the entire quote aloud with me. It was electric.

I went on to say that this great vision was traceable in large part to Jedda's remark about scarce resources and ended by saying, "Jedda was a scarce resource. Scarce resources have value. But he cannot be replaced, and his value had no upper bound." It was a magical moment.

I am proud of having thought at the last moment to add those words.**

My speech also appeared to have made some impression on some of the professionals in the audience. I began receiving letters from people who heard me that wanted to know more or wanted to tell me about their projects. That was also true after I spoke the next day at the Israel/Palestine Center for Research and Information (IPCRI).

The day ended with dinner in Jaffa with several people, after which Hillel

---

* The phrase from Isaiah 12:3 had long ago been set to music for a well-known Israeli dance, known not only to Israelis but also to many Diaspora Jews.
** Part of my speech was later broadcast on the radio in Israel, and it was published in Hebrew in an Israeli journal.[1]

Shuval and Yehuda Bachmat drove Ellen and me to Jerusalem.

Preparing for The Hague

The next morning, Ellen went off for two days to Jordan to visit Petra again. Shula and I were due at Beilin's office for a meeting—and before that we had a bit of a crisis.

Shula, coming from Tel Aviv, was stuck in bad traffic that was partly due to visits to Jerusalem by the Dutch prime minister and Vice President Gore. When she made her first of what seemed like a thousand calls to me on her cellular phone, she told me we had another problem. We had suggested to Uri Shamir that he come with us to the meeting, believing he was politically "in." But when she asked Amir Tadmor about this, she was told in no uncertain terms that his presence would be a disaster. Indeed, Tadmor said that if Uri knew about the Dutch Initiative, we might well be out of the running. Shula advised me to get to Beilin's as soon as possible to see whether I could straighten this out while she tried to head off Uri.

By the time I got there and managed to find Tadmor—no trivial feat given the confusion sparked by the state visits—the crisis was over. Uri, it turned out, had been called away to some other pressing engagement. I asked Tadmor about Uri and he told me it was simply an "inter-ministry thing." The planning minister had not yet decided what they were going to do, and until that happened, they didn't want other ministries to know about it. Once decisions were taken, he said, they would certainly turn to water experts and Uri would be fine.

Tadmor could not have been friendlier, and friendship continued in the weeks that followed. It did not hurt that we chatted about the New Israel Fund while waiting for the meeting to begin. He turned out to be a friend of Avi Armony, then the director in Israel of the Fund, and was thoroughly supportive of the Fund's efforts. He was interested that I had been the treasurer for a long time and was going to be president. I noted that my connection with Peace Now and the New Israel Fund hadn't harmed me with the Palestinians either.

The meeting began without Shula. Once she arrived, we had to crowd around a computer screen because of technical problems with the overhead projector. Also present were Amir Abramovitz, Rafi Bar-El, the engineer whose little pamphlet on water they had given me in November, and Roby

131

Nathanson from the Tel Aviv University economics department, whom I had not met previously and who had attended the lecture the previous evening.

Of course, we knew the Israelis were disposed to support our project. Beilin and Nabil Sha'ath had discussed it, and the Israelis had already been quite helpful. But the presentation got them really excited. Certainly, from that moment, they were fully behind us.

We were told that they had briefed Yossi Beilin quite thoroughly the night before and that he had decided he was fully behind the project, too. The question became a tactical one: *how* would they help us? And we did need help and advice, pretty soon.

We had assumed we would be invited to The Hague. The Israelis had given material to the Dutch ambassador, who had passed them on to Jan Pronk, the Dutch minister for development and co-operation who was leading the initiative. I had even canceled an obligation in California just before the trip, supposing the invitation would come. But it had not.

Now came the news from home that the Dutch Ministry of Planning was trying urgently to find us.

Tadmor and his colleagues suggested strongly that if we were going to go to The Hague, as the message from the Dutch indicated, we should try to do so just before the meeting of ministers on Tuesday, January 30. Even if we were not invited to that meeting, they thought it would be a good idea if we were in The Hague in case people wanted us, so to speak. It was a significant change from when Tadmor and Liel had told us that there was no way we would ever be invited to the ministers meeting.

After the meeting in Beilin's office, we had lunch with the Haddadins, who were visiting Jerusalem, and Shaul Arlosoroff. The Haddadins, accompanied by Shula, then went off to cross the bridge back into Jordan. Shaul and I spent considerable time discussing various aspects of the model.

Generally, Shaul was prepared to sign off on *Liquid Assets* at this time, even though he thought there were a number of things that obviously need to be done to improve the model. He also gave me very sensible advice about why Israeli conveyance costs may actually be lower than Jordanian: principally, because they do get electricity at special rates, since the water company agrees to interruptible service when dealing with the electricity company. But conveyance costs remained a problem.

By that evening, we began to receive requests for "terms of reference" from Tadmor. Not sure of what he meant beyond a simple proposal, we consulted Theo Panayotou back in Cambridge. The Israelis were very eager for results in six months or less, and the month of May began to be mentioned prominently. This could only be due to the timing of the Israeli elections, which looked as though they are going to happen at the end of May. Exactly why output from this project was required for political purposes is hard to say.

Shula finally succeeded in getting the appropriate person in The Hague on the phone on January 17. It was Myriam Van Den Heuvel of the Foreign Ministry, who was not very forthcoming but said they could give us an hour on January 29. Some of her colleagues might be able to stay longer than that, she added. This was the invitation we had been trying to get, and we were very excited.

After some back and forth with Theo, we produced a document of eight single-spaced pages that laid out what we wanted to do: make improvements in the model, check the data, train national personnel, and design a joint management authority. But we were told the proposal was too long, so we eventually produced a page-and-a-half-long document as well. I don't know if anyone ever read either document. (On the advice of the Israelis, we also included a proposal for building an Egyptian model. The Egyptians, though, said they were not interested—at least for the time being—but were happy to see our project sponsored by the other participants in the Dutch Initiative.)

I should mention that we were aware by then that there were other projects being discussed as possibilities for the Dutch Initiative. I don't know what most of them were, but Michael Porter of the Harvard Business School headed one that concerned which industries would tend to lead trade and development. When I heard a presentation of the project at Harvard, I thought it had essentially no analytic content whatsoever. Porter was and is well known, and perhaps this sort of thing is successful, but I couldn't tell why. Nevertheless, throughout our trip, we worried whether we were competing with Porter and whether his project might sound more seductively pleasant than ours, particularly since it would involve all four countries.

When representatives of the Porter project went to The Hague for a meeting on January 26, their project was adopted—after some skeptical discussion. They did not stay around for the ministers meeting. We had our own meeting set for January 29th, and on the advice of the Israelis, planned on staying around the next day.

Israelis and Palestinians

On the same day we were invited to The Hague, I gave my talk at the IPCRI. It was largely a well-attended professional meeting of Palestinians and Israelis interested in water and environmental problems. The talk was well received.

I was quite surprised and gratified when Nabil Sharif showed up and seemed to be quite interested. The following day, I learned from Jad Isaac that Sharif had been investigating the use of our model for some time and was particularly interested in a one-state Palestinian version in which Palestine would operate with a fixed quantity of water. Indeed, over time, all the countries asked for such versions as ways of getting used to thinking about water in our fashion.

It was clear, however, that for Palestine to operate in such a way would be a disaster, because of the interconnected nature of the Israeli and Palestinian water resources. An idea that had come to the forefront of the project was the examination of gains from trade at quantities and prices put forth by the model (to avoid monopoly problems). While these ideas had always been in the background, their importance had been somewhat neglected. I suspect this is because I (perhaps like many economists) tend to take certain things for granted and to forget that non-economists don't necessarily think in the same way we do.

In any event, following a suggestion of Zvi Eckstein's at our meeting the previous November, we performed some runs that compared trade with a situation in which Palestine has to make do with the water specified in the Oslo-II agreement signed the previous September. The differences were quite startling. Having to make do in that way, as opposed to a situation in which Palestine merely owns that water but can trade for more, would impose on Palestine a yearly cost (by 2010) of about $25 million. This comes from the need to desalinate water to supply Gaza. With trade, the Palestinians would save $25 million and the Israelis and Jordanians together would gain $15 million a year. That meant that the selling countries could compensate their consumers and still have $15 million a year left over.

Even those pretty big numbers—which would be even bigger if the costs of desalination are greater or one considers later years or higher population growth—are not the only important point. A strain like that on the quantity of water is destabilizing, and it would not be to Israel's advantage to impose such a solution on the Palestinians because it would lead to considerable and unnecessary tension within the fifteen years. That doesn't make for a stable peace treaty.

I made a point of saying this whenever I spoke on the subject. At the Fishelson memorial, Shaul Arlosoroff had gone even further, quoting Gideon as saying that the Oslo-II agreement was "shameful." I tried to impress on Nabil Sharif the necessity for trade.

When Uri Shamir arranged for me to speak at Bet Sokolov the following week, he strongly suggested to me in advance that I soft-peddle my reference to the Oslo-II agreement as "a catastrophe waiting to happen." Of course, he was one of the negotiators. I did moderate my language a bit, but I made the point substantively. It was a very important thing for both Israel and Palestine to realize.

Revson Foundation

It was about this time that a comedy of errors began to ensue.

When we met in Beilin's office that Tuesday, we told the people there that we thought we could make more progress with the Revson Foundation if Yossi Beilin would sign a letter stating his support of the project that we could fax to Bob Rifkind, a foundation board member and old and close friend of mine who was then also the president of the American Jewish Committee. He had told me that he knew Beilin personally and that such a letter would enable him to go and find out why nothing had happened with Revson since the previous July.

Tadmor said he would propose that to Beilin, and that evening he told us Beilin had agreed. He asked me to draft the letter, which I wrote out by hand and faxed to Tadmor with a cover sheet—or so I thought. I received back only the first page as a confirmation that something had been faxed.

Tadmor called the following evening saying he had only gotten the cover page and not the second. Fortunately, we could reconstruct the second page, and again sent a fax to Tadmor. This time, I fixed the incorrect reference to Rifkind as president of the American Jewish *Congress* rather than *Committee*.

A day or so later, we began to hear from the office back home about Rifkind's office having called and asked why they had received a handwritten version of a letter from Yossi Beilin that was incorrectly addressed to Rifkind as president of the American Jewish Congress. They had faxed what they had received to Anne Marie Kelly at ISEPME, who recognized my handwriting.

Apparently, the Hotel Moriah, which had just become the Radisson-Moriah, had outdone itself. Seeing that the second page of the original fax itself had a fax number on it (it was Rifkind's, put on there for Tadmor to use), hotel staff faxed the second page to Rifkind without telling me so.

Then, just before we left Israel, Tadmor informed us that Beilin had signed a typed version of the revised letter and had it faxed to Rifkind—except it hadn't been. Beilin's office was moving, and the signed version of the letter had apparently gotten lost.

When I was in The Hague later that month, I spoke with Bob Rifkind about this. He said it would be very useful to have the actual version. Shula got Beilin's office to fax us copies of his stationery so Beilin could sign a typed version.

More Discussions in Palestine and Jordan

Shula and I spent a good deal of Thursday, January 18th, with Jad Isaac at his Institute. He was wonderful. He understood this project up and down, and was as big a supporter of it as I am. He told me there had been very substantial interest among the Palestinians, but repeated what I had already heard about Nabil Sharif wanting a Palestine-specific version of the model.

I only wished it were possible to put Jad entirely in charge (as he once was) of the Palestinian part of our effort. It would be smooth. He's very smart. We liked each other a lot. We made great partners.*

Alas, things were not that simple. Jad's position in official Palestinian circles was not what it once had been. He is outspoken and very much his own man. It was not entirely clear that he was *persona grata* with the higher ups. (It was, of course, largely to provide political cover that Jad later began working through PCG.)

I discussed this situation with Jad and told him that Nabil Sha'ath was our big supporter.

---

* I was sorry to find many years later that Jad had become confused on a crucial point—that the model would not, and could not, lead to a situation in which wealthy Israel would buy all the Palestinian water. Others had already made that mistake, and still others would later. I take it up in a later chapter.

"Does he know that I'm involved?" he asked.

When I asked if that question meant he was on bad terms with Nabil Sha'ath, Jad simply said, "He wants to own me."

Nobody, I hasten to add, owns Jad.

Jad's view of the political situation was that Nabil Sha'ath would not win reelection to the Palestinian Council in the elections to be held two days later; in this, he shared a widespread view of Sha'ath's chances. Even if he did win, however, Jad thought he would not continue as minister of planning, although he might continue as minister of international cooperation and hence be in charge of the Palestinian participation in the Dutch Initiative.* Were he not to remain minister of planning, Jad suggested, he would probably be replaced by Abu Allah, his perennial rival. In that event, Jad suggested that he, Jad, might very well be put in general charge of Palestinian water.

This, would, of course be highly desirable from the project point of view, even though Nabil Sha'ath was a big supporter. So, in a sense, we would win either way. But Jad, as usual, was obviously not optimistic about Palestinian politics.

As it turned out, Sha'ath won in Khan Yunis, and he won very big.

To Amman and Back

On Friday, January 19, Shula and I went to Amman to confer with Munther Haddadin and Ali Ghezawi. It had snowed and hailed in Jerusalem and Amman the day before, and snow—although hardly any to speak of in Boston terms—was still on the ground in parts of Jordan's capital city as we approached the Royal Scientific Society.

Friday, is, of course, generally a holiday in Amman, and when we arrived at the Royal Scientific Society, we learned that the meeting was to take place at the Haddadins' home. This house being somewhat higher up the hill, our driver felt it necessary to change to a four-wheel-drive vehicle, a change which we found quite amusing, since Shula and I both live in New England.

Once we realized the King Hussein Bridge over the Jordan closes at 3:00 p.m. on Fridays, there was little time for substantive discussion. Munther later complained that we didn't spend enough time in Jordan and it was a nuisance

---

* Incidently, it is not clear to me that Jad was supposed to know about the Dutch Initiative, but I had already told him.

137

for him to gear up for discussions every few weeks that were then aborted. We had been told he would be away in Cairo for most of the time we were in the region, which explained a Friday meeting, and when he ended up not making that trip, he didn't let us know. In any case, we scheduled three solid days in early March to meet with him in Jordan.

The discussion itself was distressing in one way and hopeful in another. It was apparent that Munther had really not understood how the model works to generate prices or what those prices mean. He relied on having read the June 1994 model version with moderate care; I don't think he had read later versions. That was too bad, because the discussion of prices and their role went in as an explanation for non-economists after that version. So, in a sense, we were back at square one. But in another sense, that was not true. Munther had obviously committed to the project, probably both for substantive and political reasons, and he was putting in the effort to understand the issues and seriously discuss the problems. I believed we had reached the stage where we could, in fact, work together.

We returned to Israel, and that night Ellen and I dined at the Sheshinskis' with Yoram and Menuchah Weiss and Amira and Jesu Kolodny, also old friends. I learned that Israel radio had carried portions of my speech at the Fishelson memorial. It was plain that I was becoming widely known as the "water man."

I spent the next day producing a revised version of the terms of reference. I was almost done when I accidently erased the files from my disk. I began to rush about the hotel room, anxiously.

"This is no time for panic!" Ellen said.

"What do you mean this is no time for panic?!" I countered. "This is exactly the time for panic!"

But it ended well. I thought to contact Harshadeep, and in a frantic 45-minute-long international phone call I got the advice I needed to restore the files. It was a great relief.

Sunday, January 21, was a particularly significant day for us. We traveled to Tel Aviv because Amir Tadmor had told us that Yossi Beilin wished for us to discuss the project with Ron Pundak and Yair Hirschfeld, the two academic historians who had been instrumental in arranging the original Oslo meetings through back-channel negotiations with the PLO. In December of 1994, the *New Yorker* had run a major story on them, and I had read it with considerable interest. I regard them as heroes.

Beilin was still using them as back-channel negotiators, a fact that was widely known in Israel. Just before we went to see them, the opposition in the Knesset claimed they had been negotiating the future of Jerusalem as a prelude to giving it up as Israel's undivided capital, an allegation promptly and strongly denied by the government. I don't know whether they actually spoke with anyone about Jerusalem, but I do know now that they were used quite extensively—including in connection with our project.

They were very unassuming guys, sort of rumpled Israeli academics. When I told them toward the end of our meeting that I was going to say an embarrassing thing and expressed that it was an honor to meet them, they were genuinely embarrassed. Relations with them remained very good. They were essentially carried away with what we were doing.*

The following day, Monday, January 22, was my talk at Bet Sokolov that had been arranged by Uri Shamir. It was attended by a large number of people from the water establishment (although not the major water negotiators, who were off talking to the Syrians). Among them was Ilan Amir, an agricultural economist from the Technion. Shaul Arlosoroff sent him to us, with Uri's approval. He was the author of models of agriculture (particularly in the Jezreel Valley) that can be described as optimizing models giving farmers' actions as functions of the price, quality, and quality of water available—just what was needed to improve the agricultural demand sector of our model. I looked forward to Ilan's participation, and he did join the project, even spending a sabbatical at Harvard. (It is a comment on the financial state of the project that we found it difficult to make a commitment to Amir for expenses in exchange for six months of free work.)

The meeting at Bet Sokolov went quite well, with a number of the participants requesting continued contact. One of them, however, looking closely at the model output, suggested that our conveyance costs to bring water from the Jordan Valley up over the West Bank must be too small. He was obviously right. The shadow value in Hebron was not far enough above the shadow value in the Sea of Galilee to cover the lifting costs. The difficulty went back to the original Israeli report, which simply estimated conveyance

---

* Years later, Pundak became a valued member of the New Israel Fund's advisory council.

costs as 8 cents per cubic meter per 100 kilometers across flat country and 12 cents per cubic meter per 100 kilometers when the water had to be raised into the hills. But that can't be independent of the height to which the water was to be raised. We had tried before to get this corrected, with little success.

Did it matter? The issue had to do with whether it would be worth building the great conveyance system we were then suggesting that would bring water from the Sea of Galilee into the West Bank, from the West Bank into the Israeli national carrier (INC), from the carrier to Gaza, and then (recycled) to the Negev. That *really* mattered. The desirably of doing that appeared to be there because the new carrier would lead, through lower conveyance costs, to lower prices in southern Israel. It may very well have been that our estimate of conveyance costs to bring water to southern Israel along any route, including the existing one, was too low. If both the costs of the new conveyor and the existing costs were too low, the new conveyor might still be superior.

But perhaps it didn't matter. After the issue arose, I investigated the question of increasing (in the model) the costs of raising water out of the Jordan Valley into the West Bank by 10 cents per cubic meter, an approximate figure supplied by Uri after the lecture. The results were extremely interesting. On the one hand, it would no longer pay to take that water into the INC; it would be too expensive. On the other hand, even without doing that, it remained true that water prices in southern Israel would drop substantially, although not by as much as before. The reason appeared to be that this would provide a much less expensive way of providing water from the Sea of Galilee to Jerusalem than the existing way of bringing it around in the INC and then back up into the hills. With the INC no longer stressed to provide water to Jerusalem, relatively low-cost water in the center of the country around Revohoth and Ramla—which, with the existing system, it was optimal to send to Jerusalem—now would get sent to southern Israel. Water that enters the INC from the Sea of Galilee would not, in fact, reach southern Israel when the new system was provided, because much of the water that was then actually put into the INC would be diverted into the new carrier. As a result, there would be a break in the shadow values at approximately the point that the conveyance line to Jerusalem comes off the main line of the INC.

It took me a day or so to figure all this out. I expected to use it as a training exercise in model interpretation.

In any event, it was obviously very important to get the conveyance costs right. I once again made that a high priority item and hoped to get Shaul Arlosoroff to deal with it as soon as possible.

After the Bet Sokolov talk, I was interviewed by Sever Plotzker, the economics reporter for *Yediot Ahroniot* and probably the best economics reporter in Israel. He had been a student of Eytan Sheshinski.

The next day and a half was devoted to the meetings of the New Israel Fund. Before meeting with Nabil Sha'ath, I had agreed to become the Fund's president, despite my misgivings that having to spend so much time on the water project would make it difficult to do a good job for the NIF. Still, on Wednesday, January 24, my campaign to be defeated in an uncontested election failed and I was elected to a three-year term that would begin the following July 1.

I looked forward to it, although it was apparent that the water project would take even more of my time than I had previously expected.

It was during this time that Hillel and I again discussed his participation in the project. I had broached with Hillel the idea that there might problems on the night of the Fishelson memorial—both Uri Shamir and Shaul Arlosoroff had previously suggested this to me—and Hillel and I discussed it again when we went to a reception at the house of the outgoing New Israel Fund president the first night of the Fund meetings.

As Uri explained it, the main problem was Hillel's long-held identification with a position of cooperation with the Palestinians. Uri suggested that the water establishment simply wouldn't take seriously the recommendations of any project with which Hillel was associated.

Hillel was really upset and had grown even more so since I had first raised the issue. He believed the problem was that he was known for advocating cooperation with the Palestinians, but he didn't see any reason that should be a bar to his involvement. He argued that it was all some sort of political prejudice by the establishment"—the negotiators and *apparitchiks*. I put it to him that I hadn't had that kind of reaction directly from Beilin's office and that he shouldn't worry until we saw what happened.

The bottom line was that that I did not want to be caught in the middle of an old feud between people the project needed and had much to offer. I decided it would be morally wrong to try to read him out of project and that doing so would cause more of a problem than retaining him. Later, I

141

explained to Uri that Hillel would remain, and he accepted that with my assurance that this would not be a problem with the Israeli government.

The reception at the President's House was a great occasion—except for the spectacle of aging Israeli President Ezer Weizman, who had become an embarrassment when speaking in public. On this occasion, he chastised New Israel Fund board members for not making *Aliyah*—coming to live in Israel—and presuming to criticize Israeli society. Instead, he insisted, we should just raise money for projects. Apparently, it was the stock speech he made to all American Jews.

Earlier that day, at the New Israel Fund meeting, board member Nechamah Hillman had asked whether Weizman knew who we were. Apparently, he did not.

Jerusalem, East and West

On the afternoon of Wednesday, January 24, a Palestinian taxi from east Jerusalem came and took me to the center of Nablus, where I was to give a seminar for An-Najah University students, organized by Marwan Haddad, that took place at Marwan Haddad's consulting offices.

I was a little nervous about going to Nablus. Various Israeli and American friends expressed some concern, as did a couple of Palestinians. After Gaza, Nablus had been the city with the greatest unrest. I remembered that even in 1973, when we visited with our young children, Nablus was a place where tourists were seriously advised not to park their cars or wander about. But at this time there was no problem.

The driver found the office building for me, but when I entered, all the signs were in Arabic. I wondered what to do. Two academic-looking young men came along and asked me whether I was Professor Fisher. I said I was and that I was looking for Marwan Haddad. They were his assistants and were coming to the seminar. We all walked up several flights to Marwan's office.

The seminar itself was well attended, but the audience was quite passive with their questions. I spent a pleasant time with Marwan who invited me to come back and speak at a conference in May, and then was driven back to Jerusalem.

We had been invited for dinner that night at Jad Isaac's house in Bet Sahour along with Hillel and Judy Shuval, Sari Nusseibeh and Issa Khatar, and their wives. This was an important occasion. To be invited to an Arab's home is to be marked as his friend. I knew that, and I was honored, as was Hillel. Most of our discussion turned on the Palestinian elections that had taken place successfully five days earlier.

I also spent some time talking to Sari Nusseibeh about research on Jerusalem and other matters, a discussion we continued the next day when I joined Lenny and Shula at Orient House for a meeting with Faisal Husseini, the Palestinian's chief official in Jerusalem. That meeting was also largely to discuss ISEPME's Jerusalem project. Also present was Samih Al-Abed, who was serving as the acting Palestinian minister of planning, Nabil Sha'ath having temporarily resigned as required pending the election. He was also a member of the Palestinian delegation to The Hague, and we already had an appointment to meet with him in Ramallah the following Saturday.

I had met Faisal Husseini twice before, first in 1990 at the time of the conference to honor the retirement of my close friend Don Patinkin, the founding leader of what became a world-class economics department at Hebrew University. Husseini and other leading Palestinians, including Sari Nusseibeh, were then on a hunger strike in East Jerusalem to protest some incidents of violence. I took a delegation of the economists attending the conference to visit the hunger strikers.

There had been considerable debate among Israeli friends regarding whether I should do that. The leaders of Peace Now were all for it and came with us. Eytan Sheshinski, a Peace Now supporter who at the time had a semi-official government position, was doubtful about whether he should attend. He also suggested I take along Gur Ofer, a friend of mine who was also an economist at Hebrew University, as a "cooler head."

At first, the visit with the hunger strikers was going well. Faisal Husseini made a formal speech, and I made a formal speech. The Palestinians made much of the fact that there were two Nobel Prize winners among us, James Tobin and Lawrence Klein. I had to disabuse them of the notion that this meant they had won the Nobel Peace Prize. Everything was fine in a somewhat ritualistic manner until Gur Ofer, a Soviet expert, said that there was "just one thing": he wished the Palestinians would stop opposing the emigration to Israel of Russian Jews. It was totally inappropriate, and the meeting went wild, with angry protests against Jews who opposed the return of Palestinian refugees, and so on. It was very embarrassing.

So much for bringing a "cooler head" along.

The second time I met Faisal Husseini was at dinner at the Casablanca Conference in November 1994. We exchanged a few words, but did not really talk. He barely acknowledged us.

The meeting at Orient House went well, but was somewhat formless, probably because we didn't have a clear idea of what the Jerusalem project would do. Lenny emphasized the usefulness of thinking about joint problems such as sewer systems, police, and schools that would have to be solved by any administration in a dual city, which we think will be the ultimate solution. The Palestinians were somewhat noncommittal about this. The Israelis, when Lenny talked to them later in the day at Beilin's office, were very emphatic about how sensitive the issue was and how they could not appear even to be talking about it, especially before the anticipated Israeli elections.

Somewhat symbolic of all these sorts of difficulties was just trying to get taxis that particular day. I started out at the Hotel Moriah in West Jerusalem to get a taxi to Orient House, but could not find a driver who would take me there—a patently illegal refusal. I rushed back in to the hotel, called Shula, and asked her what to. She suggested I find a taxi to take me to the American Colony Hotel, also in East Jerusalem and literally around the corner from Orient House. Finding *that* taxi was easy; apparently, the hotel was considered a safe and usual destination by West Jerusalem drivers.

After we left Orient House, Lenny, Shula, and I needed a taxi to get to Beilin's office. We tried to get one from the American Colony Hotel, but being the month of Ramadan and with darkness falling, East Jerusalem taxis were not to be had. As Muslims, they were now permitted to eat. Finally, we managed to get a West Jerusalem driver to come down from the Hyatt Hotel on Mount Scopus.

At Beilin's office, Lenny, Shula, and I met with Amir Tadmor and Alon Liel. Most of the meeting concerned the Jerusalem and refugee projects. The Israelis were much less sensitive about the refugee project than about the Jerusalem one. The rest of the meeting concerned the Water Project. Alon Liel explained he had been at the Fishelson memorial and didn't need the entire presentation. What we wanted, I said, was tactical advice, particularly regarding the Jordanians. We had not heard from Rima Khalaf again and did not know what the Jordanian position would be. There was also the difficulty about whether I could even tell Munther Haddadin about the Dutch Initiative.

I described my relations with the Jordanians as one of having the feeling they were in another room. I had my hands inside that room and was manipulating something; something was happening; but I couldn't tell what.

"Don't worry," Alon Liel said. "It isn't you. With them, it's always like that."

Liel and Tadmor told us who would make up the delegation from Jordan to the Hague. There was a new name: Bassem Awadallah. They said it would be very desirable for us to speak to him in advance, and suggested further that we reveal a little more to Munther and that we try to get Nabil Sha'ath to speak on our behalf.

"He is your soldier in this," Liel told me, emphasizing how enthusiastic Sha'ath was about the project.

It was then time again to deal with the taxi situation. I needed to get to the Sheshinskis, where Ellen was to meet me. Despite the best efforts of Amir Tadmor with various taxi companies, there was no taxi to be had in West Jerusalem, even for the government, at approximately 6:00 p.m. Eventually, Eytan came and collected me himself.

The meeting with Liel and Tadmor thrust us into somewhat frenzied action over the next 24 hours or so. Shula spoke briefly with Munther, who was annoyed that we hadn't come to see him again.

More important, on Friday morning, January 26, she phoned Jawad Anani at his home in Amman. He asked her who was going to The Hague from Jordan, and when she came to the name Bassem Awadallah on the list, Anani said "He's very important. What is more, you can stop right there. He is coming to my house this afternoon at three o'clock. Call me back at four o'clock. I will have briefed him and he will speak with you then."

We did just that. He seemed quite sympathetic and asked that we fax him materials at his office that evening. That turned out be something of a nightmare.

The terms of reference, which we wanted to send him, existed only in the ChiWriter program on the project computer. We went to a friend of Shula's near Tel Aviv with a big computer set up, but he had only Apple-based machines and we couldn't get his printers to run for us. Eventually, we translated everything into Word, cleaned it up, printed it out, and faxed it. We got through most of the major materials before the receiving machine ran out of either memory or paper. Still, Bassem Awadallah apparently received the important material.

That same morning, Shula and I also met in Tel Aviv with Roby Nathanson, an economist who was advising the Beilin ministry. He was also going to The Hague. He gave us some sensible advice about how he thought things would play out. We agreed that we must not be seen there to be very friendly with the Israelis, although it was clear they were doing their best to push the project forward, and, of course, we felt extremely comfortable with them.

I was also beginning to feel rather comfortable with the Palestinians. With the Jordanians, although perfectly nice, we were not on such easy terms.

Ramallah

On Saturday, January 27, Shula and I went to Ramallah for what turned out to be a large meeting at Samih Al-Abed's office on the water project. There were lots of people in attendance.

Karen Assaf, whom I had met at Jad's institute in January 1995, came and brought observers from some Scandinavian countries. She held some water-related position in the Ministry of Planning and was particularly interested. She said she had spoken with Nabil Sharif and they were all looking forward to receiving copies of the model.

We also met with Samih Al-Abid himself. It was plain he did not know what our project was about. He did tell us that Nabil Sha'ath would be traveling to The Hague on Sunday, the same day as us, and going on Lufthansa by way of Frankfurt. I suggested to Shula that we try to change our reservations to get on the same plane. I had been unable to speak with Sha'ath since our meeting in Cambridge the previous November.

It turned out that either Shula had remarkable persuasive powers or Lufthansa simply lacked security. She phoned Lufthansa and was told there were three flights to Amsterdam connecting in Germany. One went through Munich, so we knew that wasn't Sha'ath's flight. The other two, one in the morning and the other in the late afternoon, did go through Frankfurt. Shula asked which one Nabil Sha'ath was on; amazingly—and *illegally*—the Lufthansa agent told her it was the late afternoon flight.

Ultimately, we decided not to change our reservations. We needed to be fresh for our meeting with the Dutch early on the morning of January 29th, and Sha'ath's flight would get us in at a very late hour.

11
The Minister's Meeting

Finally, on Sunday, January 28, Shula and I traveled to The Hague. It was an easy trip, but marred by Shula's wallet being stolen between the train station and Schiphol Airport and the hotel at The Hague.

Our travel agent had made an extremely fortunate choice by booking us in the Hotel Des Indes. Not only was it a terrific hotel, but it was also the hotel in which the various ministers and many members of their delegations were booked. Over the course of the next day, we watched security people of obviously different nationalities arrive, and we could tell who was about to come through the door next.

Shula and I spent a lot of time hanging out in the lounge, trying to talk with people and get our papers into their hands—which was not always so easy.

The next day, we went to our meeting at the Dutch Foreign Office that had been arranged by Myriam Van Den Heuvel. She had predicted that she herself would be too busy to meet with us, which turned out to be true. She gave us ten minutes and then left. We met with three other people, the most important of whom was Simone Filippini, deputy head of the Middle East Section. There was also a water expert named Becker, and someone else.

At the time, the meeting seemed to me to be, at best, anticlimactic. The participants didn't know much about the project and they seemed quite skeptical of what we told them. Filippini, in particular, kept asking the standard questions we frequently got: Isn't there a shortage of water? Why would people ever sell? Won't the rich Israelis end up using all the water? And so on.

Toward the end of the meeting, however, she suddenly woke up to the fact that this solution would depoliticize the water question. Then she became much more interested. When the meeting ended after nearly two hours, Filippini told us specifically not to expect to make a presentation at the ministers meeting the next day.

"It has been decided that there will be no presentations," she said.

Shula and I returned to the hotel. We talked about Filippini's reaction, but I had the feeling that we were just talking ourselves into believing things had gone as well as we had hoped. We spent the rest of the day sitting around and again watching people arrive.

The first one I saw was Rima Khalaf, from Jordan. I met her at the elevator, introduced myself to her again, and we chatted for a bit. She told me that she gathered I was to make a presentation the following day. I told her that I was certainly available, but that I had been told there would be no presentations.

"Well," she replied, "your project will certainly be discussed."

When I had met Khalaf the previous November, her name struck me as unusual. At the time, I had asked whether she had ever read *Green Mansions*, a semi-classic novel by William Henry Hudson once beloved by teenage girls. Its heroine is Rima, the bird girl.

She said she never had, and I said I hadn't either. So, when I got back to Cambridge, I sent her a copy.

Now, in The Hague, I asked her whether she had received the book. It turned out she had sent a thank-you letter earlier in the month that hadn't yet arrived. She told me she hadn't yet read it, but she was surprised to see it was a jungle story. I reminded her that I hadn't read the book and noted that my only guarantee had been that the heroine was named Rima.

I went back to "minister watch" with Shula, and eventually Nabil Sha'ath appeared. He saw me in the lounge and said, "Oh, there's Professor Fisher." He came over to shake my hand and made a pleasant comment about my having changed his life.

I replied that he had certainly changed mine, but that was the extent of our conversation.

So, brief a conversation was not unusual. The ministers mostly fraternized with each other. None seemed particularly interested in talking with me in person—and of course, the Israeli delegation did not want to speak with me at any length in public. I saw Yossi Beilin arrive last, coming from a meeting upstairs in the hotel with, I suppose, Dutch Jews. He did not notice us at all, probably deliberately.

Below the ministerial level, however, there was more contact, as with Ali Sha'ath and Samih Al-Abed among the Palestinians and with Bassem Awadallah from Jordan, who came over and introduced himself and said he had received our material.

Tuesday, January 30, was the day of the big ministers meeting. We sat at breakfast and watched the ministers head from the hotel to their 9:30 a.m. meeting at the Foreign Office. We thought it was politic to remain in the hotel in case something happened.

At 10:30, the phone rang in my room. It was Myriam Van Den Heuvel. With a somewhat changed tone of voice, she asked whether it would be possible for me to come to the foreign office by 11:00. I told her I thought we could probably manage that.

I raced downstairs to find Shula, who I had decided to take with me even though I knew she would not play any direct role. Her husband Shalev, who had joined us in The Hague to attend a conference, told her to change into more formal business clothes, as we were going to the meeting.

We waited for a while after arriving at the Foreign Office and then were ushered into the ministers meeting, presided over by Jan Pronk, the Dutch minister. There were perhaps forty people in the room. Hans van Mierlo, the Dutch Foreign Minister, joined later. The Egyptian delegation included a minister I had not met; she had replaced Yussuf Boutros-Ghali.

The delegations sat around an immensely long table, with Minister Pronk at its head and me at the foot. The Egyptian delegation, which included the minister I had not met, was on my left; on my right was the Israeli delegation. Nabil Sha'ath and the Palestinians were seated further up the table on my left, near Pronk. Across from them, between the Israelis and Pronk, sat the Jordanians.

Yossi Beilin and I exchanged barely perceptible nods.

Minister Pronk said it had been decided that I would make a presentation. I asked him how long I had, and he told me about twenty minutes. They would like to finish the discussion in an hour, he said.

I made my standard presentation, emphasizing the benefits that could be had through trade. I knew by then that a number of the parties wanted this model for internal planning purposes; indeed, I had been offering it with that as the bait. But my basic message was that the water situation could be solved through trade, with everyone gaining. I was not going to moderate that message, and never have since.

Pronk then opened the floor for questions. Rima Khalaf was the first to speak. She advocated caution and experiment, suggesting that they should all learn more about the model and experiment with the it as applicable to their own countries before plunging into anything as controversial as trade in water. She thought that plans for a joint management authority should come toward the end of (what they saw as) the project, which was to be in the early fall before their next meeting.

It occurred me that she had described pretty well the very work plan we had devised for the next six months. But before I could say so, Nabil Sha'ath spoke up and suggested to Pronk that it would be better if each delegation could speak and then have me comment at the end. He then launched into a speech I wish my parents had been alive to hear. He spoke of how important our meeting in Cambridge had been to him and how the project had already changed Palestinian water policy. While we had not been the only force pushing in that direction, their thinking had moved away from desalination plants and toward retreatment plants and conveyance facilities. They looked forward very much to experimenting with our model as a planning tool.

Sha'ath's support was truly effusive.

When my chance to speak came again, I remarked that Sha'ath made me go from being afraid no one would listen to being afraid people *would*. While I did not say it at the time, I was reminded of an old science fiction story in which a Berkeley mathematician publishes a paper predicting that on a certain date, at a certain time, an earthquake will strike and California will fall into the Pacific Ocean. Because the professor is so well known and respected, everyone else leaves—but he stays behind, since he thinks it's only a theoretical exercise. On the morning in question, he is reexamining his work and finds he has made a sign error in the mathematics. He hears a rumble and looks to the east. The Atlantic Ocean is coming toward him.

I hoped it wouldn't be like that.

After Sha'ath spoke, the Israelis offered support in a suitably reserved way. They emphasized that it would matter who got to work on the project. I said in reply that that was indeed very important, because it was important that the project be taken over by the parties themselves. We needed politically suitable people, nominated or approved by the various governments.

The Egyptian minister essentially said that his country had no interest in the project and, at least at this time, he did not wish to have an Egyptian model built. From my point of view, that was just fine. We had enough to do in the next six months. With a little more discussion—we had actually taken an hour-and-a-half—the meeting adjourned for lunch.

Minister Pronk asked whether Shula and I would join the ministers. I replied, as one is supposed to when offered food, "*Graag, ja.*" ("Yes, please.")

I could speak a very small bit of Dutch, having spent 1962–63 living in a Rotterdam suburb while visiting what was then the Netherlands School of Economics and is now Erasmus University.

Pronk stared at me for a moment and then said, "Well, another surprise."

I felt as though my entire life had been preparing me for that day and that project.

At lunch, I offered to sit down at the central table, but that was reserved for the ministers. I ended up at a side table with the Palestinian Ambassador to the Netherlands and a woman from the Israeli Embassy. They had become friendly with each other.

When lunch ended, the ministers went back to their meeting and we returned to the hotel, thinking we were no longer needed. Later, we learned that they wanted to continue the discussion with us and had sent someone to run after us, but we were already gone. They decided it wasn't worth calling the hotel.

The upshot of the day was that we were accepted for the Dutch Initiative. Several steps were laid out. The Dutch were to write us a letter asking for terms and conditions and a detailed outline, which we would then send to the Dutch so they could send it to all the parties.

We did not know much money would come from the Dutch Initiative. The Israelis told us, however, not to go forward with Yad HaNadiv, since that institution always wants to be the first funder. Alon Liel suggested further that we not proceed down the Revson line—something we agreed with at the time.

Further, we were to go forward with an eye to reporting at a meeting of experts in September or October, to take place before the ministers would meet again later in the fall. We were also asked to produce "countrified" models—adaptations of the model that would allow each country to experiment with what would happen without trade. We were in a position to do that very quickly. We were also to provide training in the use and interpretation of the model—something for which the Palestinians were particularly eager.

In addition, the model was to be revised, with close attention paid to data problems such as conveyance costs. We were to attempt to introduce seasonality.

Finally, we were asked to approach the question of trade and a joint management authority with great caution. Any write-up would have to be delicate in this matter; no one had committed themselves to anything.

Indeed, we were strongly cautioned to keep the entire matter secret. It would have been politically explosive within each country for it to be known that they were talking about trade in water. Later, at the end of February, Alon Liel told us that the Israelis would not even meet with us if we came between the first of April and their elections. I had become something of a public figure in Israel on the water matter and it should not be known how interested the present government was.

That evening, I was invited to a formal dinner in the hotel. I didn't get to hobnob very much with the ministers, though. Yossi Beilin and I exchanged a few sentences before dinner, and that was it.

The next morning, I took the train to Schiphol and returned home.

12
Intermezzo

With one very major exception, the month of February 1996 was something of an anticlimax. But the exception was very large, indeed.

When the Dutch foreign minister joined the ministers meeting, he had remarked that Syria and Lebanon had been invited to join the Dutch Initiative (not then focused on water) and that he had some reason to believe they might accept the invitation in the future. Then, around February 10, we received a call from Yair Hirschfeld, who asked us to prepare a proposal for a workshop that would be centered on our model and that would involve not only Israel, Jordan, and Palestine, but also Syria.

That was easy to do. Earlier, we had prepared a letter along those lines both for Uri Savir, negotiating with the Syrians, and for the Syrians themselves.

Hirschfeld warned us to make no reference to the Dutch Initiative and to emphasize in the proposal the various things that might interest Syria. So, I addressed the usefulness of building a Syrian model for domestic purposes, the possible modeling of Syria's dispute with Turkey, the advantages to joint modeling of Syria and Lebanon with the three current model partners, and what could be learned about regional cooperation and gains from trade between, say, Lebanon and Israel-Palestine-Jordan. In addition, I mentioned that it might be possible to settle the likely dispute over water in monetary terms when the Golan is given back.

Yair emphasized that that this approach to the Syrians was super secret, even more secret than the Dutch Initiative. At the time, only Lenny, Shula, Ellen, and

Theresa Benevento, my secretary, knew about it in addition to me.

Interestingly, when Lenny went to Israel in late February, Yossi Vardi, who was one of the water negotiators with Syria, and who had attended the Fishelson memorial, told Lenny that "not everyone agrees with Frank's ideas. We are worried about Israel's security." He asked that I meet with the water negotiators on my next trip.

The Dutch Initiative was secret, too. As I have already remarked, the notion of trade in water is political dynamite. Everyone believed it meant that they would be *forced* to sell their water. In fact, of course, this was wholly untrue: our system forced no one to sell, and of course not everyone can be a seller. It looked as though the Palestinians and Jordanians would end up buyers—despite the myth that rich Israel would buy all the water.

With the rather startling exception of this possible opening on the Syrian front, the rest of February and early March was devoted to administration and cleaning up loose ends.

Growing Pains

With respect to funding, it had become clear by the end of December that David Brooks and IBRD would not, in fact, fund the Palestinians for our project. He was funding a joint project between Jad and Atif Kubursi on environment and the aquifer. We had been instructed not to proceed with Yad HaNadiv, and we had received no reply from Asit Biswas as to the Japanese.

Finally, on March 1, we met with Eli Evans of the Revson Foundation. Bob Rifkind had suggested that Evans wouldn't be meeting with us if he were going to give us only $25,000. I hoped he was right. I had suggested back to Bob that it was time for the foundation to earn a little *koved* ("honor).

When we told Evans we had a governmental sponsor but could say no more, he asked whether it was because we were embarrassed by which government it was or because it was something like the CIA. We were happy to assure him that neither was the case. He agreed to give us $30,000 immediately, and said he would entertain an application for much larger funds if we put together a thoroughly serious budget.

Of course, the Dutch were to be our principal funder. They were, however, remarkably slow in getting us the formal letter, and time kept passing. Finally, we called them and then faxed them asking for permission to return to the Middle East beginning on March 8th. In response, they asked for a detailed trip budget, which we provided and they approved—imposing a fair amount of routine, but annoying, reporting details.

The positive reply from the Dutch also gave us the appropriate political cover for our visit to Jordan. Bassem Awadallah had spoken to Shula and suggested we not come to Jordan or even meet with Munther Haddadin before the Dutch paperwork had gone through. Afterward, Rima Khalaf replied to us and suggested we meet with various officials.

Even still, we needed some closure on funding with ISEPME, which was still in a stage of worrying about every couple of hundred dollars. I insisted to Shula that we never again make reservations with nonrefundable tickets. She at least followed that instruction for my ticket on the next trip; hers and that of Harshadeep were nonrefundable.

The AID proposal remained undecided. In early February, AID asked again for an extension of our proposal, which I told Shula not to grant. With the Dutch Initiative, we were already where we wanted to be; it would take another three years to get there along the AID proposal route.

The French proposal was dead. Madame Barbutte, the head of the French Environmental Agency, told us that despite her strong support the French Foreign Ministry would not proceed. They do not want to fund an American university, and they had stated that a market approach to water is not in line with French thinking. *Zut!* They had missed *le bateau.*

Madame Barbutte did send me a report on the Jordan River Basin her agency had put out.

Our preparations for the Dutch project were a little slow, or a little difficult. We got a lot of help from Theo Panayotou at HIID, but it was not always quite clear what we needed to do. Still, we did decide on a few things. First, HIID would take charge of organizing a study for the joint management authority. Second, in what seemed a very good idea, we asked Atif Kubursi, who was on sabbatical, to spend considerable time in the region. We wanted him to assist the Jordanians and, particularly, the Palestinians in revising and expanding the model and help them gain experience in model use and interpretation.

We also employed Harshadeep full time for the month of March. We wanted to employ him for as long as the Dutch Initiative lasted, getting him "seconded" to us from the World Bank where he was about to be employed. But he decided that would be impolitic. So, we also began to interview possible replacements. One was Annette Huber, who grew to be my principal colleague.

We also had to revise the World Bank paper, due on March 8th. It was mess. Moshe Syrquin, the editor, expected a polished paper. I had to leave it the hands of Bob Dorfman and Aviv Nevo—which was not a success. Bob wi not only very deaf but was also getting old and forgetful. He and Aviv wou talk, quarrel politely, and then Bob would forget the conversation. So, he wou reject Aviv's write-ups. He would claim to have worked on something for a whi and then show me what was substantially the same as the first 12 pages of th paper we had already submitted. Aviv's alternate write-up was certainly no bette Moreover, Bob—who wanted to quote results—overlooked the fact that Aviv write-up did in fact include results.

So, there had been no progress, and I had no time to deal with it. I said s(Shula never wanted to offend anyone, but worried about our obligation. A ne version was supposed to be waiting for me in Cambridge when I returned fror Philadelphia, and we tried to get a final version arranged by fax.*

All these funding and administrative matters unfolded at a time I was bein inundated with requests for papers and information and with various visits an meetings. This stemmed largely from the talks I had given in Israel in January and partly because of the appearance of a short version of the Stockholm pape in the *Stockholm Water Front*, the newsletter of the Stockholm Water Conference Even the Germans, who had promised to stay in touch with us the year befor despite deciding not to fund us, asked what was happening.

The project was having growing pains. We needed a better administrativ structure to manage all the paperwork—which had become a minor distraction

Some of these visits were more substantive than others. After the March 1 lunch with Eli Evans of the Revson Foundation, I went to a talk at MIT's Applied Sciences Division given by Jona Bargur, an economist with Tahal, the official Israeli water engineering company. He had come to visit at Habash's suggestion.** Bargur, an entertaining, friendly, and very intelligent guy, had buil a model of water systems in North China. Its key feature was something our

---

* I was playing in a bridge tournament in Philadelphia. Since retiring from teaching in 2004 (but still working on water issues), I have become fond of replying to questions about how I spend my time: "I divide my time between bridge and troubled waters."

** Not to be confused with the infamous and late terrorist George Habash, who was then the leader of the Popular Front for the Liberation of Palestine, this was the nickname for Haim Ben-Shahar.

project didn't do at the time: a policy maker with varying objectives could choose between different outcomes. Using a technique he had named after Tchebychev,** the implicit weights given to the different objectives could be worked out and the optimum solution found. Implicitly, the weights given to different objectives (employment, agriculture, industrial output, etc.) are provided by the demand curves for water, including national policies. We agreed that there might be a future place in our project for such techniques, and that we would explore them again.

A Low Moment

We continued our work, but things in the region got very bad. March 4th saw the third major terrorist bombing in Tel Aviv in a 10-day period. It was the work of an extreme faction of Hamas, in retaliation—I supposed—for the killing by the Israeli secret service a couple of months earlier of the chief bomb maker of Hamas, who was known as "The Engineer."

I thought at the time, and I think now, that it was unwise for the Israelis to retaliate. The probable effect would be that the Labor Party would lose the upcoming Israeli elections, set for May. Even before the second attack, Shimon Peres's lead in the polls had dropped and the race was about even.

It was all very foolish, tragic, and hard to understand. The Israeli public would probably erroneously believe that if only the peace process was halted, these attacks would end—because the attackers opposed the peace. I didn't believe it would. The people carrying out the attacks were not only criminals but also fools—their best chance for a Palestinian state in the foreseeable future was to refrain from such activities. The attacks were likely to get a Likud government elected, which would delay—if not derail—the peace process.

It was true that in the wake of the Palestinian election, the peace process might be irreversible, since any Israeli government would have to deal with the fact that there is a properly elected government for the Palestinians. But that wouldn't stop Likud from putting every barrier in the way of further accommodation (as they continue to do today).

** Pafnuty Lvovich Tchebychev (1821–1894) was a great Russian mathematician and statistician of many remarkable qualities beyond that there are eight ways to transliterate his name into English, German, and French. This one here is one of the French transliterations; his Wikipedia entry uses Chebyshev.

It was a very low moment. Peace and the peace process seemed dead in th water. With no pun intended, I thought that might very well also be true of ou project. I hoped that would not be the case.

(A few days later, while I was in Israel, Roby Nathanson told me that eve if Labor did lose the election—which, of course, it did—our project *wou* survive, at least in the sense of having provided a tool for Israeli wate management.)

Making Our Trip

After some hesitation, Shula, Harshadeep, and I decided to go ahead with ou trip to the region beginning on March 8th. Amir Tadmor was quite encouraging as was Jad, who in the wake of the second bombing had said that people like u must go forward and work—as the only answer to those who believe in violenc and terror.

Our purpose was to organize the three country teams that were to work under the Dutch Initiative. While we did not yet have the formal letter from the Dutch approving the project and asking for plans and a budget, we had gotten their approval for the trip.

In a number of ways, the trip was quite frustrating. To begin with, Atif Kubursi—we had arranged to hire him half time for the next six months—had kidney stones and was unable to come. More important, the trip ended up mostly involving Israel and Jordan, because following the terrorist bombings in Israel no one was allowed to go into Gaza. Not even Yossi Beilin's office could get us in. That meant we were unable to meet with any Palestinians, who wanted to meet with us. Indeed, it looked as though the most practical thing would be for the Palestinians to leave Gaza by way of Egypt and come to the United States to see us.

It was a tense time in Israel. At one point during the trip, we met with Roby Nathanson in his office in Tel Aviv. Yossi Beilin was conducting a large meeting in the next room. Suddenly there were loud sirens outside. We all moved to the window, looking out with apprehension and dreading that it was another bombing. But it was only a cavalcade of official cars bringing Shimon Peres to the Ministry of Defense up the street.

Despite the tensions, our time in Israel was particularly positive. We had gone from a situation in which we had made the least headway with Israel to one in which our project was an idea whose time had come.

On Tuesday, March 12, we had an excellent meeting with Alon Liel and Amir Tadmor. We then headed to Ben Gurion University in Beersheba for a conference on the Negev, the last part of which was to involve a number of Jordanians, including Bassem Awadallah. Liel was chairing that session, and he put me on the program at the last minute.

At that conference, Avishai Braverman, the university president, made it very clear that he and his school wanted to participate in our project. I took the view that the inclusion of Braverman and Ben Gurion University was an internal Israeli matter that had to be settled, and I discussed it the next day with Uri Shamir. His attitude was that he would consider anyone based on merit.

Uri also arranged a meeting with several people in the office of the water commissioner—chiefly Yossi Dreizin and Yehoshuah Schwartz, who had expressed interest after hearing me at the talk Uri had arranged for me at Bet Sokolov the previous January. Dreizin in particular raised a number of doubts about whether the model gave adequate attention to the need to preserve agriculture,* but as on other occasions under similar circumstances with technical people, a good deal of progress was made once we got down to brass tacks and Harshadeep presented and discussed the model. Ultimately, they decided that the water commissioner's office would assist in revising the data and exercising the model.

That evening, we dined with Uri and Yona Shamir, Shaul Arlosoroff, and Zvi Eckstein. We could finally tell all of them about the Dutch. The three of them were to head the Israeli team, with Shaul in charge of coordination, especially of data. They were quite enthusiastic.

We also discussed Braverman and Ben Gurion University, and the issue of what to do about Hillel Shuval.

The three of them made it clear that while they had no objection to Hillel continuing to be involved as an advisor to me, they could not include him on the Israeli team. They didn't believe he had anything professional to contribute that they really needed. Later, Shaul told us that Uri and Hillel had a history from when they served together on an investigatory commission. Hillel had spoken to the press before the commission had reported, which caused considerable

---

* Unfortunately, this was only the beginning of problems with Dreizin. He went on to become a bitter and biased foe of the project, partly because of his attitude towards Arabs.

political damage for the commission's other members. As a result, Uri would not work with Hillel.

We agreed this would be considered an Israeli matter and that it would be up to the Israeli team, not me, to inform Hillel of the decision. Zvi Eckstein was to do the informing. This was somewhat problematic, since Zvi was not as tactful as the others. But he had no professional issues involving him and Hillel.

I doubted very much that I had heard the last of that issue.

As for the Ben Gurion issue, a meeting was set up between Uri, Shaul, Zvi and Beilins's office for after I returned to the United States. Shula and Harsh attended. Uri apparently resisted including Ben Gurion University. More important, though, the Israeli team said they wanted to be included in expanding and improving the model, and they wanted full credit for it—something with which I had no issue problem. Surely, Zvi, in particular, as well as Uri, had a good deal to contribute to model building. Any problems that might arise would be solely political. The Israeli budget couldn't be higher than either of the other two budgets, and the model could not be seen to be an Israeli model. But I believed all that could be worked out.

We were drawing up a budget with the three country teams based on a budget for the Israelis, but there were some difficulties. Rima Khalaf had apparently insisted that the people who would work on the model going forward ought not to be the same people who had worked on it in the past. She said that would yield an objective evaluation. But that would also be impossible. We wouldn't object to evaluations by the governments and experts who had not previously worked on the model, but we couldn't tolerate totally cutting off all the people who knew about the model and had worked on the data in the three countries.

It was going to have to be worked out.

On to Jordan

In January, I had remarked to Amir Tadmor and Alon Liel that in dealing with Jordan I felt as though I had my hands extended into another room, where I was moving something, where something was happening, but I couldn't tell what it was.

"Don't worry," Alon Liel had said. "It isn't you. With them, it is always like that."

We were particularly anxious to meet with Munther Haddadin, who we had

160

missed the last two times in the region. As we prepared for the trip, we learned that Munther was quite annoyed at us. He claimed we had devoted much more attention to Israel than to Jordan and that it had been our own fault that we had not met with him—despite that we had been told he would be away on trips he ended up not taking. Apparently, Munther had forgotten—or chosen to ignore, rather conveniently—that it was his own schedule that had been responsible for us not seeing him at length in January.

This time, we offered him three days. He got annoyed at Shula for forcing him to meet with us on particular days. He was quite difficult.

Our original plan was to fly to Jordan to begin our trip, but that was not convenient for Munther, so we ended up agreeing to meet with him on Thursday, March 14, return to Israel because Friday is a Muslim holiday, and then come back again to Jordan on Saturday and Sunday. Munther warned us not to be late or he wouldn't wait for us.

We were delayed at the bridge, and phoned ahead to warn Munther. Still, we ended up arriving at the Royal Scientific Society nearly on time. But he had decided to do something else first, so we had to wait for him. When the meeting began, something very strange happened.

We knew that Jawad Anani had already told Munther about the Dutch Initiative. Munther knew that we knew. Nevertheless, I began our meeting by informing him fully about the Dutch Initiative and the extent to which the various governments were now involved, particularly his Jordanian government.

When I finished, he congratulated me, saying we were now "going in by the front door"—which, he continued, is what we should do. He then made it clear to us that he thought it was now totally inappropriate for him to talk with us, since we had some official governmental status and he did not. He plainly felt he should not participate until he was asked to do so by the appropriate ministries in Jordan. He counseled us on how to proceed, and then we spent the rest of the day largely having lunch and taking a tour of Amman.

Munther's attitude made perfect sense—except we couldn't figure out why he hadn't told us on the phone instead of absolutely insisting that we come to see him, be on time, and not make appointments with government officials until we had seen him? Munther was certainly manipulative, and liked both to exercise power and have due respect shown him, but this particular behavior baffled me. Could it have really been that he just didn't want to talk about it on the phone?

In any event, we called the Ministry of Planning and spoke with Boulos

161

Kefaya, an official who had sent us a fax about whom we should see in reply to our inquiries made to Rima Khalaf. We made an appointment with him for Saturday morning, and his office promised to make other appointments for us to see the other people Minister Khalaf had suggested.

We returned to Amman on Saturday morning and met with Kefaya and an assistant. We had another meeting with the Water Minister and his assistants the next day; I had met the minister two years earlier with Lenny.

In both cases, it was pretty clear that it was everyone's first exposure to the model. Things went less than swimmingly. For the first time in a long while, I found myself confronted with people who had just never thought about water as we did. Only when we got down to details, with Harsh talking to assistants did things really begin to move.

Their basic problem was that none of them could imagine why Israel—also water "short"—would ever sell them water. As a result, they did all their planning on the basis of a fixed quantity of water. To make the point, the Water Minister said he would be glad to pay Israel $2 per cubic meter for water in the Jordan River, but where, he asked, would Israel get it to sell to him? Someone else suggested that Israel might buy the water from Turkey for less and then resell it.

I pointed out that just because both Jordan and Israel were water "short" didn't mean they valued a marginal unit of water at the same amount. That made no impression.

After I was back in Cambridge, I realized what I should have said, and phoned Shula and Harsh to tell them for their next meeting: If Jordan would pay $2 per cubic meter, then Israel would find it profitable to sell to them, because Israel could (if nothing else) desalinate near Haifa and put water into the national carrier there. It was an example the Jordanians could understand immediately.

Much of the rest of the discussion had to do with whether the model was telling them things they already knew. The first issue concerned the model results that strongly suggested that water from the Disi aquifer should continue to go to Aqaba and not to Amman. At the time, the project to build a pipeline from Disi to Amman was in the planning stages; it has since been built. The Jordanians believed that the conclusion of our model was wrong—because of the question of whether there would be trade.

The water situation in Amman would become ever more serious unless water was brought there from elsewhere. The efficient way to do that would be to bring water from the Sea of Galilee—what the model strongly stated. With sufficient water sold to Jordan from Israel and coming from the Sea of Galilee, the crisis in Amman would be alleviated. It would negate any reason to bring water from Disi, well over 300 kilometers away.

The Jordanians didn't think like that. They insisted that they didn't have sufficient water in the north to bring to Amman, and so the crisis would not be alleviated. Hence, water must be brought from Disi.

I pointed out that the model would give results consistent with that prescription if they were to limit the amount of water available in the north and have a no-trade scenario. In other words, the model would still give (from their point of view) sensible results and would be a useful tool. I think that penetrated, somewhat.*

The second thing they thought they already knew had to do with the model's call for desalination at Elath or Aqaba. They didn't see why they needed it at Aqaba. Unless there was something wrong with the original Jordanian reports' estimates of demand in the south (as there well may have been), I didn't understand the problem.

Moreover, while on the way to the conference in Beersheba the previous Tuesday, I had asked Shaul Arlosoroff what to talk about regarding the Negev, and he specifically discussed the proposition that the desalination plant being constructed at Elath would, in the near future, be able to realize economies of scale by selling water to Aqaba where it was needed. His question was whether it would continue to be needed if Disi water were brought to Aqaba in increasing quantities in later years.

It was my experience that Shaul always knew what he was talking about when it came to such matters.

Most of the Sunday meeting at the Water Minister's office, however, was spent not on substance but on formalities. The Minister insisted very strongly that it would be his Ministry and not the Ministry of Planning or anyone else who would be in charge of this project. Boulos Kefaya, who was present, agreed, telling us that the Ministry of Planning is responsible for signing international

---

* It is interesting to note that as I write this chapter in 2013, Jordan and Israel are discussing (and may have already agreed on) an arrangement in which Jordan would sell water to Israel in the south and buy from Israel in the north (at a much lower price). Such an arrangement would be beneficial to both parties if they can agree on the prices. Much of the Disi water would continue to be conveyed to the North, but some would presumably be rerouted to southern Israel.

agreements of the sort we were involved in, but that the substance was then turned over to the responsible Ministry. Moreover, the Minister absolutely refused to have anything whatever to do with the Royal Scientific Society in this matter.

It all made sense, except that we happened to know that the Minister and Munther Haddadin were very, very close. Indeed, as the meeting began, the Minister took a phone call—apparently from Munther.

Eventually, to my vast relief, someone at the table asked whether they could have a meeting to discuss what the model was actually like and how it worked. Harsh and Shula agreed to return the following Tuesday, and that meeting also ended with the formation of a committee that would become the Jordanian team for work on the project.

We left the meeting somewhat bemused. Late that evening, just before I left for the airport for my 2:30 a.m. flight home, we saw Jawad Anani at our hotel. He listened to our description of what was going on and said that what had happened was exactly what he expected. We were in good shape, he told us. He offered the view that Munther wanted to be extremely careful, because he very much wanted to be asked to participate, and that this was the way it would happen. Jawad was quite optimistic.

I'm sure there must be many Jordanians with whom one has an open conversation and from whom one gets a clear explanation that one can understand on one's own terms. Jawad Anani was the only such Jordanian I had met like that so far. That made him extremely helpful, in addition to the fact that he was very powerful.[*]

So, it appeared that things would be moving in Jordan after all, if not quite in the way we had expected.

---

[*] Recall that the previous November, when Munther and the Royal Scientific Society agreed to work on the project, Munther had said to me, "You have a very powerful friend, and I will tell you his first initial—Jawad Anani."

13
Palestinians in Cambridge

When I returned to Cambridge, my work on the project began with a massive administrative headache—beginning with the fact that we had still not submitted a formal proposal to the Dutch. Shula stayed behind in the Middle East for an additional two weeks, and Theo did not get us a written proposal for the study of the joint management authority until the end of March. Most of all, Shula, ISEPME, and I discovered serious procedures involved in applying for a sponsored research grant at Harvard—including that only a Harvard appointee can be a principal investigator.*

Complying with those procedures, putting together the full budget, and getting it out was a major nuisance, although I recognize that the procedures really are designed to be helpful. As we put together our budget, it reached $2.5 million dollars. Then we got it down to $1.9 million. We also had the continuing problem of "overhead"—something I assumed the Dutch would object to paying. Harvard wanted to take overhead on the entire budget, including what we would be paying regional teams (that would not even be at, or part of, Harvard). The entire overhead issue was a continuing problem for ISEPME, which would get none of it and had to bill for its administrative expenses directly in the budget.

Meanwhile, delays continued. Even the Israeli team couldn't get started because no full-time people could be hired without the assurance that they'd be paid. And all this was made even more difficult by the fact that we had an implied deadline of October, the time of another ministers meeting. Further,

---

* This meant, apparently, that I was to be a Harvard faculty member—which would, of course, realize my fondest dreams! (That despite no salary.)

Ali Sha'ath was continuing to take the position urged by Rima Khalaf that people who worked on the model previously should not continue to work on it as a way to ensure an objective evaluation. As I wrote in the previous chapter, this made no sense whatsoever—including to the Dutch—when it came to revising and updating the model. Of course, it bore on whether Jad and the PCG would be permitted to participate.

Adding to my headache was the state of a paper promised to the World Bank. Harshadeep and I had produced a really good draft while we were in the region. I sent it around to all the prospective authors, telling them that I would list them unless I heard otherwise. I also had to deal with sensitivities of how the list would be ordered once the names were settled. Obviously, Gideon Fishelson and Zvi Eckstein should be near the top, but I couldn't make it appear that the Israeli contributions were so much greater than everyone else's, so I finally hit on a plan: first list the actual writers of the paper (me, Dorfman, Harshadeep, and Nevo), then the heads of the country teams, essentially in alphabetical order; and then everyone else, in alphabetical order. If I took the view that Abu-Taleb should be listed under "T," the order after the first four authors would be Eckstein, Fishelson, Haddad, Issac, Salameh, and Abu-Taleb, which was about right.

To my great surprise and disappointment, Jad faxed me that he did not want to be included. I begged him to tell me why. Bob Dorfman also told me he did not want to be listed; he didn't think it was appropriate to list him when he hadn't had a chance to read the final draft to determine whether he agreed with it, even though we had been corresponding by e-mail and I thought I had persuaded him that he would have time to do so. Clearly, the problems I described in the previous chapter with Dorfman's participation in drafting this version of the paper had taken a toll. I had to take the World Bank paper away from him entirely to get it out in anything like finite time. He did not complain about that directly.*

Even with the administrative headache, though, things were looking very promising on the substantive and personnel fronts. Harshadeep had handed

---

* As it turned out, Dorfman was once again leaving the project—and this time I was going to permit it. I am sorry to say that, after a spurt of activity and a certain amount of rejuvenation, his age was beginning to show. He bowed out very gracefully at a farewell dinner we had for Harshadeep at the beginning of April.

over his work to three younger students in his department, all very smart, very amiable, and very willing: Annette Huber, Hynd Bouhia, and Trin Mitra. They had all already been very helpful. Annette, in particular, took charge.*

They made their first work appearance over the weekend of April 5–8, when we finally managed to meet with the Palestinians.

The Palestinian Engineers

When our repeated attempts to get into Gaza had failed miserably during my last trip to the Middle East, Shula began working on the idea that three Palestinian engineers—Ghassam Abu Ju'ub, Luai Sha'at, and Khalid Qahman—should come to Harvard for training. It took some doing. They all worked for the Palestinian Ministry of Planning. We had to arrange for American visas and get them airline tickets. Two were from Gaza, and came to the United States via Egypt. The other was from the West Bank, and traveled by way of Jordan.

Getting the visas was an example of how things should go when all the governments involved cooperate. Even so, there were problems. We were asked first to send letters to the U.S ambassador in Tel Aviv, which I did— and apparently to good effect. The Palestinian coming through Jordan held a Jordanian passport, so I sent a comparable letter to the U.S. ambassador in Amman. At first, he took the position that if the man lived in the West Bank, he should get his visa in Tel Aviv. David Mullinex, the scientific attaché at the U.S. Embassy in Tel Aviv, helped us resolve this issue.

Still, though, there was the matter of how the Palestinians in Gaza would get to Tel Aviv if the crossing was closed. The answer came in the form of Rafi Moshe's taxi-driving family.

Rafi Moshe (who had now gone back to his original name of Rafi Cohen), along with his brother Avner and nephew Doron, were all very helpful taxi drivers who had worked for ISEPME for many years, providing conveyance for us whenever we were in Israel. When our needs involved Gaza, they would drive us to the checkpoints at which documents would be passed either

---

* Annette (later Annette Huber-Lee) was to become my closest colleague. Through the years, her name has appeared with mine as co-authors of numerous papers and as the two senior authors of the *Liquid Assets* book. She remains heavily involved, enthusiastic, and invaluable.

to or from Norwegian diplomats, who could take them to the other side where they would be passed to or from Palestinians.

On this particular occasion, the standard process nearly worked. In the end, however, Rafi and Avner realized it would not be possible to get the last documents to the Palestinians in time for them to get to Cairo and catch their scheduled plane. So, they took it upon themselves to go to a travel agency in Tel Aviv and remake the reservations and plane tickets for a later time. This would have worked, except that one of the Palestinians had an expired *laissez passer* and was afraid he would not be allowed to go to Europe even to change planes. So they had to remake the flights.

Eventually, they all arrived, one on Wednesday, April 3rd, and the others the next day. We began our work that Friday and worked very successfully for several days, including over the weekend.

Atif Kubursi was with us all weekend, too. He was a tower of strength. He thoroughly understood the model, was experienced in dealing with software, and was an excellent teacher. We thought he was going to be an immense asset to the project.*

Working with the Palestinians and Ilan Amir

The Palestinians became fired up by the ideas of the project and began to experiment with the model. Doing that, they made a number of valuable suggestions—some about how to make the model more user-friendly, and some more substantive. For example, the Palestinian engineers ran an experiment in which Palestine was cut off but had a lot of water in the Mountain Aquifer. They discovered that it would not be optimal to pump all that water. There could be two reasons for that. One was reasonable: it might not be cost-effective to pump all the water unless demand is high enough. The other reason, though, revealed a flaw in our thinking. In the way we were doing it, the Palestinians had an upper bound on the amount they could pump from the Mountain Aquifer. We just assumed that meant the water would be theirs to use. But with that formulation, the model permitted water not

---

* Unfortunately, Atif did not long continue in the project. Also, he later produced a paper stating that the model would have rich Israel buying all the water from poor Palestine—an oft-repeated error that Jad, unfortunately, persisted with. In fact, though, the Palestinians would sell only if it were beneficial for them to do so.

168

pumped in the Palestinian districts to be pumped in the Israeli districts. If prices in the Israeli district were high enough, the model would assign the water there rather than to the Palestinian districts (as we had intended), because it still had Israeli benefits in the objective function. So, we had not, in fact, succeeded in cutting Palestine off.

This was easy and important to fix. We simply needed to specify that when using a countrified version of the model the upper bounds on shared sources for all countries must be specified. For example, if the Palestinians are given an upper bound on Mountain Aquifer water, Israel must have an upper bound that is equal to the total amount of Mountain Aquifer water minus the Palestine upper bound. Doing so removed the problem.

As one of the Palestinians pointed out, having an upper bound on total Mountain Aquifer water was probably not sufficient, since it might produce the anomaly that Palestinians were instructed by the model to pump all the Mountain Aquifer water there was in some northern set of districts and none in the south. No peace treaty would ever permit that. Indeed, the Oslo II Treaty was quite specific about what the sources of additional Palestinian water would be. So, we made another fix.

The Palestinians realized very quickly that trade would be a major source of benefits for Palestine. They also understood very well that decisions regarding infrastructure would surely depend on whether there was trade in water permits. It was a point I had made to Nabil Sha'ath at the ministers meeting.

Ilan Amir, the agricultural economist from the Technion, arrived about the same time as the Palestinians and went immediately to work—including giving a presentation to us while the Palestinian engineers were in Cambridge. His agricultural sub-model would put constraints on water qualities and calculate the optimizing behavior of farmers. Since that would also generate shadow values for water as a function of water quality constraints, his work could be described as generating a demand curve (or inverse demand curve) for water for farmers. Moreover, when the big model would give water quantities and water prices, Ilan's model could then be used to predict cropping patterns.

But Ilan wanted to work in terms of four qualities of water—fresh, recycled, brackish (saline), and untreated surface water—and three seasons: rainy, transitional, and dry. Regarding the seasons, I convinced him that the

169

overall model should run in terms of two—rainy and other—and that he could then divide up what happens in "other" to suit himself. As for the qualities of water, things were a bit more complicated.

The model then ran in terms of two—fresh and recycled—and didn't allow agriculture to use recycled water until it was treated to make it equivalent to fresh. At first, I resisted moving to a four-quality model, but then I remembered that the reason we had chosen not to use more than two water qualities was that we lacked demand information, which was—precisely what Ilan was going to provide. He would produce, as it were, an inverse demand hypersurface, with four shadow prices a function of four quantities. Given that information, it was fairly easy to see how to incorporate it in the model. The technical relationships would permit either direct use of water or turning one kind of water into another with appropriate treatment costs. There would be four quality constraints in each district, not one.

The actual application of these methods was set to begin. The programming was to be done by early May, and we would be able to put improved agricultural sub-models into the main model, district by district, as Ilan prepared over the spring and summer. I saw this as a major advance, and we put it in our proposal to the Dutch.

Ilan raised another issue in presenting his model when the Palestinian engineers were in Cambridge, when he pointed out a result in which the behavior of maximized agricultural income as a function of water qualities as essentially the same at two different price levels for water, except for a constant gap between the two functions. He claimed this had been met with considerable skepticism and that he had had long arguments with a number of economists about it.

Kubursi and I looked at each other, and while I was fairly sure Atif knew the answer, I was the one who gave it. I used the envelope theorem—a general principle that describes how the *value* of an optimization problem changes as the problem's parameters change. In this case, the derivative of the thing being maximized with respect to the *price* parameter was just the partial derivative of the Lagrangian with respect to price. Hence, to a first order of proximation, one would expect to find the effect of a change in price to be to induce the same difference in maximized income at all levels of the constraints.

"Ilan," I said, "you've been talking to the wrong economists."

Ilan came to me a day or so later and said he wanted to write a paper with me in which he would present his results and give my envelope theorem argument. I declined, telling him my contribution didn't deserve my name on the title page (I thought at most I deserved an appropriately grateful footnote). I did offer to help him write it, though, and we would see what developed.*

In general, it looked as though Ilan's visit to Cambridge for the next six months was going to be extremely productive.

Back to the Middle East
Our full proposal to the Dutch went out in late April. They took until late May to send it to the three parties—Jordan, Israel, and Palestine—with requests for comments.

In late April, Lenny, Shula, and I went to Washington to speak with Leon Furth, who was Vice President Al Gore's national security advisor, and Caio Koch-Weser, vice president of the World Bank. Furth understood what we were doing, but was skeptical we could ever sell it politically in the Middle East. Lenny's objective with Koch-Weser was to get him to share his proposed "water summit" with us, but it was not clear it would happen. Koch-Weser did offer cooperation from the World Bank.

Meanwhile, the Dutch had approved another trip to the region by a team from ISEPME. A group of us went from May 4-16: Shula, Anne Marie Kelly of ISEPME, Annette Huber (who would be doing computer installations in Gaza, Bethlehem, and Israel), and me. Atif Kubursi was supposed to come, too, but events in the region intervened.

In April, the Israelis had launched "Operation Grapes of Wrath"—the code name for a serious, 16-day attack on Hezbollah in southern Lebanon in retaliation for the shelling of Kiryat Shmona (in the Galilee) with Katyusha rockets. Shimon Peres, in a tough election campaign in which his lead had largely disappeared after terrorist bombings in March, certainly felt pressed to do something. But the Israeli reaction turned into an over-reaction, and there were many refugees and heavy casualties in Lebanon.

Kubursi is Lebanese, and shortly before our trip was to start a large group

---

* In the end, Ilan and I published several papers together, and he remained with the project as an important figure.

171

of his family arrived unexpectedly to visit him in Canada, no doubt finding that a much safer thing to do than remaining in Lebanon. Atif's wife then put her foot down very hard on the question of whether he was going to leave her alone with all those relatives. We pressured him just as hard, and eventually he agreed he would come but a day later than planned—which he never did. Instead, we learned he had broken his ankle, which may have been true, but may also have been about saving face. Efforts to reach him by phone were very difficult; it seemed he was evading our calls. I resolved to check whether he was limping when he next came to Cambridge, which he was scheduled to do to help train Munther Haddadin in the use of the model after our return.

Kubursi's absence was a major inconvenience, since the trip's main purpose was to continue training Palestinians. It put a big strain on Annette and me. Part of that strain, however, was our own fault.

Harshadeep's replacements—Annette, Hynd Bouhia, and Trin Mitra— had undertaken a substantial revision of the program that involved a change from from Quatro Pro to Excel, which made the program much more user-friendly, and that adopted a number of suggestions made by the three Palestinian engineers during their time in Cambridge. We vastly underestimated how many bugs could creep into a seemingly simple set of changes. Moreover, it turned out there were other bugs we had not introduced, but had come from the Palestinians themselves or from Harshadeep when he was playing around with the program in early April.

All this made debugging particularly difficult. We started it only two weeks before we left Cambridge and continued until after we left on the plane. Indeed, I insisted that Annette and I had to be on the same flights, and when we changed planes in Zurich, she and I spent the entire three-hour layover period continuing to debug. Debugging went on every night throughout the first half of our trip.

We had chosen to begin our trip in El Arish, in northern Egypt—about thirty miles from Gaza—because we could be sure of meeting the Palestinians there. As it turned out, we probably could have gotten into Gaza, but that was uncertain right up to the time the trip began. Meeting in Egypt proved to be something of an adventure.

Our plan was to meet the Palestinians on the Egyptian side of the border crossing at Rafah, at the corner of the Gaza strip, and travel together to El

Arish. We entered through the Israeli side, while the Palestinians came from the Gaza side. The Palestinians were a bit late, and there were three fewer of them than we had expected. One had not shown up at all, and the Israelis had turned back two others because of problems with their ID cards— including Khalid Qahman, one of the engineers who had been with us in Cambridge and who had organized our El Arish trip.

Even though Shula had spoken to the Egyptian representative in Cairo, who assured her there would be no difficulty, the appropriate names, passport numbers, and other material had not been properly sent to Cairo and then back to the border station. So, we had to wait from noon until after 10:30 p.m. before the Palestinians were allowed to proceed into Egypt. In the interim, Khalid Qahman got the correct ID card and showed up at the Egyptian border post around 8:00 p.m.

There were some positives to all this waiting around. Most important, it gave us an opportunity to form some kind of bond with the Palestinians. We, the Americans, could have continued on to El Arish and waited comfortably in the hotel—but I immediately decided we would not. It was plain to see that my decision was greatly appreciated. Every one of the Palestinians, in one way or another during the day, came up to me and told me that now I was learning what it was like to be a Palestinian. Staying with them created a certain amount of trust.

When we first arrived at the border checkpoint, we had to sit outside the offices in the general area that dealt with transit travelers. Shula asked whether she could use a telephone, and was told there were none. Eventually, she overcame her fears and used her cell phone to call officials in Egypt and Israel—she had heard it was illegal to use an Israeli cellular phone in Egypt. (I suppose, technically, we were not yet in Egypt.) Eventually, Shula reached a member of the Egyptian delegation to The Hague with whom she had dealt before. He went out on a limb for us, telling Shula that, since she was the only one of us that he knew, she was going to be responsible for the fact that he was accepting personal responsibility for the group's *bona fides*.

From time to time as we waited, the police would call into their offices a Dr. Samir, who was one of the Palestinians in the group. They claimed they were negotiating with him as a group leader, but we suspected something else. Dr. Samir was from Islamic University in Gaza and was seriously bearded; we figured they thought he fit the profile of a radical and so kept on grilling

173

him.

Eventually, as the hours wore on, Shula and I were invited into the office of the captain in charge of security, who personally had to approve the passports of everyone who entered Egypt. He was quite affable, and discussed water and related matters for some time despite his poor English. Luai Sha'at, another of the Palestinian engineers who had been in Cambridge, provided translation; he was one of those who did arrive at the checkpoint. It was also despite my poor knowledge of actual hydrology. I put on a pretty good show of knowing something about water, nevertheless—but was certainly not about to try explaining to him the economics of water and what the project was about.[*]

Finally, in the evening, it became clear we were going to be able to proceed to El Arish. Luai Sha'at came to Shula and me and asked us for $20 dollars to give to the captain as a *baksheesh* ("tip"), assuring us it would be appropriate. He was quite wrong. It might have been appropriate for a soldier, but not for an officer—at least not for that officer. The captain became extremely angry, told Luai he would be reported to his superiors, and threatened to hold up his passport.

Meanwhile, it took until 10:30 p.m. to get the final clearance for Khalid Qahman—at which point there was general applause. A bus then took us to El Arish, with a police escort, which I suspected was because of an attack on tourists in Cairo a few weeks earlier. The Egyptian agency we dealt with had told us Egypt would be particularly safe while we were there because of the government crackdown following the incident.

Finally, we arrived in El Arish at around 11:30 p.m. Annette and I stayed up until 5:00 a.m., continuing our debugging work. It was indicative of just how hard Annette worked. She did get at least one reward, though. One evening in El Arish, we were all sitting outside when an Egyptian leading a

---

[*] The conversation reminded me of a time in 1984 when Ellen and I were in the city of Wuhan, China for three weeks. Ellen has an interest in transportation, and so we arranged an interview with the vice mayor, who wanted to talk to us about how to get more boats to stop at Wuhan on the Yangtze River. Ellen put on a great show of explaining how the Port Authority of New York works in this regard, with public docks and private boats. The Chinese were highly interested. During one of the long interruptions for translation, I asked Ellen how she knew all this. "I read the *New York Times*," she whispered.

camel came up to us. Annette was thrilled to be able to ride it.

Fruitful Work in El Arish

The sessions in El Arish were quite fruitful. We trained a large group of Palestinians, exposed more and more bugs, and learned more things users wanted to do with the program that hadn't occurred to us. For example, we had provided the facility to put in recycling plants in what we thought were the likely places, based on a list from a scenario Hillel Shuval had provided two years earlier. But the Palestinians wanted to put one in Tulkarem, so we made that possible in the program.

I thought there were two high points in our meetings. The first was when we broke the Palestinians into two groups, each given an assignment to develop a scenario and report on the results. They went busily to work, and Luai Sha'at came out and said to us, "They're hooked."

The implication, of course, was that he was already hooked and was on our side.

The second high point concerned team formation. Ali Sha'ath, the Palestinan deputy Minister of Planning joined us on the second evening and remained for a day and a half, returning to Gaza when we went to Israel. We finally had a chance to speak with him about using people from the PCG, especially Jad.

That had been a difficult issue. Ali Sha'ath had originally thought he was required to use only government personnel in this phase rather than people who had worked on the model—to "ensure an objective evaluation," as Rima Khalaf had put it. We convinced him that this would be impossible as regarded model development; the Palestinian engineers who had been in Cambridge helped us make the argument. He told us he was prepared to consider the PCG—although by then it wasn't clear the PCG was prepared to consider him.

Ali Sha'ath read me a letter from Sari Nusseibh in response to one he had written inviting the PCG to workshops and asking for information. It was quite peculiar. It didn't provide much information, but stated that the PCG's principal role in the project had been to assure that the project recognized things such as the fact that Palestinian water consumption had been suppressed, the importance of hydrology and area rainfall as opposed to prior use in assigning water rights, and so forth. In a word, the implication

175

was that the PCG had devoted itself to reserving the Palestinian bargaining position.

This was, of course, effectively untrue. It was correct that Jad, and others, had been careful never to abandon the PCG bargaining position, but the entire point of the project was to get away from arguments of this sort. Such bargaining positions were never what the project is about. I interpreted the letter as one in which Sari was covering his ass politically. Ali Sha'ath agreed.

The letter didn't make things any easier, but we had quite a good discussion with Dr. Ali. I specifically raised the question of Jad, and he said he would go ahead with him and would call him.

We left El Arish on Wednesday, May 8, returning by bus to the Rafah crossing. Along the way, the Palestinians sang Arab songs as we all clapped along.

At the crossing, the security captain's threat to Luai Sha'at about holding up his passport did materialize, briefly. Ali Sha'ath had to go in and argue on Luai's behalf, and got his passport back.

Annette, Anne Marie, and I arrived in Jerusalem very late, and again Annette and I stayed up working on the program, as we had every night up to then.

I spent the next day at the New Israel Fund, after which we all flew from Tel Aviv to Amman, Shula losing a suitcase on the way.

14
Waiting for Decisions

We arrived in Amman on Thursday evening and were met by Najeeb Fakhoury, who took us to the Regency Hotel and then to his house for dinner. I could not say no, even though I was ready to drop. The Fakhourys were very hospitable and were of considerable assistance to us, even though it could sometimes be a nuisance to work with Najeeb. Their son, Imad, was an economics attaché at the embassy in Tel Aviv and had been particularly helpful.

During this visit, Camille Fakhoury told us a bit of his father's history. Apparently, in when various airplanes were hijacked to Jordan in September 1970, Najeeb—the long-term agent in Jordan of Dutch airline KLM—commandeered a KLM plane with the appropriate air conditioning equipment and took it out on the tarmac to provide air-conditioning for the hijacked passengers. At considerable personal risk, he also led the negotiations with the hijackers. The Jordanian government awarded him a medal.

Our visit to Jordan was as wildly successful this time as our previous visit had been greatly disappointing. Whereas previously I had felt as if I didn't understand what was going on, I was now able to see that certain ritual niceties need to be satisfied in Jordan while real arrangements are made behind the scenes.

In our case, it was Jawad Anani and Munther Haddadin making the real arrangements. They came to see us at the hotel on Friday, May 10th. Anani

177

would deal with Rima Khalaf at the Ministry of Planning, who we were to see the next day, while Munther would deal with the Water Minister and propose a team to work on the project—with himself at the head. Munther informed us that he had actually been in the next room during our meeting at the Water Ministry the previous March. I wasn't surprised in the slightest.

The next day, Shula and I went to the Ministry of Planning while Annette worked with Samir Dweiri, the very smart young man from Haddadin's firm who had worked with Harsh in March and had delved deeply into the computer program issues. We had a brief meeting with Minister Khalaf and then were taken by Boulos Kefaya to his office, where we were joined by assistant Abdelghani Hijazi, with whom we had also met in March. It was a disappointment to learn that our March meeting had been for naught, at least in one respect: although we had given the Jordanians the computer program, they hadn't looked at it or at anything else since. Moreover, in both this discussion and its continuation the following day, Kefaya—unlike Hijazi—showed no interest in learning what our project could really do for Jordan.

On Friday, they sought my opinion on two reports, one by someone at the World Bank and another by a Canadian engineer. Both purported to rank different projects based on cost-benefit analysis. When I returned the next day, I told them that while the authors had some idea of what they were getting at, neither paper was very good. A discussion, of sorts, ensued.

It was scary to think that a person of so little understanding could be in charge of such projects and of project evaluation. I pointed out that both authors worked with the wrong discount rates, using nominal instead of real ones. Kefaya commented that surely the choice of a discount rate can't matter to project evaluation. I thought it an appalling statement for someone who purports to understand, let alone be in charge of, project evaluations. Hijazi corrected him sharply.

Although Kefaya had asked my opinion, he kept doing things other than listening. From time to time, he would burst out with his particular views about rules of thumb for how to do things. I simply could not get him to listen to the proposition that we were offering him a serious, systematic way to do project evaluation correctly.

We went from there to the Water Ministry, where we met with Munther Haddadin, Ali Ghezawi, and some others. Kefaya was there as well. Munther had secured the Water Minister's tentative approval of his suggested team, which was to include participants from the Ministries of Planning, Water, Agriculture, and others.

When Jawad Anani drove us to his office the next day for a meeting, I told him I felt I had comment to about Boulos Kefaya. He asked whether I really had to.

"I'm afraid he's a dope," I replied.

"I know he is," Anani answered, "and I also know what he's doing to you. But don't worry about it. I will take care of it." And I believe he did.

The meeting at Anani's office that evening was our most important one of the trip to Jordan. We got through a serious exposition of the model, a little bit of training, and there was particular interest in Ilan Amir's agricultural model. But as with the Palestinians, there was the claim that yields do not vary linearly with water quality. One Jordanian seemed very knowledgeable about this. I offered to send Ilan to Jordan to meet with them late in June, when he would next be in the region.*

That night, we dined at a garden restaurant with Jawad Anani, his wife, and some of their children. The Haddadins were invited but couldn't make it. It was a very nice experience. Anani himself is a very intelligent and likeable man; he agreed to serve as the economist on the team. The Anani children, particularly the eldest daughter,** were very pleasant. I never got to see the smallest child, who fell asleep and was put on a couch inside.

The Ananis were also well connected. While at the restaurant, they

---

* Later, Munther insisted he be present if Ilan came, and suggested strongly that Ilan represent himself as a Harvard person who happened to be an Israeli. Even 18 months after the peace treaty, professional contacts with Israelis were risky for Jordanians; professional societies had been boycotting Israeli connections. I suspected Munther felt his presence would make the Jordanians feel comfortable that this was happening under semi-governmental auspices.

** Sometime later, when the daughter flew into Ben Gurion Airport on her way to visit Shula, a female security guard stopped her and asked whether she had a return ticket to Amman. She was not returning to Amman by plane, she replied. Pressed by the guard, she eventually had to say something she did not want to say about her father, whose position in the government had changed since we had met. "My father is the foreign minister and deputy prime minister of Jordan," the daughter told the guard. "Oh sure," the guard answered. "So is mine!" Shula had to come to the airport and rescue Jawad's daughter.

179

pointed out a cousin of the royal family who was also dining. When he got up, the cousin and the Ananis greeted each other, and I was introduced to him.

Our departure for Israel the next morning was delayed, partly because Shula had picked up some young Jordanian academics who were starting a grassroots movement for contact between Israelis and Palestinians. Heaven knows how! She introduced me to them, and I told them about the New Israel Fund and Shatil.* Later, I told Avi Armony, the executive director in Israel of the New Israel Fund, and Sari Rifkin, then the head of Shatil, of the conversation. They were both very anxious to get the names involved and work with them.

On to Israel and Palestine

We returned to Israel from Jordan over the King Hussein Bridge, which gave us an opportunity to show Annette the Jordan River. Harsh had told her how small it is, but she was still surprised.

We went to meet with Alon Liel, Roby Nathanson, and Rafi Bar-El. Amir Tadmor was out of the country. Our discussion concerned three topics. First, Liel wanted us to write a letter describing the members of the Israeli team as a reply to our proposal sent out by the Dutch. The Dutch had told us they were sending it out to the parties, but had not yet arrived. (We ended up giving the proposal to each of the parties ourselves.)

Liel wanted the Israeli reply to get to the Dutch at least a week before the May 29 elections, pointing out that no one knew what the situation would be afterwards. Even assuming Labor won, Yossi Beilin might not remain in charge of us, and Liel thought it important to commit the Israeli government to the project before it might be in a panic over the election outcome. I thought that very wise.

Liel also made it clear to us again that someone from Ben Gurion University had to be included. Avishai Braverman, the university president, had already talked to Shula and pointed out he had spent the day with Shimon

---

* Shatil ("seedling") is the arm of the New Israel Fund that assists fledgling organizations to get started and learn how to become effective. It has been asked to give workshops worldwide. It has been quite successful and has grown greatly in importance over the years.

Peres, who sent his regards. Avishai was not being overly subtle in this matter. Liel gave me two CVs of proposed people from Ben Gurion. I later called Avishai and suggested he find someone who could deal with the agricultural yield problem, and we arranged for me to come to Bet Sahour and interview the people involved.

When he heard about the project's new agricultural part, Liel pointed out that the Dutch had expressed interest in having a further project involving agriculture. He said, however, that this could not be a Harvard project, since both existing projects in the Dutch Initiative* were from Harvard. He asked us to think about what such a project might consist of, but I couldn't see how it could be divorced from the overall Water Project.

We left Beilin's office and drove to Tel Aviv for our next appointment with David Mullinex, the U.S. Embassy scientific attaché who was very well informed about water, water negotiations, and water negotiators, and who already knew a fair amount about our project.

Indeed, our project was pretty well known in Israel by then. In late May, Israeli television carried a video of my speech at the Fishelson memorial. More important, the idea was in the wind. At Liel's office, I was shown a pretty favorable piece in the daily newspaper *Ha'aretz* written by someone from Tel Aviv University, who pointed out that the value of all the water that could possibly be involved could not, because of desalination, exceed the value of all the cars reported to be stolen by Palestinians. It suggested Israel should simply give Arafat all the water he wanted in exchange for a promise not to steal cars. More seriously, it observed that Israel would not go to war over stolen cars or consider it a threat to its security. Our project was named in the article as generating that sort of idea.

We spent Tuesday morning at Jad's for what turned out to be a strange and, in some ways, quite difficult meeting. After listening to a description of where things stood, Jad announced he would no longer work on the project, claiming he didn't have time. Eventually, Jad admitted that he was tired of working on projects in which he and his Institute had to do all the work without a clearly defined contract. We assured him there would be a specific contract from either the Dutch or through the Palestinian Ministry of

---

* Incidentally, the Dutch didn't like calling it the "Dutch Initiative." They regarded it as an initiative of the four parties in the region that they were "facilitating."

Planning. He came around and said that it would be fine if that ran through the PCG.

We also discussed the self-serving letter from Sari Nusseibeh to Ali Sha'at I described in the previous chapter. Jad, who should have known better, told me he had written it.

We then turned to the World Bank paper, which we had agreed to work out. Again, Jad said he was too busy to work on it, but when he learned that we were not restricted to examining the gains from trade starting with the Oslo II quantities but that we would examine them again starting with different amounts of water for the Palestinians, he became quite excited. Just as he had done in Cyprus two years earlier, he proposed that we should deal with allocations based on different principles of international law. I said such a study would be fine; indeed, it would make it a much better paper. We agreed that he would get me the appropriate quantities, and when Annette Huber went back to see Jad the following Saturday, she helped him run a bunch of these scenarios. She reported to me that he said he had managed to reproduce the entire Palestinian negotiating position. I had not yet seen the results of those runs.

All in all, it was a strange meeting, and to this day I'm not sure I understand exactly why it happened as it did. But make no mistake: Jad and I were on extremely friendly terms. I tried to make clear to him that we had fought quite hard for his inclusion.

Only in mid-June did we get news that the PCG had replied to Ali Sha'at saying the group wanted to work on the project. Shula had pushed Issa Khatar into finally writing the appropriate letter. Despite that I was friendly with the people at PCG, I did not understand their behavior. From January through May, PCG and its associates appeared to have talked themselves into the belief that we were abandoning them. By the time they understood this was not true, they had gotten themselves into a mindset that prevented them from acting. Obviously, there were also issues of internal Palestinian politics here—issues I did not understand.

After lunch that day, we went to see Hillel Shuval. Zvi Eckstein had still not talked to him about the Israeli team not wanting him involved. I felt obligated to be totally honest with him—Shula silently disapproved—and told him all about the Dutch and then about the Israeli team's attitude. He was quite upset but stated that he would do nothing to damage the project.

I asked him, instead, to become the environmental consultant to the entire project, pointing out that we had other money from the Revson Foundation. He readily agreed, but insisted that his professional pride required that he be listed at least as an environmental consultant to the Israeli team. He stated that it was all right if they never consulted him. I agreed to those terms.*

Hillel also gave me a paper on water usage in Israel and Syria that he was going to present at a conference at Lake Bellagio in late June being run by Lenore Martin from Emmanuel College and Harvard. I'm told Lake Bellagio is very beautiful, and I would have complained that I hadn't been invited, but the fact was that I was being besieged with invitations to speak on the water issue.

Hillel's paper was part history of the conflict and part discussion of water needs. It made the point quite forcefully that monetizing the dispute shows that it isn't really very big. I later gave Hillel some assistance in getting the monetary values right.

I spent the morning and part of the afternoon of May 15 in and around Beer Sheba. Much of this was on behalf of the New Israel Fund (of which I was the incoming president), looking at early education for Bedouins and speaking with some Arab students at Ben Gurion University. At best, the way Israel treats its own Arab citizens as second-class citizens and keeps them from being fully integrated into Israeli society can be characterized as self-destructive.** The New Israel Fund works hard to help change this.

---

* After we left Israel, Hillel ran into Uri Shamir. Each of them reported the facts of their encounter to me in much the same terms. Hillel told Uri he had been named an environmental consultant to the Israeli team and wanted to participate. Uri informed Hillel—coldly, I'm sure—that that they had no need of him. Hillel told me this showed that the problem had nothing to do with his previous relations with Uri and the press. But of course, it did—as Uri emphasized when we met with him, Shaul, and Zvika in the late afternoon of Wednesday, May 15.

** For example, only students who have served in the Army are eligible for scholarship and housing funds. Israeli Arabs are ineligible to serve in the Army, so it is a well-known form of discrimination. And unlike new Jewish immigrants, Arab students cannot substitute classes in Hebrew for general study classes, which prevents them from learning the language better and further hampers their integration into Israeli society.

At Ben Gurion University, I met with Avishai Braverman and two people he proposed for the project: Uri Regev, an economist, and Gideon Oron, a soil and water specialist and a friend of Ilan Amir. We agreed I would recommend Oron for participation in the project, but that it was probably impossible to get two people involved.*

Back in Tel Aviv, I met with Shaul, Zvika, and Uri. We took up the two difficult personnel-related issues: Oron and Hillel. They had no issues with Oron personally, but they had clear objections to the manner in which he was being forced on the project using political muscle. In the end, we were all realists, and we agreed to his inclusion as a member of the steering committee. It meant expanding the budget—something we weren't sure we could arrange.

As for Hillel Shuval, there was agreement on my proposed solution, but also general unhappiness. As Uri put it, "We will all live to regret this decision."

Someone suggested we propose to the Ministry that a letter be issued to all members of the Israeli team emphasizing the necessity for confidentiality and forbidding unauthorized statements by any team member regarding the project results. I drafted the letter, but I don't know what became of it.

At a late hour, we checked into a kibbutz guesthouse north of Tel Aviv. Shula and I got up at an ungodly hour the next morning and returned to Boston. Annette remained behind, doing installations for the Israelis and then going to Gaza for a full day to discuss GAMS programming and the model. Finally, she went back to see Jad.

When I told Annette how appreciative I was of the wonderful job she was doing, she said she regarded it as a privilege to work on the project. I guessed that was true.

Shula and I had a stopover in Zurich on our way back, and there we managed to get the draft of the material describing the Israeli team to Alon

---

* I gathered this recommendation might cause a problem with Ilan Amir, but he was too nice to say so. He simply said he knew Oron very well and liked him a lot. Shula suggested to me that Ilan was put out because he felt he didn't need Oron's kind of help. But I had no choice in the matter. I tried to explain to Ilan that he was, indeed, the agricultural person for the *central* project, with Oron being named (in part for political purposes) to assist with the Israeli portion, just as there would be a Palestinian and a Jordanian.

Liel and Admir Tadmor. Together with the Israeli reply to the Dutch, it was sent out a couple of days later.

Back in Cambridge

Several significant events unfolded after we returned. One was that Munther Haddadin came to Cambridge for two-and-a-half days in late May, which was extraordinarily valuable. For the first time, Munther listened carefully to the full theory of how the optimizing model worked and what shadow values are. More important, he got his hands on the computer and did a number of exercises. As Luai Sha'at put it, "He was hooked."

A number of the exercises concerned the Disi-Amman pipeline and related topics. Now that we had a chance to talk about it, Munther made a very good case for the proposition that the pipeline would probably be necessary because a good deal of fresh water would have to be retained for agriculture in the Jordan Valley to mix with recycled water. Indeed, we agreed that a very good use of the seasonal version of the model would be to show what mix should be used for that purpose. But even more important than the answer to whether the Disi pipeline should be built was Munther's appreciation that the model provided a systematic tool for evaluating that question, taking into account all the things that bore on it. Obviously, the agricultural component of the model had struck Munther's interest when we were in Amman, and now it seemed he was highly committed to going forward with the project.

There had been some developments regarding the agricultural model itself. First, Ilan Amir, who was still in the region gathering information and talking with people, told me that he was quite sure that, as a good approximation, the response of agricultural yields to blending of water qualities was, in fact, linear.* This was contrary to what we had heard on our trip, where opinions ranged from the proposition that it was not true to Shaul Arlosoroff's insistence that no one knows. Gideon Oron said he had access to the results of all the experiments.

It suddenly occurred to Annette Huber that the enterprise of running the

---

* Imagine irrigating a field with a *mixture* of half fresh water and half brackish water. Would you get the same crop as if you had irrigated half the field with fresh water and the other half with brackish water? Ilan said you would.

185

agricultural model, generating demand curves, and incorporating those demand curves into the general model would be far more complex than is necessary. She pointed out that one could simply include the revenue of farmers minus the expenditures on water in the objective function of the full model, with the various constraints of the agricultural model also included. She was absolutely right, and we anticipated that it would save a lot of work.*

Also related to the agricultural model was the question of how to handle quotas—pricing systems in which farmers pay a certain amount for a given amount of water, a higher amount for another given amount of water, and so on. The importance of dealing with this extended beyond agriculture, since, as we had discussed when in the region, we wanted to be able to accommodate particular pricing schemes for households and industry as well, whether those involved subsidies or (as it were) taxes. But there were difficulties. There was essentially one kind of water—recycled water—in the version of the model we were using. Farmers can use recycled water only after it has been turned in to something they regard as equivalent to fresh water. So, we handled quotas by figuring out in advance where the farmers' demand curve crosses the supply curve represented by the quota steps, and then imposed on the model the constraint that agriculture must receive at least that amount of water. (If there is no such intersection, with demand by agriculture greater than the limit of the last quota step, then quotas just represent a pure subsidy to farmers and don't affect demand.) But a similar technique would not work with several kinds of water, unless the quota system specified a complete set of prices and quota steps for every one of them. This is because the demand for any one type of water (say fresh water) will depend on the prices of the others.

We spent a couple of days trying to deal with this problem by altering the objective function of the model in various, apparently clever, ways. The best we seemed able to do was the following: When a user specifies quota steps for a certain type of water, the model would treat the water in the amounts

---

* However, as of the writing of this chapter in 2014, this has not been implemented. It is not quite so simple to do, since the agricultural sub-model "AGSM" uses four different types of water, while the main model uses two. We expected to provide MYWAS with a user-driven choice as to whether to incorporate the agricultural sub-model or to treat it separately.

given by the specified steps, as being removed from the system and held in reserve, should agriculture actually demand that water. For example, if agriculture is to receive up to 40 million cubic meters per year at a specified low price, the model would set aside 40 million cubic meters per year for that purpose, even though farmers only demand 30 million cubic meters per year at that price.

Obviously, this wasn't quite what was wanted. We thought it would be nice to have a system in which the quota water not demanded by agriculture could be used by the rest of the system. (In effect, in the example just given, one wants to reduce the quota step to 30 million cubic meters of water.) But given the problem of determining in advance what demand will be, we were unable to do that. We could tell the user to be careful about specifying steps that are too big. Further, the user could do another run, reducing the steps. Nevertheless, we were unable (and I now believe it is impossible) to deal with the model so that such steps are reduced automatically. In effect, you can't do that and also not have the shadow price of quota maintenance be charged to farmers in the optimal solution. Instead, users would have to realize that announcing a policy of up-to-so-much-water-at-a-low-price for agriculture itself involves a cost if agriculture does not in fact take that water.

Another thing we dealt with when we were back in Cambridge was to prepare for a trip to Kuwait that Shula and I were to take at the end of June. It had been arranged by Lenny Hausman, principally with the Kuwait Institute for Scientific Research (KISR). Lenny traveled around advertising our way of thinking about water, and was right to do so, but my willingness to do projects elsewhere in the region or elsewhere in the world was somewhat limited. If the Kuwaitis wished to mount such a project, I thought, I'd give them advice, but only if after this first trip they paid my consulting rate.

Our meeting with Imad Mahzoumi at the end of May was also connected to Lenny's "proselytizing." I had met Mahzoumi, a Lebanese member of Lenny's board, previously. He was also a manufacturer of pipelines in the Gulf, and had contacts with both the Syrian and Lebanese governments. Earlier, he had sent along some of our materials to the Syrians. He asked me to write a two-page piece for each of the two governments explaining why they would benefit from participation in the project. He also was arranging for me to speak at a desalination conference in Madrid that was scheduled to take place a year-and-a-half later.

Mahzoumi was absolutely frank about the Syrian and Lebanese governments' attitude toward the Institute and our project. He told us they were well aware of our project, but that they regarded the Institute as operating primarily in the context of Israel and branching out to its neighbors only when it suited Israel's interest to do so. That perception was, no doubt, fueled by the fact that many of the participants in the Institute were American Jews.

In some sense, the perception was absolutely right. It described my own interests precisely. But in another sense, it was wrong. Not all the participants were Jewish, and I think that all of us—Jews and non-Jews alike—had learned to move in a wider circle. Still, the perception Mahzoumi reported did match my attitude about working on water systems in Kuwait or the Gulf. I wouldn't do that for free, whereas I regarded my work on a project that involved water conflicts between Israel and its neighbors as my contribution to *Tikkun Olam*, the repair of the world that is a central tenet of Judaism.

By the way, neither the trip to Kuwait nor to Madrid ever took place.

The Israeli Elections

All of this unfolded as we were back in Cambridge and the Israeli election campaign was happening. The vote was on May 29: Benjamin Netanyahu defeated Shimon Peres in an extremely close election and became prime minister for the first time. Had 20,000 to 30,000 Israeli Arabs not cast blank ballots, Peres would probably have won.

The result did not make me feel particularly cheerful. For several weeks, I even stopped writing what has become this memoir.

It was too early to tell what the effects of the election would be either on the project or in wider terms. Netanyahu had not yet formed his government, and there were even some signs that he might be more moderate than one might have supposed. Overall, though, the general outlook was not good. And while the peace process was important—it would surely continue, I though, but at a much slower rate—it was far from the only problem. Under the Labor Party, the West Bank was already being divided into what Jad called "ghettos"—Palestinian cities ringed by roads to protect the Jewish settlers and that cut off these cities from each other. The issue of Jerusalem was yet to be decided.

There were certainly people of tremendous good will toward the peace process in the Labor government, Yossi Beilin one of the chiefs among them. I believed there would have been an accommodation on Jerusalem, eventually—as there would need to be. Further, there were discussions with Syria that held out the prospect of giving back the Golan. The new Netanyahu government seemed determined not to do that.

However, the real problem for Israel seemed likely to be the accession to power of a large number of right-wing religious politicians. This put civil rights, the rule of law, women's rights, religious toleration for non-orthodox Jews, and the position of Israeli Arabs all in considerable danger. Alas, it was an important opportunity for the New Israeli Fund, the presidency of which I was to assume at the end of June.*

What did this all mean for the Water Project? I didn't know. It would surely make matters with the Syrians more difficult. The new government, though, seemed to be suggesting that water was one of the subjects it was prepared to go on talking about.

Meanwhile, on the Dutch front, there was very good news. Shula was informed that the Dutch had no intention of waiting for Israel's new government to comment on the project; they were going ahead. I believe the outgoing government, which had already given its comments to the Dutch proposal, had urged them on. The Palestinian and Jordanian comments—the latter largely drafted by Munther Haddadin—were, we were told, about to be sent.

Still, though, I did not know what would happen to the project in Israel. We did not yet know who or what Ministry would be in charge of us. For a short time, we had enjoyed enthusiastic support from a part of the Israeli government and at least heavy interest from the other two governments. All that was now uncertain.

Before hearing the news about the Dutch, I was particularly depressed. I

---

* Indeed, the issue of religious freedom for Jews and the associated issue of "who is a Jew" became a major issue for the New Israel Fund—one that remains important today. In 1999, when I gave a farewell address to an Israeli audience, I commented that I felt I had not put enough energy into those issues because of the Water Project. However, I said, "I comforted myself by the realization that, in future years, NIF members would recall that at this time they had served under the best fundraiser the Fund had ever had. Unfortunately, that man's name was Bibi Netanyahu."

didn't realize how much so until I totally lost my temper at Shula and knew I was doing it for no particular reason. But while not as optimistic as before, I remained convinced that this project had a serious future, including a future as part of peace of the region.

Two Stories from Jawad Anani

In the aftermath of the elections, I spent a fair amount of time in June in Cambridge with Jawad Anani. He came to ISEPME on business other than water, and we went out to dinner with him and the Gilads.

Jawad, as I have already remarked, was a very powerful man in Jordan without actually being in the government. He is a very nice man and, I believe, quite straightforward in his dealings. He is also an economist—which, I guess, is another recommendation. He had been one of Jordan's chief negotiators in the overall negotiations with Israel. He told me two stories about that, both of which were quite revealing and worth remembering.

At a break during the last days of the negotiations, at Bet Gavriel on the Sea of Galilee just south of Tiberias, Jawad put a question to Yitzhak Rabin.

"Mr. Prime Minister," he asked, "why is it that many people say you drink too much?"

Jawad told me that King Hussein's face went white.

Rabin answered in his deep, slow, gravely voice. "Well, when I was an army officer, I only drank water. Then I had to negotiate with [here Jawad named a French general whose name I have forgotten], who was reporting to General de Gaulle. When I asked for water, he said 'in France, water is for horses. Wine is for men'."

Rabin, Jawad told me, then paused for a long time before he added, "I have taken that advice very seriously ever since."

That story is revealing about both Anani and Rabin. The second story reveals a good deal about King Hussein.

When the negotiations were over, and the document had been prepared for signature, Rabin said to the King, "Your majesty, you have gotten a lot out of this treaty—water, land, and so forth. I need something to take back to my people, something tangible. Can you not provide it?"

The King looked at Anani and indicated that Jawad was to answer. Anani told me that he knew perfectly well that the Israelis had something in mind or they would not have brought this up, so he suggested to Rabin that Rabin should say what would be suitable. Rabin replied that they would like complete and open trade or at least trade in relatively large amounts.

Anani told me the question of trade had been amply discussed during the negotiations. He replied to Rabin that Jordan would be willing to have a great deal of open trade with Israel—and he stated a fairly large dollar amount—provided that half of it could be with the West Bank and Gaza.

Rabin said that was impossible. "You are opening Pandora's Box."

So Anani offered a much larger amount of trade, if 30 percent of it could be with the West Bank and Gaza. Rabin gave the same reply.

"Mr. Prime Minister," said Jawad, "you seem fond of Greek mythology. I know something about Greek mythology, also."

At that point Shimon Peres apparently interjected something about Anani being a learned diplomat.

"You say that we are opening Pandora's Box," Jawad continued, "but you are attempting to press us into Procrustes' Bed."

While the legend of Pandora's Box is well known, that of Procrustes' Bed is much less familiar. Procrustes was an innkeeper who would invite travelers to his lodgings and then cut off their limbs when they were too large to fit into the beds he provided. He lent his name to describe situations in which sizes or properties are fitted to an arbitrary ("Procrustean") standard.

After a little bit more talking, they settled on immediately permitting a relatively small number of tourists to come from Israel to Jordan—the number being half that suggested by Rabin.

The King then told Anani to come with him, and the two of them left the room. In the next room, the King began laughing uncontrollably. Even when he asked Anani to light his cigarette for him, the King went on laughing.

Finally, Jawad began to laugh as well, but he said to the King, "Your majesty, I would feel better if I knew why we are laughing."

The King replied, "Don't you remember who won the last three wars?"

15
Problems and Progress

In April, I had received an unsolicited invitation from the James A. Baker Institute at Rice University in Houston, Texas, to attend a conference in late June on Israeli–Syrian relations. The institute's director, Edward Djerejian, was a former U.S. ambassador to Syria and also to Israel, and the plan was to discuss the Syrian track of the peace negotiations. As water did not seem to be featured, and as it appeared that I would be just be one of many attendees, I saw no great reason to attend—unless Yossi Beilin's office and associated people thought it important.

Things changed after the Israeli elections. The start date was pushed back two days. Beilin, who had originally planned to be there, canceled. So, too, did the Syrians—at least any "official" representatives. One Syrian living in France, Bassma Kodmani-Darwish, did attend. The focus changed to how to make progress in the peace process given the election results.

As it turned out, Yair Hirschfeld, who had largely arranged the conference, would be there, and very much wanted me to attend and talk about water. Yair sent me a Syrian-authored paper on Syria's water claims and problems. I ended up writing a paper, too.

It was a very small but very high-powered conference—actually, more like a workshop. There were three people, in particular, who could be said to represent the parties in one sense or another: Jawad Anani from Jordan, Yair Hirschfeld from Israel (who, of course, did not represent the new government), and and Maher El-Kurd, an economist who was the economic advisor to Arafat. There were also a few academics, and a Dutch investment banker. One evening, James Baker, the former U.S. secretary of state, joined us for dinner.

The plan was to issue a report that would address a number of items. It ultimately became a book with the various papers.[1] There was general agreement that the new Israeli government could not afford to stop making progress on the Palestinian track, even though progress would be much slower. To do so would be to risk an implosion in the West Bank and Gaza— a return to violence and *intifada*—with the disappearance of the Arafat government. That was a situation no reasonable Israeli government should want to face, and there were signs that, despite the presence of Ariel Sharon and Rafael Eitan, the new government would be reasonable. This might, therefore, be the time when progress on subjects such as water could be made. Indeed, the two subjects suggested were water and natural gas pipelines.

There was also general agreement that we were in a dangerous period before the U.S. elections, and general apprehension that the new Israeli government would take action on the ground that the United States would feel it could not oppose—such as strengthening settlements and expropriating land in East Jerusalem. There was some discussion of the role of American Jews, and I spoke, mentioning the New Israel Fund.

Everyone felt further that the boat had been missed in the Syrian negotiations. Hafez Al-Assad was, in part, responsible, and he could have had a deal for the Golan a year earlier. But responsibility was shared. For instance, the U.S. government had failed to push hard enough for that kind of deal. Former Secretary Baker was very critical of Secretary of State Warren Christopher's repeated visits to Damascus, which turned up nothing at all.

These were the more general items that came out of the gathering. As for the water project, a great deal of the first day was devoted to my presentation (which was by far the most elaborate paper) and discussion. I think it is fair to say that I bowled everyone over. They became quite taken with the idea of water as a cooperative venture rather than as a source of tension. I was sure it would be pursued further.

The most important thing for the water project, though, occurred on the sidelines. Maher El-Kurd took me aside and opened a conversation that continued the next morning at breakfast.

El-Kurd began the conversation by asking me what I thought to be a somewhat strange question, given that he obviously already knew the answer: what Palestinians had worked with me on the project? I told him.

He suggested that Jad Issac and Marwan Haddad had "feathered their own nests." But he added, "And why shouldn't they?"

Then he expressed his view that that the Palestinians needed two things. One was to put together a relatively united negotiating team. In the past, Riyad El-Khoudary had led the team, but it had been somewhat disunited. His description matched what I knew about Marwan being kicked out of the negotiations over Oslo II before they were completed. The other was that the Palestinians wanted someone to write a background paper to give them a consistent position on the use of the model in negotiations and related subjects.

There were only two Palestinians I was quite sure really understood this matter, and I told him they were Jad Issac and Nabil Sha'ath. El-Kurd told me that Sha'ath would not participate, because he would be saved for "mega negotiations." If a background paper had to be written by a Palestinian, then, I said it would have to be done by Jad.

My response prompted El-Kurd to bring up what I suspect was his intention in the first place: to ask me whether I would be willing to assist the Palestinian negotiating team in the use of the model and in preparing the negotiating position. He revealed to me that he had been instructed to do this by Abu Mazen, Arafat's number-two person in the PLO and the Palestinian in overall charge of the negotiations (also known as Mahmoud Abbas, and who is now the Palestinian president). Some of the details of his conversation with Abu Mazen revealed that there had been an intelligent discussion of what was involved in this approach—in particular, on the question of how this would all work if the amount of water under dispute was itself in dispute (I explained that was a fairly trivial issue monetarily and that there were ways of handling it). El-Kurd also told me he had discussed this matter with Ali Sha'at.

I told El-Kurd that I would be delighted to do what he asked, but that he must understand that while anything the Palestinians told me would remain confidential, I would provide exactly the same service to the Israelis if asked—and that I hoped to be asked. He promised to get back in touch with me in a couple of weeks, and I told him that I planned to be in the region at the end of July.

With El-Kurd's permission, I told all this to Yair Hirschfeld (who had a very friendly relationship with El-Kurd). Yair told me that El-Khoudary and Nabil Sha'ath didn't get on well and that maybe I could bring them together. I raised with Yair the idea of my also working for the Israelis, and he was quite intrigued. He and Yossi Beilin were off to see the Dutch at the end of June and both he and El-Kurd spoke of getting the Dutch to finance this phase. I told him I would take no money for this.

Yair also suggested that it might be possible to bring Yossi Beilin back into this for the Israelis, which I thought would be wonderful. But wouldn't it be the kiss of death from the new administration?

Yair said it would not, and indeed suggested that it might be possible to approach David Levy, the new Foreign Minister. He was also interested in getting the Germans to do a feasibility study of the West Bank connector that the model and project had proposed—something that would greatly promote cooperation in water were it to become a reality.

Shula was doubtful about my conversation with El-Kurd when I first told her about it. I explained to her that this was not a matter of helping the Palestinians negotiate with the Israelis for Palestinian advantage, but about bringing them both into a win-win situation in which both sides would benefit from using the model and the associated way of thinking about water.

My wife Ellen reminded me that there was still danger here. Whatever the substance, she said, it would be important that I not appear to be helping one side against the other.

Shula became quite enthusiastic. She informed me that the people in and around Beilin's office thought there was a good chance Beilin would be used by the new government for back-door negotiations. In that case, we might very well have been off and running. Ideally, it would be wonderful if both sides would negotiate with the help of the model and with me mediating the discussions or otherwise assisting (I thought I might try to enlist Howard Raiffa and the Conflict Management Group).

Shula wanted me to go see Minister Pronk to discuss such a possibility, and perhaps that would happen. Unfortunately, none of these suggested meetings ever occurred. I never heard from El-Kurd again.

Intellectual Progress

After the workshop at Rice University, I returned to Cambridge, anxious to continue the work on water. A couple of weeks later, on July 8, 1996, Ariel Sharon was named Minister of Infrastructure in the new Israeli government. The *New York Times* reported that he would be in charge of negotiations over water with Israel's neighbors and with the Palestinians. That was not going to make my work easier.

I kept thinking about my experience at the Egyptian security post at Rafah. Each Palestinian there came up to me and said, in effect, "Now you know what it's like to be a Palestinian." It would now be my task to teach Ariel Sharon "to do justly." I guess that's what it's like to be a Jew.

The summer unfolded as a time of political frustration and considerable intellectual progress. We created a really easy-to-use version of the existing model. Our interface (WAS 3.2) let the user choose all sorts of things and get the results quite conveniently. We also seemed well on our way to a version (WAS 4.0) that would incorporate several types of water as well as Ilan Amir's agricultural model all at once. It proved quite a difficult enterprise, and we were still debugging the program, but we succeeded in going from a situation in which convergence did not take place all, to it taking twenty-five minutes, to it taking well under five. A combination of the agricultural model with the main model, I thought, would make a very attractive planning tool.

We were attracting attention as well. I spoke at a conference of the World Bank and was invited to speak at the Society of Water Engineers Meeting in Memory of Yitzhak Rabin in Israel at the end of November. Moreover, Uri Shamir visited Cambridge in September. He was contemplating doing a project on the Aral Sea and the five former Soviet Republics that border it, and wanted to build a model like ours to deal with their water problems.

When I spoke to ISEPME's Board of Directors, there was similar interest in applying such methods elsewhere. One board member had a possible project in Kazakhstan. Another director, Gil Shiva, was insistent that we pass on to Israeli negotiators with Syria some of our conclusions, and remarked that the Syrians had proposed to the Israelis negotiating on the basis of our project when they had met the previous February.

On another front, Eytan Sheshinski and Dagobert (Bob) Brito, who had been working on natural gas and cooperation in the region, asked us to see what difference it would make if water could be imported to Gaza from Egypt. They proposed that the Egyptians could sell it at relatively low prices

(25 to 40 cents per cubic meter). It turned out, we found, that this would bring a benefit of at least $16 million per year by 2010 and more than $60 million per year by 2020—which seemed conservative, since it was assumed that everything else would be handled efficiently and that the water does not flow beyond Gaza.

There was a potentially important further conclusion to be drawn there. Were there to be trade in water permits, the Palestinians would have to consider the fact that, unless they built a desalination plant in Gaza and did not use it, they would have a problem if the agreement broke down. It was the only realistic case I could think of in which reliance on trade really wouldn't produce a hostage to fortune. But with a pipeline from the Nile, such a desalination plant would not be necessary in either case, and the Palestinians would not have to rely on the agreement with Israel for their water supply. They would have two suppliers.

The fly in this ointment, of course, was that the very notion Egypt would sell precious Nile water seemed extremely unrealistic. So far as we could see, they wouldn't even begin to talk about such a possibility.

Political Frustrations

Meanwhile, over the summer, we were still waiting for the Dutch to sign the contract and give the project the go-ahead. The delay was not a problem of Dutch attitude about our work. Indeed, they had already approved our budget and an additional $300,000 for work on agriculture. Further, we knew that Minister Pronk and others were extremely enthusiastic about the project. The difficulty was that they didn't feel they could proceed without the assent of the Netanyahu government.

Would that assent be forthcoming? I thought it would, but we were still waiting. All through the summer, the Dutch moved quite timidly. They were attempting to get Dan Meridor, the new Finance Minister, to take charge of cooperation in the Dutch Initiative, which included our project and another Harvard project led by Michael Porter in the Business School. Porter's project, which seemed to me to have very little analytic content, was much less politically sensitive than ours.

There was a general reluctance to ask straight out about the water project in part, no doubt, because of the possibility that it would lead to Ariel Sharon as the person in charge of water. At different times, there were approaches to Meridor. Yossi Beilin was supposed to see him and tell him that there was this second Porter project. That meeting took place early in the summer, but the information wasn't transferred very directly, if at all.

In mid-summer, Meridor sent the Dutch a letter saying he was happy to cooperate with them in their initiative. Minister Pronk was advised by his staff to read the language in the letter as sufficiently broad to include the water project. He then wrote a masterfully phrased response expressing pleasure that Meridor was going to be involved in all their projects.

Still, though, this was not considered quite enough. The Dutch wanted, as it were, a wink and a nod. So, a meeting was set up between the Dutch Ambassador and Meridor for late September. The Dutch Ambassador's mother died, and the meeting was replaced by one between Pronk and Meridor when they would both be attending meetings of the International Monetary Fund in Washington, D.C. in early October.

Several sources gave us essentially the same account of the meeting. Apparently, it was quite positive, but Pronk eventually felt he had to ask directly about the inclusion of the water project.

Meridor asked, "What water project?"

Meridor's response showed that while his staff knew about the water project, the view that he had really known about it all along was quite wrong. Still, he reacted favorably and promised to look into it and get back to Pronk.

I believed this would go reasonably well. Shula had been speaking with Dan Katarivas, who was on Meridor's staff and quite favorably disposed, stating that they wanted to look into it. Shula offered immediately to send me to Israel to talk to Meridor, to which Katarivas replied that they wanted to look at the written materials at this time and talk to the Israeli team, meaning Uri Shamir. Apparently, as we knew from an earlier conversation with Katarivas, they were very pleased that Uri was involved in the project. Certainly, if they were going to see an Israeli, that person ought to be Uri. He was one of the two best people in the world—me being the other—suited to persuading them this was a good thing. In some respects, I thought, having him do it might be better.

Meanwhile, work kept being put off and time kept passing. The Dutch had been planning a meeting in The Hague for November 7, possibly followed by something at a Cairo conference the week after. They then reported that they were thinking of having something at the Cairo conference only. But it wasn't clear to me that even that would happen. Still, so long as the project continued to go forward, there was no real reason why a big conference had to happen quickly.

I believed the Netanyahu government would wish to go forward with the project if it was understood. The Israeli government needed very badly to make progress in areas of cooperation that were not wildly sensitive.

After summer's end, around the beginning of October, the West Bank and Gaza exploded and pent-up Palestinian frustration was becoming violent. The triggering incident was the opening of a tunnel in the Old City of Jerusalem—a matter of no real consequence, if viewed rationally. Netanyahu and Arafat came to Washington at President Clinton's insistence and negotiated for two days about the tunnel without results. However, they then started negotiations on evacuation from Hebron.

Unless Netanyahu was very foolish, I thought; he and his government should be seizing on opportunities to study modes of cooperation such as ours. But it was difficult to remain cheerful. I generally began talks about water by commenting that I wasn't there to tell the audience that it is not true that the next war in the Middle East will be about water. During this period, I felt I had to reassure audiences by saying that that was not because the next war in the Middle East was imminent for lots of other reasons. I hoped that was true.

Other political events during this time concerned Jordan. In mid-September, we learned that Munther Haddadin had been having political difficulties with the Jordanian government, as happened from time to time. We were told his battles had nothing whatsoever to do with our project. But then, Munther withdrew from the project without telling us. That triggered his friend the Water Minister to write to Rima Khalaf, the Planning Minister, saying that Jordan should withdraw from the project. Apparently, Khalaf had not quite gotten around to writing to the Dutch when the news finally came to us.

Shula and I were alarmed. The first thing we did was to try to convince Munther, I think successfully, that he wanted to change his mind. We mentioned that the Dutch had approved the budget, with a lot of extra money for his team. He made it clear to us, however, that while he could get the Water Minister to change his mind, the Planning Minister would now have to ask the Water Minister about it—something we set out to try arranging.

Jawad Anani was in China, so we spoke with Bassem Awadallah, who said he would attempt to fix things—but he wanted his name kept out of it. I don't know how far he got, but eventually Anani returned and informed us that he was not worried and he would take care of it.

As always on matters like this, Jawad was correct. When he came to ISEPME's board meeting and informed us that the matter was entirely under control and that Rima Khalaf would ask Munther to rejoin the project. It was plain, once again, that Anani was very powerful. He had the ear of the Crown Prince. He was also always a pleasure to deal with. Apart from being a charming companion and a wise man, his yea is yea and his nay is nay.

Still, an issue persisted with the Jordanians: their extreme reluctance to have us put out a preliminary report or publicize widely that we were recommending trade in water. As a result, when we produced a preliminary report, we restricted its circulation.* According to Anani, the worry in Jordan was that the project would recommend that certain projects in the 1994 Jordan-Israel treaty not be built and that Jordan wouldn't get certain water to which that treaty entitled it because the project would recommend a "better" solution that would be seized on by the Israelis. I thought the fear was baseless, and that perhaps we could manage to clear it all up if we could ever get to work again.

Meanwhile, we put together a proposal for negotiating workshops at which the Israelis and Palestinians would participate with Howard Raiffa and myself. That proposal appeared stalled in Yair Hirschfeld's firm, because, after pressing us to hurry, they had not completed their part. We needed to get back to that as well.

At the end of October, whether the project would go forward still remained up in the air. Meridor's office said they would go ahead and seemed poised to do so, but we then hit a possible snag. Uri Shamir had spoken with people in Meridor's office, and it appeared that the Israelis were seriously worried about confidentiality. In other words, they worried that participation in this project might force them to reveal either real data they didn't want others to have or to reveal what exercises they would make with the model.

---

* Shaul Arlosoroff, noting in the report our quotation from Isaiah—"With gladness shall you draw waters from the wells of salvation"—suggested that it would be appropriate to find a quote from the Qu'ran, which we did.

The concern appeared to come largely from Yossi Dreizin, the head of the Water Commission planning office. He had expressed them earlier, in March, when we had met at his office. I had attempted to allay them then, and I did so again. I wrote a letter to Dan Katarivas, who was handling all this, and assured him of confidentiality. Of course, I thought there should have been nothing to hide. In terms of operating the model, it would not be in anyone's interest to dissemble. But they had to learn that, and they were obviously still thinking in terms of negotiating over water quantities only rather than dealing with the value of water. I hoped my letter would overcome the problem and that we would learn the outcome in a few days.

Meanwhile, I held up sending my paper for the Rabin memorial conference. It would be a serious hot potato and draw a lot of attention if there was opposition to the project in Israel. In it, I pointed out the reasons for believing that water should not be an obstacle to a return of the Golan Heights. Israel's relations with Syria had become very bad. Since there was some reason to believe there might also be opposition from people who thought the model would give results "contrary to Israel's policy," I was reluctant to release such a paper until the contracts between the Dutch and the three countries were signed.

In any event, the various ministers were due to meet in Cairo on November 12. Foolishly, I couldn't imagine that we wouldn't have an answer by then.

A Busy November

A great deal happened in early November. On Sunday, November 3, Katarivas met with a group of water experts, including Yossi Dreizin and Rafi Ben-Venisti. Meir Ben-Meir, who was about to be, once again, the Water Commisioner, was also on the scene, but not at the meeting. According to Uri Shamir, he was said to be calling the shots.

I had two reports of that meeting. The first came from Shula, who spoke to Katarivas. She described him as saying "Don't worry. It's not all 'ay, ay, ay, ay', but there are serious problems." Apparently, the opinion of the water experts was very negative. Katarivas expressed it to Shula this way: "We have to find a way to say yes." Shula, on hearing this, suggested to the Dutch that they get back to work on it and that we be permitted to say that the Dutch government was really quite interested in this project.

Shula's report made me think that while the problems were with the people heavily invested in water only, at least Meridor's office was smart enough to realize there were larger political issues at stake.

The second report came to me directly from Uri Shamir. I called him after the Cairo meeting, because Katarivas said he wanted to speak with him when he returned to Israel. Uri had spoken with Yossi Dreizin. He told me the project would probably not go forward—which was, of course, a report from Dreizin, who was very negative about the project. Apparently, confidentiality was only a minor problem. Their real worry was about being identified publicly with anything that might run counter to their current negotiating position. Apparently, they did not realize that the project would give them an even better negotiating position! Uri said he thought the prospect was very bleak, particularly because of the change of power in the Water Commission office. Later in the week, Uri told me he was resigning from any official position in that office.

I was very upset.

Some Divine Intervention?

Around this time, there was an episode of the sort that would have seemed natural to my ancestors. But it was something modern academics do not speak of without some embarrassment.

For years, I had prayed for the success of the project and that it would contribute to peace in the Middle East. I pray pretty hard. Indeed, at the conclusion of one *Yom Kippur* a year or so earlier, a stranger standing next to me told me it had been quite an experience for him because I obviously was praying so hard and well.

On Tuesday morning, I was descending the stairs between the two floors of our home when a shaft of sunlight lit up the staircase. I found myself singing the psalm *"hodo l'adonai kee tov, kee l'olam chasdo."* ("Give thanks to the Lord, for He is good, for His mercy endureth forever.") My heart was uplifted and I suddenly felt quite confident the project would go forward.

Half an hour later, the phone rang. It was Munther Haddadin, who was in New York for ten days. He told me he had called Jordan that morning (I think the water minister's office) and, as chance would have it, interrupted a meeting with Rafi Ben-Venisti and Meir Ben-Meir. Munther knew both of them quite well, and spoke with Ben-Venisti; I had even considered asking him to call Ben-Venisti.

Munther told me he asked Ben-Venisti what was going on with the Harvard project, to which Ben-Venisti replied, "Oh, that project is very dangerous."

"You think everything is dangerous," Munther had responded, and went on to explain that Jordan also was being very cautious about the project and that it had been decided that each country would work quite separately. Agreements to cooperate would come, if at all, after all the parties had gained confidence in the model. Meanwhile, it was just an academic exercise.

Munther then told me that Ben-Venisti promised him that on his return to Israel later in the day he would be back in touch with Meridor's office and attempt to change the decision in order to go forward.

And that is exactly what happened. Apparently, the Israelis took great comfort in the fact that Jordan was also being very cautious. They agreed, as we heard from Katarivas the next day, that they would go forward, although with a great deal of caution.

So far as I can tell, Munther was on the phone with Ben-Venisti just about the time I was descending the stairs. But any reader who gives some thought to the various roles Munther Haddadin had played in the course of the project so far would find it remarkable that it should have been his intervention that turned things around. If, as I believe, God had a hand in what transpired, then it is truly said that "God moves in mysterious ways, God's wonders to perform."

When I spoke with Shula the next day, she confirmed that Ben-Venisti had reversed himself and that Meridor's office was now prepared to go forward. Uri did the same. Shula also raised with me quite strongly the question of whether I should go to Israel and present my paper at the Rabin Conference. She had spoken with both Shaul Arlosoroff and Uri about this. I considered the question, spoke with Uri over the weekend, and decided to withdraw. The Israeli assent—assuming it would be given—was still a very nervous one, and there were a number of people who would be looking for an excuse to reverse it. For me to give a provocative paper in public would provide exactly such an excuse. It would be better if I kept a low profile.

There was no question that the paper would have been provocative. It made the point, carefully phrased, that, while there could be many reasons for Israel to hold on to the Golan Heights, the value of the water involved was certainly not one of them. That would have run exactly opposite to the commonly believed view and (presumably) to the position being taken by Israel (not that there were any current negotiations with the Syrians going on, so far as I knew). To give such a paper at a big conference (attended by Ariel Sharon) could reasonably be described as an "in your face" thing to do.

Further, even if I were to remove the part about the Golan, my paper was supposed to be on the gains from trade in water in the Middle East. I therefore couldn't remove the part about the gains from trade with the Palestinians, putting forward the proposition that cooperation was far more important than water rights. That, too, was not the Israeli negotiating position. Indeed, the water negotiators were said to be frightened lest the fact that it would be efficient to let others use more water undermine Israel's position on negotiating water rights. It was also said they believed the project would have results that would force Israel to sell water.

What a curious phenomenon, I should note. Every country in which I would speak appeared to be frightened lest they be forced to sell. Not only can it literally not be true that they would all be forced to sell, but if they took the trouble to understand the project, they would understand that *no* participant would be forced to sell. The sales involved in the project solution were all voluntary, with gains for both buyers and sellers. Typically, people who take the trouble to listen to me at any length about the project understand that. If they trouble also to understand the model, they become quite enthusiastic. But most were saying yes largely because they wanted the model as an internal tool. Uri had made the case that they couldn't produce such a tool themselves and that they would never get it if they sabotaged us. And partly, I assume, because Meridor's ministry wanted to cooperate with the Dutch, they might also have seen that it would look very bad for Israel not to participate in an academic study of how cooperation might take place.

In any event, the risks were too high for me to go and give a paper that would make news all over Israel. I have never understood people who would rather speak in public for a cause than win that cause, and I was not going to be one of them. I did, indeed, need to say the things in the paper, but there was no particular reason why I needed to say them to that particular audience or at that particular time. The issues were not going away, nor were negotiations on the immediate horizon. There was therefore no need to get one's back up over censorship, even the implicit kind.*

Shaul assured us he could handle any embarrassment. I called Munther and thanked him profusely for his role in the turn of events with Ben-Venisti.

The Ministers' Meeting

On November 12, at 2 p.m. Eastern Standard Time, I began to wonder about the ministers' meeting in Cairo. It was supposed to have taken place from 5:00 p.m. to 7:00 p.m. in Cairo. If so, it had been over for about two hours—and as yet I had no word on what had happened.

Minister Pronk had not gone to Cairo, traveling instead to Zaire because of the refugee crisis there. As a result, the Egyptians refused to convene a ministerial meeting. Louise Anten, the Dutch Foreign Ministry officer in charge of us and the project, spoke with both Nabil Sha'ath and Rima Khalaf, and that went off well. Unfortunately, she did not see Meridor. Instead, she spoke with Dore Gold, a close advisor of Netanyahu's who had heard about the project and reacted favorably, and with Dan Katarivas, whose answer was also favorable. He asked Louise what he should do, and she told him send us a letter—which he said he thought he could get out within a week.

It took until the 14ᵗʰ to find all this out, Shula having left Cairo in a hurry to go to Italy for a short vacation. The afternoon of the 13ᵗʰ was not a happy one for me.

Shula had been pressing Louise to get the Dutch to take a somewhat more proactive stance. They arranged for a ministers' meeting in The Hague on December 3, to which our project might get attached. Meanwhile, we continued to keep a low profile and didn't send out copies of the report.

Meanwhile, Jad emailed me saying he didn't want to be listed as a

---

* Speaking of censorship, I had recently heard a perfectly appalling story from Rob Swanenburg, a reporter from Dutch television who came and interviewed me, who told me that Jad Isaac was stopped at Ben Gurion Airport as he tried to leave the country for a water conference in Scandinavia. The Israeli authorities confiscated the paper he was to have given. Swanenburg gave me a copy: it was a joint paper with Atif Kubursi and hardly provocative at all. I e-mailed Jad saying that such an act was the hallmark of a stupid tyranny. But of course, the water establishment in Israel was correct in one sense: the project really was "dangerous"—at least to established ways of thinking and narrow-mindedness, by forcing people to contemplate the world of water in new ways.

205

participant since he had done nothing but the supply and demand material. Of course, that was not what he meant at all; there must have been some political problem. I e-mailed him back, expressed my outrage about the censorship incident at the airport (see footnote above), and reassured him that the question was pretty well moot since we weren't distributing the material widely anyway and had no present intention of publishing it.

I learned from Shula, though, that Issa Khatar had said Jad was "angry" because he had asked for a number of things to be included in the report that weren't there. I didn't know what those were, and Jad's e-mail was neither angry nor mentioned such things. It was true that Jad had asked me to put in the Palestinian position on the suppression of their water use. I accommodated this as best I could by observing that it was a Palestinian position and put it as an offset to the Israeli position on their "right of prior use." Beyond that, I of course could not go. I hoped it would satisfy him.

Substantive work on the model proceeded, too. WAS 4.0 included several water types, such as fresh and retreated water, and the agricultural model proceeded. It was difficult, but we were making progress. We appeared to have reached the point where the model would give us the same results as either the earlier version (WAS 3.2) or the agricultural model run separately. In the case of the agricultural model being run separately, this meant taking the water allocations that WAS 4.0 gives to a district (Beit She'an, for example), using them as constraints in the agricultural model .and seeing whether one would get the same acreage allocation.

Ilan Amir, Annette Huber, and I—and occasionally M. Daniel Paserman, who had succeeded Aviv Nevo as an assistant on the project—whiled away our time doing all this as our stomachs churned with nervousness over the fate of the entire project.

16

The Israelis Decide

The nervous churning of our stomachs continued. In late November, we learned that Meir Ben-Meir was appointed Israel's water commissioner. It was perfectly understandable, therefore, when Dan Katarivas told us the Israelis could not go forward with the project without discussing it with Ben-Meir. That discussion was to take place on November 25 at a meeting called for other purposes that I hoped would be followed by a meeting with the entire Israeli team.

The news of Ben-Meir threw us briefly into a tizzy. He was reported to be smart but anti-intellectual and anti-American. When Ilan Amir described him as "a farmer," I hoped that might offer considerable hope of interest in the agricultural model.

Two other events did nothing to help our down mood. First, Louise Anten reported that Myriam Van Den Heuvel had met Katarivas at the donors meeting in Paris the preceding week, and he had not seemed optimistic. Second, we spoke with Shaul Arlosoroff. As always in those days, he was extremely pessimistic, believing the project would fall victim to intra-governmental wrangling and power struggles.

My mood picked up, though, when we managed to talk with Uri Shamir. He seemed much more matter-of-fact about the situation: one could reasonably take the view that the Finance Ministry wanted to go forward and merely had to touch the appropriate bases for approval. So, we sent a long fax to Katarivas stressing the advantages to Israel of getting this model and also stressing that by going forward, Israel wouldn't be committing to anything other than an academic study.

On another front, we had received a fax from Issa Khatar with his objections to the preliminary report. It was obviously authored by Jad. The objections were very annoying, consisting of things he could perfectly well have brought up the first time around and that could then have been cured. One objection was to the use of the Oslo II quantities as a basis for experimentation. Jad had had the last nine months to supply other suggested quantities, but had not done so. Moreover, not only did the report state quite clearly that we were only using that as an example, but intelligent reading would have revealed that one of the points of the example was that the Oslo II quantities would be a great mistake for both parties. But since we were not distributing the preliminary report, I supposed it didn't really matter.

Finally, a decision to proceed came on December 17, when the Israeli Finance Ministry held its meeting. Meir Ben-Meir had not been able to attend, but had sent Dreizin at the last minute. There were a number of other people there, including Baruch Levi, a former general in the Israeli Defense Forces and a friend of ISEPME, and an observer from the Prime Minister's office. Lenny, who had talked to Levi and called me seven minutes after the meeting ended, said that he thought Shula probably had strings she could pull on 70 percent of the people present.

I also had news of the meeting from Shula, who had spoken with Shaul Arlosoroff, and from Uri, Zvi, and later from Munther Haddadin, who had spoken with Rafi Ben-Venisti.

The Israeli decision was communicated to the Dutch. Participation was subject to provisions on confidentiality and with the understanding that this in no way committed Israel to "cooperation"—a word that plainly scared them.

We were planning a major trip in mid-January. I told Lenny that even if the Dutch had not yet approved the contract by then, the trip had to take place and that I would take responsibility for it. It was the last time I was available to make such a trip until May.

The news was the best we had had since the Israeli elections. On hearing it, Ellen presented me with a Zuni artifact known as a "water fetish" she had bought the previous summer and had been holding for a special occasion. She told me frankly that she firmly believed she would never have the opportunity to give me the gift.

In late December, we received a copy of Dan Meridor's letter to the Dutch. It was far less restrictive than we had feared, and simply restated that the study was to be an academic exercise, with each country working separately, and that Israel reserved the right to keep the data and runs confidential if it so chose. Other than that, it was quite upbeat.

Back to the Region: Jordan

On the night of January 13, Annette Huber and I flew to Israel. In Tel Aviv, we were joined by Shula Gilad, Stacy Whittle, a helpful staff member married to a Jordanian, and Patricia Tucker, a contracting officer for Harvard University. The trip's principal purpose was to arrange sub-contracts between Harvard and the teams in the region that would do the work under the Dutch contract and, of course, to organize the work.

During the trip, we paid two visits to Jordan—both at the insistence of Munther Haddadin. Getting in and out and back into the country was an adventure.

Americans—and all of us held U.S. passports, including Shula—could obtain visas upon entry into Jordan except when going over the King Hussein (a.k.a. Allenby) Bridge. Annette and I both had multiple entry visas to Jordan that were good for a couple of years. So did Shula, but when she changed passports to be able to go to Kuwait—a trip that ended up not happening— she gave up the passport with the Jordanian visa. Neither Patricia nor Stacy had visas.

Typically, Americans expecting to travel to Jordan—especially if they expected to travel to Jordan frequently—would get visas through a commercial visa service. But we did things differently at ISEPME. Despite my urging, there was a tendency to try to work directly with embassies and consulates. The result was often that things that needed to happen did not.

On this occasion, we planned to fly into Amman from Tel Aviv on Friday or early Saturday. But alas, all the flights were full. By the time we found out, offices in Jordan were closed, so Shula phoned Jawad Anani and asked him to help arrange things for us at the bridge.

Jawad had considerable difficulty. Munther told us he had to spend his own time at the appropriate office fixing it so we could cross the bridge. So partly as a result, we decided to fly out of Jordan early Sunday morning and

209

fly back into Jordan on Monday. Najeeb Fakhoury produced the tickets for us and gave them to Shula, who looked at them and said that we had reservations to fly back at 10:40 p.m. on Monday evening.

At 8:40 on Monday evening, we all piled into a cab in Tel Aviv to go to the airport. It then occurred to me that 10:40 was a very strange time for a flight to Jordan and I asked to see the tickets. The time, in fact, was 20:40. Our flight had left a couple of minutes earlier.

Shula was, of course, very embarrassed. Her first instinct was to call El Al to see whether there was some chance of our making the plane, but even her connections could not return the plane to the ground. We then proceeded toward Jerusalem by taxi with the intent of trying to cross the bridge. I pointed out that there would be a visa problem. If it were not solved, Annette and I would be able to go over the bridge, but the other three would probably have to drive up the Jordan Valley to Bet She'an, cross the Sheik Hussein Bridge—not to be confused with the *King* Hussein Bridge—early the next morning and try to get to Amman in time for a 9:00 a.m. meeting. That did not seem a feasible solution.

The first question was how late the bridges stayed open? Here Shula, ever active on the cellular telephone, did what she always did in times of need. Having first called our hotel and found no one who could give her an answer, she phoned Mordechai (Moti) Cristal, an aid to Prime Minister Netanyahu she had met. Shula always used her connections, and as was often the case, it paid off this time: Cristal told her the bridges stayed open until midnight.

But of course, three of our party had no visas. As we approached Jerusalem, Shula placed a call to Jawad Anani. Jawad was not at home, and she spoke with his adult daughter Dima, who had a highly developed sense of humor and, as Shula told us rather plaintively, was laughing at our predicament.

Dima wanted a fur coat in exchange, but eventually settled for a small stuffed animal. Dima gave Shula the number at which her father could be reached and assured her he was at a private party and could be interrupted.

Jawad took the call and said, "Shula. How nice to hear from you. Just don't ask me to get you visas for the bridge."

Shula responded that she merely wished to ask his advice and then explained the situation in part. She then asked Jawad whether she should call General Mansour, a high official in the Jordanian Army Shula had previously met. Jawad assured her that would be a very good idea.

Unfortunately, General Mansour didn't answer the phone, and we were about to descend into the Jordan Valley, where cellular phone reception starts being problematic. But she finally did reach him and he said he would fix things and call us back. Shula also reached Najeeb Fakhoury's son Kamil and asked him to arrange for cars to pick us up at the bridge.

Just as we were turning towards the bridge, General Mansour called back, having made arrangements. Indeed, when we got to the bridge, the authorities had a large note that said "Shula Gilad—five passports," and Kamil was waiting on the other side of the bridge with two cars. We arrived in Amman at about the same time as if the plane had actually taken off at 10:40 p.m.

As our adventure unfolded, Shula asked me not to recount any of it to Munther.

"Shula," I replied, "what makes you think I'm planning to tell anybody in Jordan?"

She also told me she hoped I wouldn't tell the story in my memoirs.

"In your dreams," I responded.

As for the substance of our time in Jordan, the Saturday visit was spent mostly in discussion with Munther over the time allotted for the work and contractual matters. To understand this required a bit of discussion on where the project stood.

I had written a "Scope of Work" that described the project: the model at the time, WAS 3.2, needed a substantial review of data and assumptions, which would be done first. At the end of that task, the revised program, WAS 3.3, would be ready for use, and various government officials were to be trained in using it. Like WAS 3.2, though, WAS 3.3 would remain an annual, steady state model with only two types of water—potable and recycled—and no seasonal variation. It would also have a very rudimentary treatment of agricultural demand.

The second task was to implement AGSM, Ilan Amir's agricultural sub-model. This would require data on crops, land, water requirements, income from crops, and so forth. It would be useful in its own right as a district-by-district model of agricultural response.

The final task for the first year was to be WAS 4.0, a model that would incorporate AGSM, have seasonal variation (as did AGSM), and include several different water quality types. In the previous fall, Annette, Daniel

Paserman, Ilan, and I had spent considerable time putting together a program that would actually do that, but it could only be described as "user hostile" and a lot of work was still needed. We hoped to have WAS 4.0 ready in the late summer.

Munther was very interested in WAS 4.0 and wanted a serious part in its development, particularly as regards Jordan. We suggested he should send us someone to spend a fair amount of time assisting in its development and programming.

Things became difficult when we got into a discussion of what publications would be permitted. Munther's opening position was that any publication would require the permission of the Jordanian Water Minister. That, of course, was unacceptable, but he and I worked out language that involved no premature publication (during the life of the project) and publication either with permission or acknowledging that permission had not been given and giving the Water Minister (if he wished) room to express reasons for disagreement (assuming that could be worked out with editors).

Munther emphasized that we were having great difficulties maintaining project approval from the Water Minister to whom Rima Khalaf had sent us. He also said that premature publication and, especially, media attention, would kill the project in Jordan. After consultation with Uri Shamir, who assured us that the same thing would be true in Israel, we agreed to suppress distribution of our preliminary report "Liquid Assets," which reported on the project as it was a year earlier and emphasized the value of cooperation and trade.

Patricia Tucker, the Harvard contracting officer, was present during this negotiation. Then and for the rest of the trip, she could charitably be described as a mitigated pain in the ass. Of course, she was very sensitive to issues regarding what Harvard would or would not approve, confidentiality and restrictions on publication being chief among them. However, she seemed totally unable to understand when those issues were relevant and when we were talking about something else. She dipped her oar on every possible occasion, including in the technical discussions, and was thoroughly undiplomatic. Instead of noticing that Munther and I were working out a perfectly acceptable form of language, she intervened at length, explaining how she, at Harvard, had had contracting experience all over the world—and so on. Several times in the various meetings, I had to ask her to be quiet.

212

I bawled Patricia out after the meeting with Munther and told her I did not want her coming to the extremely sensitive meeting we had scheduled with Water Minister Samir Kawar. She absolutely refused not to come. Trying to make the best of a bad situation, I later apologized and told her I thought the "bad cop–good cop" act she and I were putting on was probably helpful—which was definitely not true.

Patricia was incapable of distinguishing her role on contracting issues from matters of considerably delicacy and intellectual diplomacy. I was very sorry she had come on the trip, but I had no choice in the matter.

The meeting with Minister Kawar took place late on Wednesday afternoon, after Munther had spoken with him early in the morning. Patricia Tucker did attend, but somehow finally had the good sense to shut up completely.

Meetings in the afternoon were difficult because of Ramadan, which meant that people were hungry, tired, and possibly cross.* It was clear to me and to Munther that the objective of the meeting was to be polite and get out with the project still alive. Hence, I refrained from speaking my mind, although the epithet *schmuck* did come to my mind several times to describe Kawar. Nevertheless, I was happy that the meeting was not only attended by Zafer El-Alem, one of the people with whom we had spent considerable time in Munther's office going through the model and who had been at Harvard the previous summer, but also by Ahmed Mungo, the advisor to the Crown Prince I had last seen when the Prince broke his appointment with me in 1994. The latter suggested this was not to be merely a Water Ministry affair.

Minister Kawar began the meeting by making the same speech—more accurately, tirade—he had made roughly a year before when I first met him. He fulminated about the World Bank and its insistence on water markets, stating that it was somehow the fault of our project. He again brought up how the World Bank thought that the marginal product of water in industry was many times higher than that in agriculture and must have been thinking only of a particular type of industry (medical supplies) rather than industry in general. He showed essentially no understanding of what we were doing. It was a remarkably stupid performance.

---

* I found Ramadan to be a personal problem, too. Breaking the Ramadan fast leads to meals at irregular times, which wreaks havoc on my diet.

I assured Kawar that we were not the World Bank, that the World Bank's errors were its own, and that we were not recommending a free system of water markets. I emphasized that we were building an important decision-support model for Jordan for the management of its own system and that Jordan was not being committed to trade or cooperation in any way.

The meeting's objective was achieved, and we left with the project alive.

I was extremely gratified to find that Munther was now saying clearly that he believed WAS 4.0 to be a valuable decision tool *and* that the outcome of the project would, in fact, be trade and cooperation in water. He had undergone a serious intellectual conversion. Apparently, it had taken quite some arguing with the Water Minister to get him to continue with the project.

The rest of our time in Jordan was spent in training sessions in Munther's office. The people were very smart, and it went extremely well. There was some entertainment in the evenings. On Saturday evening, the Haddadins took us to an old-time restaurant outside of Amman where we had a pleasant dinner, joined by Stacy Whittle's in-laws, Robert and Therese El-Haj. They were bridge players and invited me to their house to play bridge with them on the Wednesday night of our second visit. One of my opponents at bridge referred to Stacy having traveled to Palestine, by which she meant anywhere west of the Jordan River. She said she could not bring herself to say "Israel."

We spent Tuesday evening at dinner with the Ananis. Jawad Anani is a fascinating man, full of humor, and told some more stories about his role in the peace process. When our conversation turned to the Arabs and Jews having lived side by side in pre-fifteenth century Spain, he told me he had been assigned the job of defending the Israeli-Jordan Peace Treaty in the Jordanian Senate. He was opposed by someone who referred to the story of the Jews of Medina breaking their agreement with the prophet Mohammed. When Anani challenged the speaker to give a single example of any Muslim sage who in later years had recommended not trusting the Jews because of this, his opponent replied that he would not debate Anani because Anani was neither a professor of religion nor of Muslim history.

"I replied," said Anani, "that I knew more about it than he did, and if he was not prepared to cite chapter and verse, he should shut up and go home." The opponent shut up.

Overall, our time in Jordan went quite well. We were essentially ready to contract with the Jordanians, and the team appeared to be pretty much in place, with a budget that was more or less agreed upon.

I should add that we met with the Dutch Ambassador to Jordan, as we had been encouraged to do in all the countries. Of all the Dutch representatives with whom we met, this was the one who knew the least about the project. So, it was principally a courtesy call.

On to Palestine

On January 15 and 16, our entire group went to Gaza, commuting each time from Tel Aviv. In addition, Annette and I returned for a technical training session on Saturday, January 25, and Shula went to discuss contracting matters the following week. The meetings were held at the Ministry of Planning and International Cooperation, of which Nabil Sha'ath was the Minister.

On Thursday, our second day in Gaza, Shula and I saw Nabil Sha'ath in person for the first time since the meeting in the Hague a year before. "As you know," he assured us, "I am totally behind this." He asked what he could do to help.

Moreover, when he had met with Shula, Lenny Hausman, and others from ISEPME a couple of days earlier on a completely different matter, Sha'ath had turned, unprompted, to praise of the water project and its importance. He had also gone to see Dan Meridor and expressed the same view very strongly.

Unfortunately, Water Minister Kawar from Jordan had gone to see Meridor after that and had made Meridor nervous about the project.

Things were not so simple. The question remained as to who was going to do the work. We met to discuss this with Ali Sha'at and Said Abu Jalallah, the engineer Shula and I had met on our first visit to the Planning Ministry and who had spent the intervening year in Europe. He had now returned and was obviously going to play an important role in this project, second only to Ali Sha'at. Other relatively young staff members of the Planning Ministry were also present, including Luai Sha'at, who had been to Harvard and El Arish the previous spring. Abu Jalallah was the engineer.

The attitude at the ministry could not have been more positive; the model had been determined to be a very valuable tool. The ministry was anxious to have staff trained and using it, and wanted ministry people heavily involved, along with people from the Ministry of Agriculture and the Water

215

Commission. But they agreed with us that the work would probably have to be done by private experts. The money allocated in the budget did not seem adequate to them, so we suggested and they agreed to make an extra proposal to the Dutch for capacity building funds—that is, to train, as they put it, the next generation of ministry officials for work with the model. Louise Anten seemed quite receptive to this when we asked her about it, and told us that the Palestinian Ministry applies to the Dutch for capacity building "all the time."

It was quite unclear who the private experts would be. I mentioned Jad Isaac several times without receiving more than a polite brush off. Moreover, it appeared that the Ministry very much wanted to have the project concentrated in Gaza. We agreed that a group of private people would be brought in for a briefing on Saturday, January 25, and the group would then form a private unit that would contract with Harvard.

As it turned out, that did not come to pass. A large group of people, including some of the people who had been in El Arish, convened on that Saturday, but there was no organizational progress toward a private unit. Shula and Ali Sha'at, meeting a few days later, agreed that the ministry itself would be the subcontractor, but that specific people such as Mustafa Nusseibeh and Taher Nassereddine—the two very well-informed older men I had met at the Cyprus meeting—as well as Karen Assaf could also be involved. I hoped it would happen.

Certainly, the Palestinians were the party most supportive of the project and, perhaps, the party with the most to gain. But here, as in the other countries, there were serious matters of internal politics. We encountered some explicit exposure to this on the afternoon of Monday, January 20th, when Shula, Annette, and I went to Ramallah to meet with two of the Dutch representatives to the Palestinians, Malika Kruis-Voorberg and Michel Rentenaar. The junior of the two had been completely won over to the Palestinian point of view. He referred to the water of the Mountain Aquifer as Palestinian and talked about the Israelis "sucking it out from under them." But in discussion of internal Palestinian affairs, they were both quite reasonable and very helpful. They described Ali Sha'at as an empire builder; their view was that he would want to retain all the money in the Ministry of Planning—which I thought was probably true.

They also discussed Jad's role. They said that a few years earlier, Jad was to become head of the Palestinian Environmental Agency (PEA). But Nabil Sha'ath introduced an environmental agency into the Ministry of Planning, thus removing the *raison d'être* for Jad's agency, which caused Jad to become quite bitter. He and Sha'ath had been very much on the outs ever since, they said, but it was possible that some resolution of this was now happening with the creation, under a third party, of a new group to deal with the environment.

Moreover, the Dutch representatives told us, Jad was a recognized expert who was getting grants from the outside. This made the people at the Planning Ministry quite jealous of him. In any event, Jad was now a non-person as far as the Planning Ministry was concerned.

I was sure that much of what they said was true, although I was not certain it was a totally unbiased view. It was reinforced by a lengthy discussion that I had later with Jad himself, who was also, doubtless, not completely unbiased.

On Saturday night, January 25, Ellen and I had a wonderful time when we took Jad and his wife Ghada to dinner at a Bethlehem fish restaurant. Jad was not surprised as I began to break it to him that there was no way he was going to be permitted to work further on the project. He already knew that perfectly well, he said. I shared his view that he was out of favor because he had preserved a fiercely independent stance. He had refused to put himself in a position of owing favors to Arafat and, by implication, receiving slush money. He even described an incident in which Arafat essentially told him to ask freely for anything he wanted, but he refused.

Jad was quite cheerful during our conversation, but he was not particularly happy. He had been contemplating emigrating to Canada, and what he saw as a deteriorating situation in Bethlehem only fueled those thoughts. He and his family are Christians, and Bethlehem was changing from a Christian to a Fundamentalist Muslim town.

The model, Jad told me, had become both very well known. Earlier that day, Christof Bosch, an MIT engineering student from Switzerland who had been tangentially involved in the project, had met us at the ministry in Gaza and showed me a book with the proceedings of a conference held the previous September in which there were several mentions of the model, including one that called it "the best known model of this type." Jad pointed out that the World Bank was working on the development of a WAS-like model for the Ganges. Mrs. Koch-Weser, the World Bank's vice president, was promoting this heavily. Of course, this was Harshadeep, then at the World Bank, working along these lines. Jad also mentioned some other similar developments by others, which raised issues of intellectual property that I realized we would have to resolve.

Jad said he would like to go on working on WAS for Palestine and suggested he should do so independently and we should exchange notes. I did not tell Shula about this, but I agreed in general terms. I managed to get him a copy of the manual and disks for WAS 3.2 that was delivered after we left for the United States on January 31.

There also appeared to be also substantial controversy around the model, Jad said. Apparently, a conference was being organized at Brown University to criticize our model and another model put forward by Tony Allen of England; its organizer was someone by the name of David Miller.* Within Palestine, Jad also said, there was great opposition, and mentioned in particular Dr. Samir from Gaza University. That surprised me greatly, since Dr. Samir had been in El Arish and I had seen him at the ministry earlier that day as one of the large number of people who had come to prepare for work on the model.

Perhaps most important, Jad said the model was being opposed at the Water Commission and that the Water Commission people the Planning Ministry was involving were "too low down to matter."

Apparently, the general—and *incorrect*—view in Palestine was that the model would force the Palestinians to sell water.* We all knew that *all three* participating countries held the same view: the model would force them to sell. The view persisted despite that the three countries could not possibly *all* be net buyers and, more importantly, that the model does not force *any* country to sell water that it would rather retain.

A few days later, I found out that it wasn't the only view that made the model controversial in Palestine. We had dinner with the Law Fellows of the New Israel Fund. Gidon Bromberg, who headed an organization called

---

* Lenny knew Miller. We inquired about the conference after we returned to the United States, and his assistant said we would be on the list of invitees to the conference, which was to be held in April, and that perhaps it would be appropriate if I were to speak about the model. I never saw an actual invitation.

* In fact, years later, both Jad and Atif Kubursi fell into this error on the supposition that the rich Israelis would buy all the water.

EcoPeace, was at our table. It turned out we had met a year earlier in connection with my talks about water. He said there were people among the Palestinians who would absolutely refuse ever to work on our model because it would make the Palestinians *buy* water that is rightfully theirs.

That position is also not correct, but it's perhaps more interesting than the other one. On the one hand, it overlooks the fact that, given a property rights settlement, it has to be to the advantage of the Palestinians to be able to buy water. On the other, it may be that those who believe it think that such a possibility will reduce the intensity with which the Palestinians insist upon the property rights they think they should have. At bottom, of course, such lack of insistence would simply reflect that water is not as important as the people making this objection believe it is if they haven't studied it.

Ellen pointed out afterwards that I had not conveyed the depth of feeling in our discussions with Jad over these issues, especially about his role vis-à-vis Arafat and his government. She was probably right. I should say that Jad told me that the real reason he wanted his name removed from *Liquid Assets*[1] (the draft, not the ultimate book) was to spare the project political difficulty.

I have noted that when political difficulties lead Jad to say in one way or another that he must withdraw, he tends afterwards to suggest ways in which he might continue. At dinner's end, he suggested that there was likely to be a ministerial shake up soon and that Abu Allah was going to rise again in power from what was at the time largely a parliamentary position. Jad suggested he would arrange a meeting for me with Abu Allah, who lived in a Jerusalem suburb called Abu Dis.

After consulting with Yair Hirschfeld, I had to call Jad and tell him that such a visit would not be a good idea at the time. Hirschfeld was emphatic that such a meeting might come, but that it would risk a lot to try to end run Nabil Sha'ath, who was our biggest supporter. And however I intended it, such a visit would be seen as just such a move.

Another feature of our visit with the Palestinians led to a meeting with Yasser Arafat. The long-awaited agreement on Hebron had been signed the night before our first meeting in Gaza and signed at the military base at Erez just outside Gaza. Shula had become friendly with the commander of that base at a previous meeting on ISEPME's trade and security project, and we were taken to meet him and to see the room in which the great event had taken place.

It had apparently been a great event, brought about both the exertions of Dennis Ross and finally by King Hussein's intervention. Quite remarkably, Netanyahu not only agreed to redeploy from Hebron but also to a schedule in which further actions would be carried out. The signing of the agreement brought a wave of optimism, particularly among the Palestinians (including Jad!), but also among many of the Israelis (of course, already in the peace camp) with whom I spoke in the following week.

The trade and security meeting that had taken place in the preceding week was supposed to have included both Netanyahu and Arafat. Netanyahu did attend, but Arafat was unable to come, in part because of the pressures of the Hebron negotiations. Bisharah Bahbah, who had left ISEPME, arranged an audience with Arafat for Shula and me on trade and security as well as water.

The rains had come at last and the streets of Gaza were under water. We were driven to Arafat's office through minor floods, afraid we would be late and certain we would have to wait. In fact, the entire meeting took place on time. The only person we had to wait for was Bisharah.

From the point of view of the water project, I regarded the meeting as pleasant but perfunctory. I had a few minutes to explain the project, about which Arafat obviously didn't know much. Bisharah praised it to the skies and we left it at that. There was a picture-taking session.

We were invited to Arafat's house for dinner that evening—but for a reason unrelated to the project. It turned out that Suha Arafat, Yasser Arafat's wife, had a sister who was at Harvard's Kennedy School and had some connection with ISEPME. She had sent a letter to Mrs. Arafat and a gift for the Arafat baby, which Shula had brought and had been delivered. Our party was asked to dinner as a sort of "thank you."

Located just across the street from the Ministry of Planning, the Arafat house was a nice one, but not grand. Unfortunately, Yasser Arafat did not come to dinner, although he had been expected. The Palestinians present were his wife, their 18-month-old daughter Zahwa Arafat, an assistant to the president we had met earlier that day, and the Arafat's English nanny, who largely dominated the conversation.

There were several amusing incidents. For instance, when Mrs. Arafat asked the baby "show us how papa looks," Zahwa stuck out her lower lip. It would have been a better imitation had the baby produced a three-day beard, but given that limitation, it was pretty good.

Inevitably, during dinner, Shula's cellphone rang. She was very embarrassed and rushed to turn it off.

"Why turn it off?" asked the nanny. "Just answer it 'Arafat household' and see what happens on the other end."

There were also more serious aspects to our conversation. The nanny described being quite mistreated going through security at Ben Gurion Airport. Apparently, it happened with lower-level security people. She is now officially treated quite well, but it was very unpleasant.

We asked Mrs. Arafat whether things were improving in Gaza. She gave what I thought was a fairly perceptive reply. She said there had been a lot of building in the past few years, and so things looked better on the surface, but that as far as she could see the average Gazan was no better off than had been the case when the Arafats arrived from Tunis a couple of years earlier. She referred specifically to the floods in the streets and the lack of an adequate sewage system.

A Difficult Situation in Israel

Our situation with respect to Israel was, in some ways, perhaps the most difficult during this trip.

There were several meetings. We met on January 19 with Uri Shamir and then again with him, Zvi Eckstein, and two new players—Menachem Perlman and Elisha Kali—the following day. Perlman was a partner of Zvi's in a small consulting firm; Kali, was a hydrologist and water systems person I had heard of before. All six of us met with Shaul Arlosoroff on Friday, January 24. Finally, Uri, Shula, Annette, Stacy and I met with various government officials at the Finance Ministry on Thursday, January 30.

The Israeli team continued to be the most capable of all three of the country teams. Of this there was no doubt, despite that these meetings were not without problems. One had to do with the budget; the other concerned the role of Ilan Amir. The preceding spring, the Israeli team had proposed a budget of about $300,000 and, substantially more than had been allocated to each regional team at the time. Since then, the budget had grown to include a considerable addition for work on the agricultural model. The Israeli position was that most of Ilan's payment should come out of the central team's budget. The central team, not surprisingly, took the view that much of what was to be paid to Ilan should come from his work as an Israeli economist. But this would reduce what the Israeli team thought it ought to get. Further, there was a difficulty because we were contracting with the Technion, and no provision had been made for the university's overhead. No doubt this would all be ironed out, but Uri and Zvi were difficult and tough negotiators. Further, Ilan wanted to have his payments increased because he had stayed longer in Cambridge than the others and had made a number of substantive contributions.

The issue of Avishai Braverman and Ben Gurion University also persisted as a problem. We had gotten to a point where Uri would write to him that we were perfectly willing to talk but couldn't make any promises. We did have certain ideas as to what Ben Gurion University might do, including the building of a hydrological model in the next stage of the project. But we would see.

Another problem had been solved, at least for the time being: Hillel Shuval. We had made a separate application to the Cummings Foundation for a grant to study environmental issues in connection with the model. One possibility would have been to treat such issues in a single-year model with a system of changes; the alternative was to build a multi-year model.[*] We proposed to study the feasibility of the latter alternative in light of the possibility of doing the first. Were the money to come through, we saw it as paying both for Hillel as a general environmental consultant on the model and for some part of Annette Huber's time as she worked on her proposed thesis on environmental aspects of the model.

The governmental issues were trickier—and Uri was invaluable. For example, we had told the Palestinians that the Israelis might insist upon confidentiality for certain types of data and certain runs. Ali Sha'at and Said Abu Jalallah were quite concerned about the data issue, claimed to have great difficulty getting any data from Israel, and wanted assurances that the data available to them would be sufficient for them to build their own model.

On this issue, Uri said it was important to realize that the Israeli team could not be the conduit of data to the Palestinians. The Israeli team could, however, point the Palestinians to publicly available data, and he said he

---

[*] This was the beginning of MYWAS, the multiyear version of WAS that was completed many years later and was much more powerful than WAS itself.

222

would be willing to reassure the Palestinians on that point. He stressed that most data the Palestinians would need would be available and that issues of confidentiality probably stemmed more from an unwillingness of particular officials to give up control of information even to other Israelis than from any concerted effort to keep the information from the Palestinians.

The two-hour meeting at the Finance Ministry at the end of January was particularly interesting and went surprisingly well. I met Dan Katarivas for the first time; he chaired the meeting. Minister Meridor was unable to attend. Meir Ben-Meir, the new Water commissioner, sent Yossi Dreizin. Among other attendees, a Dutch diplomat was present.

Dreizin had a number of comments about what he needed in terms of a model. In particular, he wanted to be able to experiment with different prices for different types of water quality and declining prices as consumers buy more. We went to work to provide this facility soon thereafter.*

After the main meeting, Dreizin stuck around to discuss things further. "You know," he said to Uri, "it's going to be hard to sell this to Meir."

He was probably right, but it was notable that he said it in a way that suggested he *wanted* to sell it to Meir Ben-Meir.

Rafi Ben-Venisti made the most significant statement during the meeting: in effect, that the Israelis wanted to be really sure they had the model right before going further in the direction of water cooperation because water was something that could "start a war." The very clear implication was that despite the official cover that our model was for internal decision-making purposes only, they knew quite well that to participate in the project was to get ready to cooperate in water. That was very important.

It was Mordechai Cristal, an assistant to Dore Gold in the prime minister's office, who had told me that they all understood that implication. We had met at the Dan Hotel on the evening of January 16th. He had become quite excited about the prospects the project presented for negotiations not only with the Palestinians but also on the Syrian and Jordanian fronts—the

---

* Doing so required a good deal of time and effort and requirements on what the model user had to specify. When we next met with Dreizin and discussed it, he said of the use of declining prices, "Did I really ask for this?" Based on later events, I came to suspect strongly that his objective had really been to delay us by putting a foolish obstacle in our path.

latter particularly important because Israel was searching for ways to supply the 50 million cubic meters per year called for in the Israel-Jordan treaty. Cristal had wanted me to go to see Ariel Sharon right away, and possibly even see Netanyahu. He had also asked for a one-page memo describing the usefulness of the project.

After thinking about it, we decided to slow him down, preferring to keep a low profile until the contracts were all signed. That was definitely the advice of Yair Hirschfeld.

We had a similar, and welcome, problem when Shula and I met with Christiaan Kroner, the Dutch Ambassador to Israel on Friday, January 31st. The Dutch diplomat who had been at the Finance Ministry meeting the previous day joined us.

Ambassador Kroner worked very closely not only with Minister Pronk but also with Yair Hirschfeld and Yossi Beilin, and knew all about our project and its implications. He regarded it as immensely important for the peace process, and thought it might just be possible that progress on the water issue could break the logjam with the Syrians. So, we raised with him the idea of second-track negotiation workshops, despite that Hirschfeld had suggested that we go slowly. Kroner was extremely receptive. We had to impress on him that we were in a delicate position, and so he urged us to try to see Meridor privately.

When we tried to get an appointment with Meridor, Katarivas also slowed us down. He and others, and Meridor himself, had gone out of their way to make the project go forward with the understanding that the negotiation question would not be involved at that time. We could not now turn around and push it to the front, he said.

Meanwhile, Yair Hirschfeld was urging us to keep second-track negotiations on the second track and not try to go directly through the government. Cristal, while he understood Hirschfeld's point, still wanted to keep Dore Gold informed about our project since it might prove to be important. We sent him the one-pager he had requested in Tel Aviv.

Returning Home

After the trip, we continued to try to settle the various contracting issues, and were getting substantially closer. But still no contract had been signed as March began. We hoped for resolution by the middle of the month, but realized that further delays might occur.

Meanwhile, we had a different set of problems. The news about the project was out, and I was being invited to speak in various places. More important, Lenny had suggested that the new desalination institute in Oman, which was funded by the United States and Israel, among others, hold its first conference that fall, with our project as the central feature. Lenny had also been talking with the Italian company ENI about a similar conference in Italy. And Uri Shamir's wife Yona was running a UNESCO conference in Haifa in May on conflict resolution in water. I agreed to speak, as did Howard Raiffa.

The difficulty with all this was that we were quite limited in what we were now permitted to say. We had basically agreed with the Jordanians to be silent for the time being about anything that affected them. Further, it would be seriously impolitic to emphasize the importance of cooperation from an Israeli perspective, at least for a few more months or even a year.

Yona Shamir understood perfectly well the awkward situation. I was to speak about the theory and present results as though they were for a hypothetical set of countries. There would be actual data, but the names would be changed and no one would know. There was no way, of course, that I could prevent people from realizing what countries it was that I had been working on. Everyone would know that anyway.

One might think it would be even easier to do the same for the Oman conference, since it was to be about the model as a decision-support tool rather than about conflict resolution, I was not sure. To have a two-day conference about that would require making potential model builders and users believe there is something real there. I wasn't sure that could be done well by presenting the theory and made-up examples. I wanted very much to talk about infrastructure west of the Jordan on the assumption of regional cooperation, but even November, when the conference was likely to be held, might still be too early for that to be anything other than a very sensitive subject.

We discussed how to handle things. One possibility that emerged was to have Harshadeep come and talk about his work on the Ganges.

One more matter had arisen that at the moment was secret. The government of Yemen, which had been moving toward cooperation with Israel in various ways, had apparently indicated to Minister Pronk that the Yemenis want to join the Dutch Initiative—perhaps only in order to have their own water model. Yemen was the first of what might be many countries that would want to have such a model.

I had told Lenny earlier that I was not prepared to repeat the process at anything like the same level of effort without being paid. We agreed it might be possible to do this by involving Charles River Associates, my consulting firm, as a subcontractor on future projects.

In any case, it remained very difficult to keep relatively silent about the project, which had generated immense interest and was becoming increasingly well known. I would have liked to be able to speak freely about it in all its aspects. But that was going to have to wait. It was not made easier by Lenny going around the world, and especially throughout the Middle East, claiming I could spin straw into gold.

I began to worry that Rumpelstiltskin would make an unplanned appearance at some point.

IV. March 1997 through June 1998

17

Movement at Last

We were on the verge of signing sub-contracts in all three countries and then contracting with the Dutch—all the more remarkable because of the considerable increase in Middle East tensions. The Israelis were building a settlement in Har Homa, south of East Jerusalem near Beit Sahour; their withdrawal from the West Bank, where there were general disturbances unfolding, was quite limited; and Hamas terrorism was on the rise in Tel Aviv. But Shula said the Porter project was experiencing considerably more problems than ours. Perhaps that was because we had positioned ourselves as helping each party individually.

With a little more than a week left in March, Shula called from Israel to inform us of events in Jordan. There had been a government shakeup that had resulted in the appointment of Munther Haddadin as Water Minister and Jawad Anani as Deputy Prime Minister! Both appointments were particularly remarkable given that relations with Israel were very stressed and both men were associated with the peace treaty. Nevertheless, their new roles could only be good for the water project.

Later, Shula reported that Munther had told her the project would go forward with his consulting company doing the work—the Jordanians were simply awaiting the arrival of the sub-contract documents—but that he would have to be replaced as leader. Obviously, I regarded his elevation to Water Minister in place of Mr. Kadar to be a big plus.

As the Jordan contract was settled, I found it amusing that the solution to an obvious conflict of interest was for Munther to "sell" his firm to Samir Dweiri, his young and very capable assistant. Of course, the transaction was like religious Jews who, in the sabbatical year in which a field is to remain fallow, sell it to a non-Jew for a nominal sum and buy it back the following year. It appeared to be a bit more complicated, though, because Munther reversed his earlier position and began to indicate that he wanted to remain the team leader.

Meanwhile, Ilan Amir's relationship with the Israeli team seemed to have been straightened out, more or less. Zvi, Uri, and Shaul had been quite resentful of Amir's participation and had forced him to come see them in person, and then gave him only a little time. Generally, they behaved badly. Someone's nose was certainly out of joint, but I couldn't quite tell for sure. I suspected it was primarily Zvi's, but wasn't certain.

Nevertheless, despite difficult negotiations, the contract with Israel was nearly ready to go. The situation with the Palestinians, as always, was more complex, and it was where we had the most serious contractual problems. Apparently, they had been working seriously, at least, on the agricultural model. Ilan and Shula, along with two Dutch representatives, had gone to Gaza. Nabil Sha'ath was set to make another attempt to get extra money for capacity building, something the Palestinians were very anxious to procure.

Louise Anten, our Dutch contact, discussed with Shula the possibility of taking some of the money that was to have gone into exploring how a joint management authority would actually work (a study to be done by Theo Panayotou and HIID), and giving it to the Palestinians with the high expectation that the Dutch would restore it in their next fiscal year. This and other adjustments would allow us to provide the Palestinians with rather less money than they had asked for, but still with a substantial sum. It was an issue that we would surely have to discuss when we visited the Palestinians some days later.

Unfortunately, the Dutch representative in Ramallah, Malika Kruis-Voorberg, was much less than helpful. She harbored deep suspicions of Ali Sha'at and the Planning Ministry and was said to regard our entire project as a Zionist plot! Further, the worsening of the peace process, due in large part to the Israeli building in Har Homa, was not helping.

We couldn't get a reply from the Palestinians as to whether our proposals were satisfactory, and absent that input the Dutch would not release the money. Shula was going to try to contact Nabil Sha'ath directly, but I don't believe that happened. Fortunately, the Palestinians had agreed that if they couldn't get additional funds, they would still go forward with what they had. But there were other implications: I was supposed to go to Haifa for Yona Shamir's UNESCO Conference on water negotiations at the end of May. We needed to meet in the field to facilitate the work, but without word from the Palestinians, it would be impossible to bill such a trip directly to the project—and ISEPME couldn't cover the cost.

A Visit to Holland

On Tuesday, May 20, Patricia Tucker and I arrived in Holland on the overnight plane. Shula had arrived the day before. We took the train to The Hague, left our bags at the hotel, and Patricia went immediately to the foreign office with Shula. I followed after a shower.

At the foreign office, I finally met Louise Anten in person; at the January 1996 meeting, she had been one in a crowd. Louise was the Dutch officer who had been our principal contact and tower of strength.

Over the next 36 hours, Shula and Patricia did a tremendous job straightening out the legal aspects and budgetary aspects of the contract. The latter required dealing with officials new to the project. Shula's detailed work paid off, and Patricia, when dealing with matters properly within her ambit, was very good. Ultimately, the project was backdated and the Dutch agreed to release about $150,000 in addition to the funds already agreed upon to reimburse ISEPME for prior expenses.

Shula and I also met with Dick Van Ginhoven, who was handling Yemeni affairs for the Dutch Foreign Ministry. Apparently, Lenny Hausman had managed to interest the Yemeni government in possibly participating in the Dutch Initiative—at least to the extent of having a water model—and the Yemenis had sent an inquiry to Holland. The Yemeni interest was interesting and slightly odd. It was important that Yemen wanted to participate in a project that involved Israel, but there was no connection between the Yemeni water system and the systems we were modeling, so any joint participation would be nominal.

I had yet to decide whether adding Yemen would be something I was willing to do *pro bono* when they involved peace. As it turned out, I never had to make that decision, because nothing ever came of it.

231

On the afternoon of the 20, I made a presentation to several people in the Foreign Office and also to Hans Wesseling of Delft Hydraulics, who was to be one of the experts who would work on the project. The audience included Gerben de Jong, Louise's superior. We spoke about the possibility of having a major kickoff event for the project in Holland that summer.

That night, Shula, Patricia, and I took Louise and her husband Piet to dinner. We brought my nephew, David Paradise, and his wife Karen, who were spending the year in Holland. It was a very pleasant occasion. At one point, I turned to Louise and said I hoped I wasn't going to embarrass her.

"Are you going to sing?" she inquired.

"As a matter of fact, I am," I said, and then sang two temperance songs. I had sung the first at the Cyprus meetings and on other occasions, including in Stockholm. Its lyrics follow:

Come let us sing of fountain spring
Of brooklet stream and river,
And tune our praise to Him always,
That great and glorious giver.

What drink with water doth compare
That nature loves so dearly?
The sweetest draft that can be quaffed
Is water that sparkles so clearly.

I then said that I had thought of a companion song that seemed particularly appropriate given our long negotiations. The lyrics that follow are only the song's beginning:

Sign tonight! Oh sign tonight! Sign tonight!
Why stand ye longer waiting?
The pledge is here within your reach.
Why linger hesitating?

Sign tonight! Oh sign tonight! Sign tonight!
Your heart will be the lighter

'Twill cheer and gladden others too
And make your life the brighter.

232

We were hoping at the time that the signatures would come before we left Holland, but that did not happen.

The next day, May 21, I had a little time off and went to the Mauritshuis, the great small museum in The Hague. It had been 30 years since my previous visit, and I had forgotten how truly beautiful the Vermeers are. I consider the "Girl with Pearl Earring" perhaps the most beautiful painting ever made, and the "View of Delft" a close second. Of course, these feelings also involve nostalgia for our year in the Netherlands early in my career.

On to the Middle East

That evening, Shula and I took the KLM flight to Tel Aviv. We were to see Nabil Sha'ath and other Palestinians the next day—a meeting we had asked for urgently. Patricia went to Amman to see Munther Haddadin.

We arrived at a time when the situation between Israel and the Palestinians was rather tenuous, with all sorts of rumors and accusations flying around. Some were serious; others were rather silly. For instance, the *Jerusalem Post* had reported that Palestinians were claiming Israel was deliberately shipping chewing gum with aphrodisiacal properties into Gaza. It reminded me of the teenage myth of some 20 years earlier that green M&Ms had similar properties. Presumably, teenagers all over Gaza were singing, "My mother gave me a nickel to buy a pickle. I didn't buy a pickle; I bought some chewing gum. Chew chew, chew, chew, chewing gum—oh, I love chewing gum. Chew chew, chew, chew, chewing gum—oh, I love chewing gum."

We arrived at our hotel in Tel Aviv well after 2:00 a.m., and got up in time to leave the hotel by 9:00 a.m. Just as we were about to leave, however, Shula got a call from Luai Sha'at informing us that Dr. Nabil had been summoned to Cairo and could not keep his appointment. It was very upsetting, since the matters that had to be settled with the Palestinians were the most important still outstanding. Shula began working to reinstate the appointment for either the next day or Saturday, a task made difficult by the fact that people in Gaza become relatively unreachable on Fridays.

Meanwhile, we juggled our schedule and drove to the King Hussein (Allenby) Bridge and then to Amman to meet with Munther Haddadin— where I was pleased to find essentially nothing to meet about. Contract negotiations with Jordan had ended successfully, and the Jordanians would sign as soon as the Dutch did. Munther had reconsidered who should buy his firm and had sold it. We met with the buyer and then had a very late lunch and then dinner with Munther, joined by his wife Lexi, Nora and Dima Anani, and Howard and Estelle Raiffa. The latter had just arrived in Amman, and were going to Petra the next day before going on to Haifa to the UNESCO conference.

We had a pleasant social meeting with Jawad Anani the next morning.

Between dinner and meeting with Jawad, I learned why members of the peace party had gone into the government. In March, there had been a terrible incident in which a Jordanian soldier fired on and killed seven Israeli girls on an island at the border. King Hussein then paid a condolence call in Israel to the families. The previous Prime Minister had indicated his disagreement with the King for this gesture, which led the King to replace him and his government and introduce into the government people who had been involved in the negotiations with Israel, including Jawad and Munther.

I remarked to Dima Anani that I was glad to see that wisdom and intelligence were not bars to high office in Jordan.

Shula, Patricia, and I returned over the bridge on Friday, May 23, barely making it before the 3:00 p.m. closing.

The next morning, after repeated phone calls, we were informed that we had secured our meeting with Nabil Sha'ath for that afternoon. We were late, since getting the necessary permissions to go into Gaza took some time, especially on Shabbat. Moreover, for the first time, we couldn't get in touch with Rafi, king of the taxi drivers, and had to secure a ride on our own. The result was a short meeting, insofar as it included Nabil Sha'ath, but a very productive one.

The problem to resolve was the funding for capacity building I mentioned earlier. The Palestinians wanted a grant for capacity building, for several reasons. The most obvious and charitable one was that they wanted to feel absolutely certain that they owned and understood the model, its programming, and its code. They wanted their government officials to be thoroughly familiar with what was going on. The less charitable view, presumably held by the Dutch representative in Ramallah, was that this was just another empire-building grab by the Planning Ministry.

We had accommodated the Palestinians largely by postponing the

234

beginning of the HIID study of how cooperative management operations would work. In response, they suggested they would sign the contract directly with the Dutch and employ the Harvard team as consultants. But that was unacceptable. First, in both appearance and reality, I had to remain in control of the project. It would not do to have the Palestinian government participating officially. Further, and Shula felt very strongly about this, we presumed that the extra capacity-building money we had found for the Palestinians would be used not only to training government personnel but also private-sector people. There was no point in providing extra money if it was all to train government personnel. These objections were why I had written to Nabil Sha'ath urgently asking for the meeting.

It turned out settling these matters was pretty easy. Almost at once, we agreed on an arrangement that would have the Ministry of Planning sign directly with the Dutch for the capacity-building piece and hire us as consultants. The Ministry would also hire private people, with our approval, to work in parallel with the other teams. Ali Sha'at from the Ministry of Planning, Khairy El-Jamal from the Water Commissioner's office, and a representative from the Ministry of Agriculture would constitute a steering committee. Ali Sha'at, of course, had been the person generally in charge of us; Khairy had been one of the absolutely best people at the El Arish seminar the year before; and the man from the Ministry of Agriculture had met with Ilan Amir in March (this was my first direct introduction to him). We met at length with the steering committee to work out details.*

We left Gaza at about 6 p.m., getting a taxi from Ashkelon to take us to Tel Aviv. There, we were met by Mordechai Cristal and drove to Haifa,

---

* Before leaving, I asked Nabil Sha'ath for a private moment. I pointed out that Mariam Maari was a good friend we had in common. "Oh, yes," he said. "I almost married her." This led me to understanding something I had not figured out previously. In Chapter 1, I described the 1990 London conference held by ISEPME, at which two young men from the PLO burst in and insisted we adopt a resolution supporting a Palestinian state. I had managed to cool things down. Nabil Sha'ath was at that meeting. A few weeks later, when Ellen and I were staying with Mariam in Akko, I told her about the incident. "Oh, that was you?" she had said. I now realize that it was Nabil Sha'ath who had told her about it. I was quite sure at the time that she had some connection to the PLO, but I hadn't realized that her connections ran so high.

arriving quite late, but not too late to join the reception at the Shamir home, magnificently overlooking the bay. It was the kickoff to the UNESCO conference on negotiations on water run by Yona Shamir and chaired by Uri Shamir.

The conference lasted for the next three days. On the whole, the intellectual content of the papers was not very high—which reminded me of the Stockholm Water Conference. The better papers came on the last day, with some good discussion of negotiations in theory and empirical cases, although I thought much of the negotiation discussion was fairly empty. I note, however, that my specific opinion of various papers may not be entirely accurate, since I found myself sleeping through most of the proceedings.

There was one respect in which the conference was definitely *not* a success: no Jordanians came, and official Palestinians had been told, at the last minute, to stay away. Ali Sha'at had told us he was coming, but never showed; nor did anyone else from the Ministry of Planning. There were, though, a number of Palestinians in attendance unofficially.

I was the featured speaker on the afternoon of the first day. Keith Hipel, the speaker before me who is a professor at the University of Waterloo in Ontario, began his presentation with a small joke about the difference between Americans and Canadians. I began my talk by reminding the audience of what Hipel had said and offering the difference between MIT and Harvard professors.

"An MIT professor," I said, "is someone who chairs a Harvard project."

The audience, particularly Howard Raiffa, Harvard professor emeritus, thought that was pretty funny.

Preparation of my paper had presented a problem. Because of the political sensitivity of the project, I had agreed with Munther Haddadin's request not to talk about any actual results. With the agreement of the Shamirs, my paper would present the theory and then I presented a fictional example about three countries I took from the Land of Oz: Munchkin, Winkie, and Quadling. I told the audience the example was not real, but also made it obvious that it was close enough to a real example to have some pertinence. I showed a map of Oz; even though I had changed the geography of the Middle East, the three actual countries I was talking about were obvious to almost everyone.*

When the time came for questions, the first questioner identified himself as "Lt. Colonel Daniel Reisner, Winkie Defense Forces." Only one of the Palestinians seemed unable to get the joke, complaining that he didn't want to talk about fictional countries.

In another incident at the conference, I narrowly escaped embarrassment. At the Stockholm Water Conference in 1995, a Japanese engineer named Murakami talked about a Red Sea–Dead Sea desalination project (a canal) to provide fresh water to the cities and large towns on the coast of the Dead Sea at 48 cents per cubic meter. I had taken to giving this as an example of mindless engineering, since there are no cities or large towns on the coast of the Dead Sea. Our earliest model suggested that the value of water there in 2010 would only be about 30 cents per cubic meter. I used the example this time, and the Middle East audience got the joke. But Murakami himself showed up that evening. Had I known he was coming, I would not have told the story.

My paper excited sufficient interest that three people asked to know more: Frank Hartvelt of the United Nations Development Program (UNDP) in New York; Le Huu Ti of the United Nations Economic and Social Commission for Asia and the Pacific, who expressed some interest in the model for the Mekong; and Hans-Peter Nachtnebel from Austria, who wanted his students to experiment with the model.

Not everything went well, however. While we were at the conference, Ariel Sharon said he wanted Israel to have sovereignty over all the water on the West Bank and to give the Palestinians whatever they needed (admittedly the same per capita consumption as Israelis). There were also accusations that the Palestinians were deliberately polluting the aquifer. In fact, Hillel Shuval assured me that for 50 years, the major pollution had come from the holy city of Jerusalem.** Despite the fact that the Israeli contract was about ready to be signed, and that Meir Ben-Meir from the Finance Ministry had come and

---

* I realized we would also likely have to use made-up examples at the Amman Conference scheduled for the coming November, where we were supposed to talk about our approach to water as a decision-support model. I was hoping to get Harshadeep to come and talk about the Ganges, which, of course he was free to do directly.

** "Holy shit!" I commented, not able to help myself.

asserted his support, it was not clear that the higher-ups in the Israeli government really understood that this project was going on or that the general political situation would allow us eventually to succeed.

That was brought home to me on the last day of the conference, when someone from the Tel Aviv University Political Science Department—apparently, he was the advisor on water to Rafael Eitan, the very right-wing Minister of Agriculture—came and talked about how it was essential that Israel always own the entire aquifer, essentially since one could not trust the Palestinians. He had the political, economic, and hydrological facts all wrong. In particular, he certainly didn't understand that even in the worst case, Israel could desalinate an amount of water equivalent to that "stopped" for the Palestinians and do so for a very low percentage of its GDP. It was (and is) really hard to make people think of water in such terms.

But apart from such warning signals, the trip was a great success. If the Dutch were going sign as promised, we could finally go to work.

A Long Time Coming

On the first of June, although I was nervous about saying so, it appeared that the contracts for the project would be signed within days. It seemed our trip was about to bear fruit.

It had been a long time coming. Some three-and-a-half years had transpired since the project began in earnest. It had been a year-and-a-half since my meeting with Nabil Sha'ath, at which he had suggested we should be sponsored by the Dutch Initiative. Sixteen months had gone by since the parties and the Dutch decided to do that in principle. And we had survived the Israeli elections a year earlier.

I thought of Winston Churchill's statement after the battle of El Alamein in October 1942: "This is not the end. It is not even the beginning of the end. But I think we may say that it is the end of the beginning."

I hoped so. Few people, if any, would ever really know how much I had already put into the project. Over the preceding six months, I had felt my zeal and enthusiasm flagging considerably, not only because of the long wait, but also because of outside political events. I was hopeful that I would be able to turn those feelings around and see this thing through.

Minister Pronk had told Nabil Sha'ath as late as June 6 that the contract was about to be signed, but by then it was already being delayed. At June's midpoint, the project was still held up. We were planning to go forward with a large meeting in the Netherlands in mid-July, but only four of seven people had signed for the Dutch, and we were told that the contract was being held up for reasons that had nothing to do with us but generally with Dutch policy and, possibly, bureaucracy.

So, we took action. On June 20, Shula and I wrote to Pronk suggesting rather strongly that we could not keep the teams together much longer and that the project might very well go down the tubes were the contract not signed in the next few days. Our letter contained a generous offer from the Office of Sponsored Research at Harvard to postpone a good deal of the payments until 1998. Meanwhile, Shula secured promises from Nabil Sha'ath to call Minister Pronk. Dan Meridor and Jawad Anani were supposed to contact him when they were both in Washington during the third week of June. Munther Haddadin wanted first to talk to Rima Khalaf before taking action, since she had recently been in Holland.

"Whom the gods would destroy, they first make mad!"[1]

But Meridor didn't go to Washington. Instead, he resigned as Finance Minister. We were trying for an appropriate fax out of Netanyahu's office, but there was no certainty that we could achieve this even with the help of Mordechai Cristal. Meanwhile, the Jordanians would certainly not act except through Rima Khalaf, who said she had done what she could.

We kept looking for help. During a recent trip to Jordan, Lenny had met with the Crown Prince, who had apparently become extremely interested in the project—although it was not clear he knew exactly what it was. Still, he wanted to have a major conference on it in Jordan. That was all very well, but, as I rather wryly remarked, what the project needed was "CPR"—Crown Prince Recommendation—and fast.

Further, it was not clear that Nabil Sha'ath had succeeded in reaching Pronk.

When Pronk got our letter, Gerben de Jong and Louise Anten of his staff had gone to see him and pushed him pretty hard on our behalf. Pronk said he would think about it very seriously and took the letter home. But then he left for New York without getting back in touch with his office about the water project.

During the last week of June, I awoke every day expecting to hear something. De Jong called and begged us to hold off pulling the plug until the following Monday, when Pronk was set to return. We did not know whether we were simply not high on Pronk's agenda and he had forgotten about it or whether he was simply avoiding action. To say we were stressed out does not begin to describe our state of mind.

Finally, on July 2, all the Dutch signatures were accounted for: *Langzaal het leben in den Gloria!* ("Long may it live in glory!") The news came to me at a break in a deposition I was giving in an antitrust case involving the National Football League. I said the *Shehecheyanu*—a 2000-year-old Jewish blessing from the *Talmud* used to celebrate special occasions—and then left a message for Ellen on our answering machine, singing a slightly edited version of the Dutch birthday song quoted above. She broke the code and bought champagne.

In fact, the signature (or at least the news of it) came at the last possible minute at the close of the Dutch business day on the last day we could possibly keep going without pulling the plug—at least on the conference to be held in Holland. We had already waited beyond the time we thought we should.

I now permitted Ellen to buy a Dutch guidebook, which I had superstitiously prevented her from doing earlier in the week. We made plans to go to Holland in advance of the conference, which was scheduled to begin on the evening of July 14 in Delft.

18
Work Begins Again in Earnest

By July 19, we had completed the opening conference of this phase of the project, in Delft. It had gone splendidly.

It had looked at first as though there would be more hitches and delays. The Palestinian government team wasn't going to come, but Shula took action. Through argument, complaining, and influence, Shula got the Palestinians permits to attend. But airplane tickets had to be bought with money personally laid out by the Palestinians, including Ali Sha'at. Shula managed to persuade Nabil Sharif, head of the Palestinian Water Commission who said "I don't like Harvard," to permit attendance by Khairy El-Jamal, the Water Commission employee who had attended the El Arish workshop more than a year earlier and was plainly the Palestinian who really understood how models like ours worked.

Jordan was also well represented, partly by officials from the Water Ministry. Israel was represented only privately, save that General Baruch Spiegel—the head of the Israeli Defense Forces Liaison Unit—came for the last day and a half. The Dutch government also sent representatives.

On the evening of July 14, I realized that I had previously met almost everyone in attendance at one point or another in my various travels, and some of them many times. What I had not expected was that, by and large, they all knew each other. In particular, the Israelis already had pretty good working relations with the Palestinians and the Jordanians. That was important, because the conference was marked by an almost total absence of political maneuvering. Said Abu Jalallah provided the mild exception.

Said had several times expressed dissatisfaction with the model. H didn't think it protected Palestinian water rights, didn't handle environmental issues, and didn't handle the social value of getting water to poor people. We spent some time assuring him that none of these things were true and that the model was basically his to use. There was also some confusion about the shadow values and the prices that get charged to consumers as a matter of policy.

One other problem was easily settled. The Palestinians argued that they did not have access to the same hydrological data as the Israelis, to which the Israelis responded that it was all available from public sources. But the Palestinians said that they could not access the public sources. While the Israelis felt they could not officially turn over documents to the Palestinians, they did agree to provide the locations of these publicly available documents to me so that I could provide them to the Palestinians.

There was a considerable sense of the project moving forward. In these matters, Uri Shamir was one tower of strength. He not only understood thoroughly how ours and other models work, but he was used to negotiating and assisted in clarifying issues and developing relationships. The others were people I brought with me: Annette Huber, Stacy Whittle, and Shula Gilad. Annette impressed everyone not only with her friendly manner but also with her skill and helpfulness. At one point, I took her aside to tell her that she had made a particularly strong impression on the most important person in the room to impress. She knew right away that I meant Uri, who she referred to as "one of the gods of water."

At dinner on the first evening, I made a point of thanking Shula as most responsible for making the event happen. She had an incredible ability to develop contacts and friends in all countries, and had put in an immense amount of effort. I'm not sure anyone else could have made it happen.

The Israeli team, as expected, was unquestionably a strong one. However, I could not always be sure that two of its members— Elisha Kali and Shaul Arlosoroff—could distinguish modeling issues from reality. Elisha in particular had developed a somewhat annoying habit of pressing with questions that were not relevant to what was being discussed. Sometimes he referred to things discussed the day before or scheduled for the next day. Still, those discussions were useful for the group to have.

The Jordanian team also looked quite good. The three Jordanian attendees were Faisal Hassan and Hazim El-Naser from the Water Ministry (El-Naser was later to become Water Minister), and Salem Hamati, the young man who had, at least nominally, bought Munther's consulting company. They were all quite capable, understood how the model worked, and knew what they were doing. I had no worries about them.

The worrisome part seemed likely to be the Palestinians. It was not always clear that Said Jalallah, one of the Palestinian government representatives, really understood how the model worked, but he was very bright and that would work out. The Palestinian private team, though, seemed relatively weak and was going to need a substantial amount of help. Most of them did not speak at all during the conference; that may have been only a language limitation, but I was not convinced they understood what was happening.

Anan Jayyousi from Nablus was definitely an exception. When we were introduced, he reminded me that we had already met at the meeting held in January 1995 in Jad's office, at which the original Palestinian team was assembled. He turned out to have worked with Jad for years.

As mentioned earlier, Baruch Spiegel arrived when the conference was already underway. He said negotiations were taking place between the Israelis and Palestinians to resume talks and that it was very possible that final status negotiations would begin very soon. He also said that water was near the top of everyone's agenda and that, in Israel, water was being handled by four different ministries, none of which knew what the others were doing and none of which had a clear idea of what their objective should be. This was believed to be bound up in all sorts of things not directly about water itself. For instance, Ariel Sharon had announced that Israel should remain in control of all water sources, and that was a view about controlling the land.*

Spiegel believed our project offered serious hope for Israeli negotiation policy and for a fair agreement. He was arranging to have a presentation made

---

* Incidentally, Sharon did not replace Meridor as Finance Minister as I suggested in the previous chapter. Ya'akov Ne'eman, who was said to be a reasonable and intelligent person, instead replaced Meridor. That was probably good for the project, although it made it even more likely that were Sharon to get wind of the project, he would think it in his interest to kill it.

to Defense Minister Yitzhak Mordechai, and planned on taking Uri Shamir and Zvi Eckstein to do that. Shula was trying to arrange for me to be present as well. I didn't know whether that would happen.

All sides told me that the level of cooperation at our conference was far above what was currently being exhibited in other international meetings. The level of enthusiasm for continued cooperation was very high. The Israelis were all quite excited about what was happening.

One point of discussion at the conference was whether I should continue to accept invitations to speak or we should be issuing papers. Everyone, particularly Spiegel, was positive that we should absolutely shut both down and that it would be dangerous even for me to speak about theory only, since what we were working on was well known and there was no way we could avoid reporters. This meant we had to withdraw from the Oman Conference scheduled for the beginning of November and from the ENI Conference that Lenny was trying to arrange for the following March. I undertook to carry that out, and stated that this was the most important thing that I had done in my life and I didn't intend to let anything stop it. I was going to have to tell Lenny that I could not speak at either conference and try to convince him that it would be to ISEPME's long-run benefit—not to mention that of the project and possibly Middle East peace—that we not speak publicly in the coming months.

Elisha Kali came up to me afterwards to say he was moved by what I had said. I found that a little surprising, probably because I was so heavily committed to the project that I couldn't imagine people being surprised by the depth of that commitment. Perhaps that seems most extraordinary when the commitment is from someone from outside the region.

Of course, the speaking-and-writing blackout caused problems. Lenny was very upset—not surprisingly—when I phoned him. He saw big gains for ISEPME in such conferences, but I expressed my strong view that any gains must be postponed lest everything be risked. It was not enjoyable to ride over any friends or participants, but it was necessary for the project, the future of which seemed potentially very bright.

Strengthening the Teams
By the end of August, a bit more than a month after the meeting in Delft, a number of things had happened. In the field, everything appeared to be going

wonderfully well. As planned in Delft, Annette Huber and Hans Wesseling from Delft Hydraulics were spending two weeks in the field assisting the teams to organize data reports and provide training. From Annette's reports, things could not have been going better, particularly with the Palestinians. There was great excitement about the model. In Annette's opinion, very good people were being added to the private team, and Said Jalallah and Khairy El-Jamal were taking a very active role. Great progress was being made in convincing Nabil Sharif, the Water Commissioner, that the project was something in which he wanted to participate.

One of the new people, a young economist in Gaza whose name I do not recall, played with the model and exclaimed, "In forty-five minutes with this model, I shall know more about water than the experts!"

The level of cooperation remained high. Despite that the peace process was at a very low point following suicide bombings in Jerusalem at the end of July, and despite the closing of access to the West Bank and Gaza, joint meetings took place and continued to be planned. Baruch Spiegel was particularly active. He was very impressed both by the project and by the level of cooperation shown in Delft. He arranged permits for the Palestinians to travel. A joint meeting on agriculture was soon to take place in Amman.

The project for "code cleaning" also looked as though it would be quite successful. The teams were to send people to Harvard at the end of September, where they would learn GAMS programming and we would take them step-by-step through the code of the existing model—which had come to resemble an archeological tel. The purpose was two-fold. First, the code needed to be cleaned up, with pieces no longer active removed; in the course of that, various modeling decisions would be revisited. Second, the people who were coming would end up understanding not only the program but also how the model works. That would be a major step in the transfer of this technology to the field.

Both Said and Khairy were planning to come for the Palestinians.

I wished I could be so cheerful about events at home, which could be charitably described as a pain in the ass. They were characterized by a continuing war between Lenny and Shula that was both substantive and also a personality and status conflict. Shula felt, with some justification, that Lenny put extra burdens on her and did not really understand what it actually took to do the work on a contract such as the Dutch one. But Shula also fiercely resented any attempt to take work away from her. She had nearly left ISEPME the preceding fall when she was not given the title of "associate director," as had been given to her friend Tamar Miller. Lenny had risked losing her over that title, and the incident had provoked a continual storm of scenes and letters. Both of them, particularly Lenny, had spoken to me at length about the matter, and it was like dealing with squabbling high school kids.

More recently, there had been episodes of terrible tension. Shula accused Lenny of being an "abusive boss." Her husband, Shalev, came and threatened to charge Lenny openly with misuse of Institute funds—a charge for which I knew of no foundation—if he did not give Shula what she wanted. He even went so far as to suggest that Lenny was protecting and favoring Anni Karasik (who had become considerably less active at ISEPME), not because of her history in the development of the Institute but because, he said, Lenny had previously had an affair with her. Shalev threatened to go public with it all, which upset Lenny very much. He was unwilling to say "publish, and be damned!" because he feared the Institute would suffer very substantially. So, at least for a time, he made peace and calm was restored. But later, new tensions arose about how many resources Shula did or did not get to command.

It was very difficult. There seemed to be something to be said on both sides, but Shula behaved impossibly in these matters—including making another public threat to resign. Whether she would follow through remained to be seen.

I hated trying to be the guidance counselor.

Shula remained worried about her status and her feeling that Lenny was not treating her appropriately. For his part, Lenny saw her as not permitting him to act as director of the Institute, and he played status games with her. All this caused Stacy Whittle to threaten to resign as well, and even put Annette's status in question. Annette's resignation—which I did not believe would happen—would have been a catastrophe.

The issues that troubled Shula most were substantive, and related. The first was the decision taken in Delft not to have conferences or publications over the next several months, but rather to keep a very low profile. This had upset Lenny very much. I did not believe he really understood how complicated the model had become or, more to the point, what it could do

246

besides valuing the water in dispute—in particular how it could be used to allow all parties to benefit by trading water among them.

Of course, I had run the meeting in Delft in part so I would be forced to make a commitment not to do things such as the ENI Conference during that period. Lenny probably thought Shula had arranged that, and he was furious at her. The provoking incident for Shula's resignation threat was a meeting at which I was not present. Lenny and the other ISEPME senior staff discussed whether it would be possible to have the ENI Conference without my participation, since it was my participation that was so sensitive. The idea was floated of having Harshadeep or Zvi Eckstein do the presentation.

Of course, that idea simply wouldn't fly. So far as I could gather, it didn't even get very far. But the very discussion pushed Shula over the edge. She was particularly aggrieved when she insisted that I be brought into the discussion, and that was refused.

The other substantive issue concerned whether Shula was getting the appropriate resources for her work in general as opposed to other members of ISEPME. Shula also believed everyone must always be consulted about everything—in her eyes, the project (and perhaps all projects) were being done by a *kibbutz* or commune. So, the fact that Stacy was not included in every meeting also aggrieved her. It was true that Stacy was more than clerical help, but she was also young and, while I valued her opinion, she was not part of top management. But sometimes Shula was wrong. For instance, this belief of Shula's once went so far that an hour into a presentation on readiness for Delft I realized the *temporary* secretary who was supposed to be answering the phone and typing memos was still in the meeting—and I had to point out that there must be something more useful she could be doing.

The situation was not helped by Lenny's demand that he be given ten copies of the suppressed report "Liquid Assets." He told me he did that because he wanted to prove he was the boss. Indeed, Shula gave them to him, but she and Stacy believed Lenny had wanted them simply to distribute them. I got him to give them back to me, returned them to Shula, and told her I had convinced Lenny he must no longer distribute such things. It was quite difficult to deal with the real issues when the two of them were playing such games.

The deep suspicion that he would continue to do so anyway remained, because he kept pressing the issue. I kept telling Shula, Stacy, and Annette that I was prepared to put up with repeated annoying conversations on the subject so long as nothing material happened. I pointed out to each of them that, if necessary, I would go to the Dean of the Kennedy School and yank the project right out of ISEPME's hands. I didn't tell that to Lenny, though, and I didn't expect to have to do so.

The most serious issue we had to concern ourselves with at the time concerned making the project politically acceptable. Lenny seemed not to understand that we had changed the way in which at least Palestine and Jordan thought about water and that we had good prospects for doing so in Israel. He kept referring to the Delft Conference as though it had been held solely to discussing what he called the "policy of silence." He showed no awareness of what the project was doing, and in saying such things, he was treading on Shula's political accomplishments.

Part of this arose because Lenny went to see Aryeh Genger, a close friend of Ariel Sharon's, to interest him in ISEPME's work and persuade him to be on the board. Without speaking with either Shula or me about it in advance, though, he told Genger about the water project and they discussed taking it directly to Minister Sharon.

This was typical of Lenny, who was fascinated by closeness to power. From various reports, however, Genger was a man with more than a hint of public scandal around him. It was not at all clear he was someone with whom ISEPME should wish to be associated.

The question of whether to approach Sharon at that time was a complex one. At some point, the Israeli government would have to deal with the question of using the project as part of its policy in the region. Of course, we wanted the Israelis to make that transition, but at this time, the level of official Israeli involvement was relatively low compared with Jordan and Palestine.

Shula insisted that we basically had to ask everyone's advice about this, but I was able to get Lenny to agree that we would take advice from three people on the question who I believed had political information and to whom Lenny might actually listen: Moti Cristal, who was then visiting the Kennedy School; General Baruch Spiegel; and Uri Shamir. Our conversations with each of them were quite interesting. Each thought that going to Sharon at the time would be a terrible idea, but each gave different, but complementary, reasons.

Moti confirmed what Lenny argued: that Sharon already knew about this project. He said that every important Minister knew about this project, and told us that he himself had spoken to Meir Ben-Meir, the Water Commissioner, who was close to (and under) Sharon on such matters. But he also said there was a difference between knowing and *knowing*. The fact that Genger was saying Sharon was really well disposed to the Palestinians did not make it likely to be Sharon's official position. If he wanted to, Sharon might very well use the project as an example of cooperation that he wanted to have stopped for his own politically expedient purposes—and he could kill it with a word.

Moti also said that the fact that the project presented a solution to the water problem was already known. But that was not something we wanted to be acknowledged publicly because the Israeli negotiating position to start was not going to be that the water problem should be solved in the way the model suggested. Israel would only "retreat" to that position in exchange for other concessions. He used the example of Israeli–Syrian negotiations: everyone knew Israel would eventually have to give up the Golan Heights, but that was certainly not the position Israel took going in to negotiations.

Baruch Spiegel's attitude was somewhat different: he warned us that it would be dangerous if Sharon were to adopt the project. It could then be seen as the possession of a minister responsible for a particular sector and with a particular agenda, and because it was Sharon, it might offend the other negotiating parties and, a year or so later when the political situation changed in Israel, it could be a disaster to have the project known as Sharon's baby.

Spiegel believed the project should be brought to the attention of the higher-ups slowly. He arranged a meeting for the Israeli team with Uzi Arad, Netanyahu's advisor, but the Israeli team decided to postpone the meeting, believing the time was not yet right.

The logic behind Spiegel's position was well articulated by Uri, whose view was that one could not discuss the project only in theory but that getting into results was inevitable. The results obtained thus far would not be convincing to people who worked in the water sector, he argued, because they involved data not familiar to them.

Uri recommended very strongly that we avoid any publicity or any approach to higher-ups until the model would produce results that could be reconciled with what water-sector people in Israel believed to be correct. In his view, it would be imperative to convince Meir Ben-Meir that the model was a useful tool that had something to do with reality. Otherwise, it would simply be dismissed as an academic exercise.

We hoped to review this situation in a few months when WAS 3.3 would be ready to go.

Meanwhile, when the last week of August came around, Shula withdrew her resignation.

A Workshop at Harvard

In the last two-and-a-half weeks of September, we held a major workshop at Harvard. Two Jordanians from Munther's former firm, Salem Hamati and Samir Dweiri, attended, as did a young Israeli named Yossi Yakhin, who was going to be the computer assistant. There was also a large group of Palestinians headed by Said Abu Jalallah and Khairy El-Jamal that included new members of the private team, the quality of whom seemed to me to be somewhat higher than those I had met in Delft—although that sense might have been about their better English.

It was a remarkable success for Shula in making it possible for the Palestinians to come. The overall situation in the region had continued to deteriorate, and the Palestinian areas were closed. With the help of General Baruch Spiegel, who had recently retired, she managed to get permission for the Palestinians to attend and personally shepherded them out of Israel. Perhaps more significant was that the Palestinians were experiencing political difficulties at home, too, and it was problematic for them to be seen to be participating in any cooperative venture. Nabil Sharif in particular was opposed to the meeting because of the general state of relations.

Those feelings continued to grow after the workshop. We had to convince the Palestinians we were doing something good for them, and that participation remained in their interest.

Khairy and Said brought their wives, and Said's young son—perhaps three or four years old—came as well. We gave a reception, and the young boy slept in his mother's arms. We were told it was useless to try putting him down; he would just wake up. Ellen suggested pretty strongly—and I think somewhat subversively—that his father ought to take a turn. She did not think that was received warmly.

The workshop itself was quite successful. It began with a fairly intensive introduction to GAMS programming, and went on with a line-by-line discussion of what the model does. I cannot say that all the participants learned to do GAMS programming, but they did become considerably more familiar with how the model operated. In the course of the training, we also found a number of small mistakes and raised a number of issues. We issued a memorandum summarizing those issues in an attempt to involve everyone in general modeling decisions.

The most interesting issues were political. One had to do with how we would treat water for Israeli settlements. Another had to do with our treatment of Jerusalem. So far, we had been careful to treat Jerusalem as a separate district assigned to no one. There were some minor problems as to how to treat the connections between the water systems when an Israeli model or a Palestinian model was run in isolation, but by and large we had successfully taken the view that there are people in the Jerusalem area and they will need water, and that we were not going to take any stand on the disposition of Jerusalem.

The Palestinians informed us that, from their point of view, this was no longer satisfactory. Those who criticized their work would see it as political, they said, because treating Jerusalem as a single unit was an Israeli position. They also wanted to know the borders of the "Jerusalem district."

I had given some thought to this made suggestions in writing to the Palestinians. These included that we didn't have to be exact as to the borders, because it didn't really matter for our purposes. Further, I suggested we should add to the model a user choice as to how Jerusalem was to be treated. The user would first see a screen indicating that the user, not us, is making the "political" decision. Then the user would have to specify the Israeli and Palestinian populations of the general Jerusalem area and whether these two populations' water systems were to be connected or separate. Dividing them would, in effect, divide Jerusalem.

The Palestinians didn't quite seem to understand my suggestions. I hoped to clear things up when I would be in the Middle East in November.*

---

* In our major book *Liquid Assets*[1] (not to be confused with papers of the same name), the Palestinians in their chapter simply specified the water situation and users from their perspective, and the Israelis did the same in their chapter. We added a short discussion of this at the start of the book's second part, which included the actual

251

Despite the political issues, all the workshop participants—including the Palestinians--referred to the project as "our" project and spoke of difficulties such as the Jerusalem one as about selling "our" project to others. I thought that was an extremely good sign.

We were planning a trip with a workshop on agriculture and modeling in Amman, but it was probable that we would have to visit each of the three countries separately and hold separate discussions in addition to one big one because of political problems, particularly for the Palestinians.

During this period, something occurred that had the potential to make life harder or easier. Lenny Hausman met with Netanyahu's chief assistant, Uzi Arad, who told him the Israelis expected to move toward final negotiations in 1998. Lenny asked what the Institute could do to help. Arad said he knew about the water project. Were he to ask for a meeting about the project, we would have to agree to go, even though the Israeli team thought it was still premature.

Shula was very nervous about it. I was not. But she may have been right.

Difficulties at ISEPME

We were also having a certain amount of administrative difficulties. In particular, we needed to figure out how to work it out so Harvard could pay the MIT Economics Department for my released time without a contract in which MIT would have to take on substantial amounts of overhead. Harvard was being very strict in trying to follow exact procedures, which I strongly suspected was related to a very unpleasant affair that was brewing.

Someone had complained that there were all sorts of conflicts of interest going on at ISEPME, that Lenny was abusive, and that there had been human resource violations. None of this appeared to be true, but an investigation was under way. I was sure it would end up exonerating Lenny, and soon, but it was a time of extreme tension for Lenny and, less so, for those of us who knew about it.

Clearly, someone had brought the charges, and it was also clear that whoever did so had been exposed to a lot of what Shula had gone around saying, often rather wildly. I didn't want to believe it was Shula herself, but to suspect her was natural. Perhaps, though, it had been Camilla, who had been

---

model results.

fired as Lenny's secretary a few months earlier and was known to be litigious. I also thought it might have been Stacey Whittle, the manager of the water project, and her husband, Rajai Haj. I would have been sorry about that, because I liked them both—but they did behave rather strangely.

For instance, when we scheduled a reception for the workshop participants and I mentioned to Stacey and Rajai that we had invited Lenny, they indicated immediately that they wouldn't attend. It was at the reception that Lenny told me there was an investigation.

Stacey certainly listened to Shula and her various complaints about Lenny and about the Institute, over and over again. She also struck me as a rather rigid in terms of what she thought was appropriate or inappropriate behavior. A few days after the reception, Stacey announced her resignation quite suddenly. Indeed, she departed ISEPME promptly, leaving us quite high and dry until we could hire a replacement. And when Rajai also resigned, he told the human resources people at the Kennedy School that ISEPME was a terrible place that destroyed its employees, and said the same about his boss, Tamar Miller.

While the situation was deteriorating at ISEPME, the general political situation continued to deteriorate in the Middle East. I remained convinced that if things did not fall apart for other reasons, we could solve the water problem with our project. But the "if" kept growing and growing.

Meanwhile, Shula had persuaded the major players among the Palestinians to participate in the Amman workshop and meet with us in Gaza. "We have a lot of political difficulties," Said Abu Jalallah had said to her, "but, no matter what the political situation, we are determined to go on with this project."

Jallalah's comment was very heartening, although not for publication at the time. It was my hope that we could persuade Water Commissioner Nabil Sharif, in particular, that the project could assist him substantially in water management and negotiations, as, indeed, it could assist everyone.

19

To the Region Again

Lenny and I flew to Tel Aviv on November 10, and along with Shula met with Uzi Arad, Netanyahu's foreign policy advisor. It was a meeting we could not refuse. When Arad had seen Lenny in The Hague, he expressed interest in how ISEPME might help in final status negotiations, said he knew about the water project, and specifically asked to talk with me.

The discussion of water went pretty well. Arad was quite interested. I cautioned him about the need to go slow, told him we were not quite ready, and then talked about what the project could do. He observed that water values were so small, which made the gains from cooperation relatively small. He pointed out that both parties, Labor and Likud, were talking about complete separation from any Palestinian entity, although that didn't seem feasible with the water systems. The party line—also his own position—was that the right solution would be for Israel to remain in control of the water sources and manage them for the benefit of both peoples.

Arad also told me he thought Maher El-Kurd had never gotten back to me because I had told him I would offer to help Israel as well as Palestine, which was not the answer he was hoping for. I doubted that very much.

Our conversation continued with ISEPME's work on trade and especially on refugees. Arad's ideological bent persisted: for him, it was solely up to the Arab countries to solve the refugee problem. If that was the Israeli starting point, any academic research seemed useless.

The next two days were spent in Gaza meeting with the Palestinian steering committee. On the first of those days, our focus was on convincing Nabil Sharif, the Water Commissioner, that our water model would be useful to him and that he would not be surrendering any claims if he used it. I believe we accomplished that end.

Sharif, who seemed not to remember being exposed to the model two years earlier, had been talking extensively with Khairy El-Jamal, the one Palestinian I was sure understood thoroughly how it all worked. The others, especially Said Jalallah, were continuing to make considerable progress in that regard.

I was able to demonstrate how the model could answer some of Sharif's specific questions. For instance, he began by asking from where on the seacoast—including Israeli ports—he should supply Hebron. Then he asked whether it would pay to supply Hebron from Gaza if the Palestinians could buy water from Turkey?* He also asked about putting another well in Jenin as opposed to buying water from the Israelis. It was a question that could not be answered at that precise moment, but I was able to assure him that a question of that sort would be easily handled going forward.

Sharif said he would be interested in practicing using the model for negotiations, possibly with me negotiating for the Israeli side. I suggested it might be wiser and more productive to have Uri Shamir use the model for the Israeli side. Uri would be joining us the next day, along with Shaul Arlosoroff and General Baruch Spiegel.

The next day, questioned by Uri, Sharif stated that he was satisfied that use of the model would not compromise any negotiating position or claim he wanted to make. I called that real progress.

Most discussion on that second day in Gaza, November 13th, revolved around issues of Jerusalem and settlements and how they were treated in the model. I began by saying that I had always wanted to be able to say, "I know how to settle the Jerusalem issue." We reached several conclusions.

---

* I knew what would result if we ran the model giving the Palestinians half the water in the Mountain Aquifer: the right answer would be that the Palestinians would not want to supply Hebron from any place on the coast even if they could buy water from Turkey at a landed cost of 50 cents per meter. The shadow value in Hebron would simply be lower than 50 cents plus the conveyance cost. When we went to a run in which only 10 percent of the Mountain Aquifer was owned by Palestine, it would become worthwhile.

With respect to Jerusalem, the Palestinians pointed out that even a model with the city treated as belonging to no one could still be considered an "Israeli-oriented model" because it would treat Jerusalem as a single, undivided entity. To address that, we planned to provide a button on the model's interface marked "Treatment of Jerusalem"; when the user clicked on that button, he or she would see a screen explicitly disclaiming any political intentions and would be able to choose one or two Jerusalems and whether their water systems were to be connected. Results would be presented similarly, and also with the reminder that how to treat Jerusalem had been the user's choice.

That solution was quite satisfactory to the Palestinians and to the Israeli team, but it worried General Spiegel. His concerns may have been partially because I had spent most of my time explaining how a two-Jerusalem version would work, and even to suggest that there could be two views regarding Jerusalem, which would make the model politically difficult in Israel. I couldn't see any other approach, though. We weren't specifying exact boundaries of what was being called Jerusalem; the user would simply have to specify the populations and demands involved.

As for settlements, the time had clearly come to stop ignoring the use of water by Israeli settlers. I proposed a solution in which the Israeli team would supply the demand functions for the settlements and the user would then be asked to specify how the settlements in each district get their water: from an Israeli line, a Palestinian line, or directly from a source. The user could then also specify the size of the settlement. That, in particular, would enable the Palestinians to specify that the settlement didn't exist and see what happened with the model.

Workshop in Amman
Late that day, we traveled to Jordan with Uri and Shaul and a significant group of Palestinians, including Khairy El-Jamal, Said Abu Jalallah, and Dr. Ali Sha'ath—a process that took an extraordinarily long time even with the very helpful arrangements Shaul and Spiegel had made. There was a well-attended dinner at the Kan Zeman, a restaurant south of Amman that features local cuisine, tourist-attracting décor, and extremely loud Arabic music and dancing.

The next morning, we began our workshop in Amman by talking about agriculture and how to predict agricultural demand curves. The discussion went back and forth on the use of WAS, AGSM (Ilan Amir's agricultural model), and other methods for those purposes.

A lot happened behind the scene. The news with respect to the Jordanians seemed good in all respects. Salem Hamati, the leader of the Jordanian private team, made a point of telling me more than once that Munther Haddadin had convened a meeting of all the people involved in the project when he and Samir Dweiri had returned from Harvard. Munther had stated that he wanted the Jordanian model to be a "regional model." I thought Munther meant a model attuned to regional cooperation, and Shula thought he meant a model attuned for the other models in the region, but in either case it was pretty good.

Apparently, Munther had also told the team, "Some day, you will be proud to have been associated with this project."

Late on Friday evening, November 14th, Shula and I visited Jawad Anani at his home. He told me that in formulating a Jordanian long-run plan for water, he had suggested to Munther that the idea of water as a tradable commodity be included—and that had been done. Further, Jawad expressed his view that by 2010 treating water in such a way would be a necessity. I assumed he meant that Jordan would have to buy water.[*]

Thinking back to Munther's initial opposition and all the distance we have come, these developments were really great. That is not to say that there was no internal opposition, but we had totally influenced the way in which Jordan thought about water. Still, we were instructed that all of what we had learned needed to be kept very quiet.

Matters with respect to the Israelis and Palestinians were more vexing. We had yet to meet with the Israeli steering committee, and Israeli governmental participation was, of course, not exactly wholehearted. In addition, it seemed the Israeli team had been inactive. They made some attempt to disguise that by claiming to have done a lot of data revision for

---

[*] In late summer 2013, the Jordanians proposed to Israel a system in which the Jordanians would sell water to Israel in the South and buy from Israel in the North. Annette and I were pleased to see that this arrangement had been suggested under Hazim El-Naser, *who had worked with us and was then the Jordanian water minister. In principle,* our model could be used to set the appropriate prices.

what was not even the latest version of the model, but so far as we could discover, had done essentially no work at all. That had unfortunate consequences, both in terms of the schedule (they were quite late with their data reports), and in terms of the reaction of the Palestinians.

None of this kept Shaul Arlosoroff from talking. He was very nervous about any statement and any model paper that suggested ours might be a regional model, lest the wrong people in Israel get hold of it. His caution was probably warranted—although he seemed to have become afraid of his own shadow after the Israeli elections—but I was not sure it helped to express it publicly in front of the other teams.

As for the Palestinians, I saw trouble brewing from quite early in the workshop, and even before. It first manifested itself with an insistence by Ali Sha'ath and Said Abu Jalallah that they needed historical data on pumping so they could estimate the size of the water resources. Those data were available in Israel, they insisted, but they weren't getting them. They asked Uri to ask his government for them; Uri said he couldn't do that because he didn't represent the government, and to ask the government for such data would not only fail but ran the risk of killing the project. He offered to get them publicly available data and help them sort it out.

Uri had made that same promise four months earlier in Delft, but nothing had happened. Again, he said that the Palestinians should tell him what they wanted and he would make an effort to get it. The Palestinians regarded this as quite unsatisfactory.

The Palestinians were the only team that came to the workshop equipped with a full set of draft reports, on schedule. They gave a quick presentation about them the morning of Sunday, November 16, but absolutely refused to give them to us for comments, arguing that they didn't want us to have their data officially approved. More importantly, I suspected, they were quite reticent about giving out data when the Israelis had provided nothing.

As we learned later, the Palestinians were also quite upset that we kept pressing them to send larger numbers of people to our meetings than they first proposed, while the Israelis and Jordanians did not send large delegations and, in the case of the Jordanians, delegations in which the governmental people made only cameo appearances. They were particularly annoyed that

Israel had sent only Yossi Yakhin* to the Harvard workshop; he was smart and capable, but he was only a graduate student.

In the end, they wouldn't even give us their reports *without* the data so we could comment on the methodology. They insisted on waiting until the other reports were ready.

So far as I could tell from their short presentation, the methodology seemed okay—except in one respect. Said Abu Jalallah presented the demand projections for households simply in terms of its being Palestinian policy that household consumption per capita should be a number comparable to Israel's. Not surprisingly, he regarded that as fair and right. I tried very hard to make him understand that the Palestinians could impose that policy on the model. Khairy El-Jamal obviously understood.

All of this created some tension, which I did manage to break a bit. When it was Said's turn to present, the overhead projector just would not work. I suggested that operatives from Israel's internal security service, *Shin Bet,* using Canadian passports, had been seen near the overhead projector earlier in the day. It was a reference to an incident that had taken place the previous September, when two *Shin Bet* operatives using Canadian passports attempted to assassinate a Hamas leader in Amman—an incredibly stupid act that very much strained relations between Israel and Jordan.

After a pause—during which I assured Said I was joking—he and the other Palestinians said it was very funny.

Still, the Palestinian problems kept coming to the fore, and three days later the issues erupted when our team—Shula, Annette, Hans Wesseling, Dawn Opstead (Shula's new assistant, who had replaced Stacey), and I—met with Ms. Nellicka and her assistant Michel Rentenaar, the Dutch representatives in Ramallah. Also present were Ali Sha'ath and Said Abu Jalallah and a third Palestinian from the West Bank.

We had met with Rentenaar the previous January, at which time I had remarked that the Dutch representatives in Ramallah seemed entirely captured by the Palestinian point of view (although they seemed distrustful of Ali Sha'ath and the Ministry of Planning and International Cooperation and later opposed any money for them for capacity building). This meeting

---

* Yossi, an excellent programmer, had joined the project while a graduate student. He went on to a distinguished career as an academic economist.

certainly reinforced that view.

Ali and Said discussed the grievances they had with us and with the Israelis, which had a sound base—but there was no question in my mind but that they were playing to what they knew to be a quite receptive audience. Indeed, after the meeting, Nellicka informed the Dutch Foreign Ministry that there was a major crisis in the project, which was simply not true.

Ali and Said made the same three points they had raised at the workshop. We forced them to do a great deal of work, they said, and the Palestinians were the only ones who had been sufficiently responsible (they noted in particular that the Israelis had done nothing. Shula, they complained, kept pressing them about who and how many people to send to meetings (largely brushing aside the fact that they had two contracts; Ali Sha'ath specifically noted that his counterpart from the Israeli government never came. They were most upset about the data.

Harvard, they claimed (although not pressing the point very hard), was being unfair. They had, in fact, already privately asked Hans Wesseling whether the project could be transferred to Delft Hydraulics—which he correctly said was out of the question.

After the meeting, we took the Palestinians to a big dinner at the American Colony Hotel that was partly in honor of Rafi Cohen, our chief driver, whose birthday was the next day. There, away from the Dutch, Ali and Said become more willing to listen to us, and we managed to deal with some of the issues from the day's meeting—at least for the time being. We agreed that the Palestinians would not hand in their reports before the Israelis did. We agreed to set up a meeting between Ali Sha'ath and Dani Katarivas, his counterpart in the Israeli government in charge of the project. And we agreed that a couple of Palestinians would come to Tel Aviv in early December to spend a few days working with Uri and Shaul on data collection from public sources and to discuss generally their data needs. I suggested that Hans Wesseling assist in that meeting as a neutral party. He could do that far better than I, since he, at least, knew something about hydrology and could, presumably, distinguish between data the Palestinians genuinely needed and data that they might simply ask for to bolster their negotiating position outside the project. General Spiegel, who had been in command of the West Bank, apparently also had a good deal of data, and moreover, he was a friend of Meir Ben-Meir, who he could encourage to make needed data available.

It was left to Shula to broker all of this and get the Palestinians the appropriate permits. It was the sort of thing at which she was really marvelous.

A heartening incident that occurred just before our dinner bears mentioning. In the hotel bar, we ran into Mohammad Nashishibi, who was still the Palestinian Finance Minister. He had been a reasonably sympathetic audience when I had spoken with him at length three years earlier. He remembered me, greeted me pleasantly, and then proceeded to insist quite strenuously that I was totally wrong in my attitude toward water. Rights had to be settled, he said, and the Palestinians would never sell them.

When I mentioned his to Ali Sha'ath, he said we would have to spend a morning with Nashishibi in Gaza and convince him that he was wrong. Ali's use of the first-person plural, I thought, was important.

Meetings in Israel

On Monday, November 17, we met with the Israeli private team on both methodological issues and the state of the work. The treatment of capital costs became central to our discussion.

For years before the Delft meeting, I had maintained the position that capital costs were to be treated separately and considered as being recovered in hook-up charges rather than in the price of the water. This was so as not to discourage the use of facilities that were not yet being used to capacity. Over the years, there had been repeated arguments about this, first from Hillel Shuval and later from Shaul Arlosoroff. According to Annette, engineers had only recently learned the lesson that capital costs had to be recovered. At Delft, I had finally allowed myself to be persuaded, against my better judgement, by the argument that the prices the model puts out had to look familiar to the water authorities or they would discard the model. Zvi Eckstein had also suggested that we should treat the model as a long-run planning one in which everything was used to capacity.

Following the meeting with the Israeli team, and based on some issues brought up by Elisha Kali concerning the recovery of historical costs, my better judgment regained control. I wrote a very forceful memorandum explaining that if we included capital costs as a per-cubic-meter charge, the model would simply give the wrong results for the use of existing facilities and those whose use had not yet reached capacity. The latter would make

261

cost-benefit analysis extraordinarily difficult, and the two points together would mean that the model would keep producing wrong results. I regarded that as intellectually dishonest, and I said so. We sent the memo out to the teams asking for comments; I expected a lot of noise.

A second meeting with Israelis was scheduled for the following Sunday, November 23. In the interim, Shula and I met with Como van Hellenberg Hubar, the relatively new Dutch Ambassador to Israel. He gave us considerably more than the scheduled hour, was very supportive and, like his predecessor, expressed enthusiasm about the project in terms of the peace negotiations. No matter what we read in the newspapers, he told us, we should know that there was a good deal going on behind the scenes and that serious negotiations would begin pretty soon. He also told us that based on his United Nations experience, he regarded Netanyahu as an unscrupulous negotiator.

On Tuesday, November 18th, Hans Wesseling and I had wonderful dinner with Ilan and Sara Amir in Haifa, after which their daughter and her boyfriend came over and pretty quickly engaged me in a serious discussion about my religious beliefs. They wanted to know what difference religious beliefs made to me. It was not the kind of discussion I was used to having. I didn't want to lie, but it was a bit much for me to come right out and say I thought there was divine intervention in terms of the water project. So, I only hinted around it.

Another meeting took place on Thursday, November 20th, with Yair Hirschfeld, who believed no movement towards useful negotiations on any issue until the government changed, and that such a change was not imminent. When that happened, I said, I would send Yossi Beilin a telegram that would simply say, "Hello, Dolly!" My point came from the song's lyrics: "It's so nice to see you back where you belong." Yair said Beilin would like that, and I told him he could quote me.

Yair also asked me to make a presentation to some people he had been talking to about water the next time I was in Israel. The group included Avraham Katz-Oz, the former Minister of Agriculture. Evidently, I learned, there were several groups that had been meeting to prepare the way for resuming negotiations on all the different topics if and when the government did change.

Incidentally, Yair wanted the New Israel Fund to assist his Institute with its "people to people" program. Yair also assisted Shatil (the New Israel Fund's subsidiary) in getting a grant from the European Union.

Among my other Fund business, I met with Yona Shamir, Uri's wife, who was heading a mediation center and wanted to apply to the New Israel Fund for grants. I was very responsive, as this is the kind of work that I believe the NIF does at its best. How we are all tied together in different ways!

Back to Jordan

We had agreed at the Amman workshop to return to Jordan for further discussions, and on Saturday, November 22, the Harvard team did so— arriving quite late, as usual. Even with the formalities at the bridge eased considerably thanks to Shula and her contacts, the crossing still took a while, and we had also, as usual, started out late. The return trip that night, too, involved a considerable delay. It was the period of a minor Muslim holiday, and many pilgrims were returning from Mecca through Jordan.

In between those two crossings, we spent the afternoon in Amman with Salem Hamati and Samir Dweiri discussing possible experiments that might be presented to a meeting of Jordanian ministers we were trying to arrange for the coming January. We were told that Munther Haddadin had made a big presentation in favor of the Red Sea-Dead Sea canal, so we tried to figure out experiments that would fit with that. These involved the question of how many more people would have to settle in the Dead Sea region to make the shadow value of water without the canal high enough to justify the cost of producing desalinated water there. The results were interesting, but it turned out that the Jordanian plan was not to use such water in the Dead Sea region but rather to pump it to Amman and Jerusalem. It looked to me as though their cost estimates for doing that were sufficiently high as to be above the shadow values that would otherwise obtain in the presence of other projects, but a serious evaluation effort would be required to find out.

That evening, we were among Faisal Hassan's guests at a very good restaurant in Amman. Munther Haddadin came, wearing the headdress. I didn't remember having seen it since the first day we met (see Chapter 2), despite that he said that he is one of two ministers who wear the headdress constantly.

Some of the conversation was quite amusing. At one point, Hazim El-

Nasir passed a salad dish to Munther, who asked what it was.

"Whatever you want it to be, sir," replied Hazim, who at the time worked in the Water Ministry under Munther, who was the Minister then (Hazim was later himself to become the Water Minister, twice).

"You're learning," deadpanned Munther, who was evidently a dedicated and, as one would suppose, an intimidating boss who worked himself and his staff extremely long hours.

Munther told us two entertaining stories at dinner. One was about peace negotiations between Jordan and Israel in July 1994 that took place in a tent in the Arava, the area south of the Dead Sea Basin that forms part of the border between the two countries. Water negotiations began with Munther speaking for a short time; the chief Israeli negotiator then spoke for an equally short time and then turned the floor over to Uri Shamir, who lectured for twenty minutes or so in what I surmised to be his fairly dramatic and distinctively didactic lecture style.

When Uri was done, Munther apparently put on a great show of indignation. "You have behaved like the professor," he told Uri, "but you seem unaware of who is the professor and who is the student." He then headed for the door of the tent. It was very fortunate, he told us, that they called him back before he quite got there, since he would have no idea of where to go once he got outside, as the tent was in the middle of the desert.

Munther also told a story about Uri that Uri did not know when I repeated it to him later. In January 1996, there was a joint Jordanian-Israeli water meeting in Israel, chaired by Uri. I don't know whether the Palestinians were included. Munther's driver begged Munther to be allowed to come with him, since he was engaged to a woman on the West Bank, and Munther agreed. The driver arrived in Israel with the rest of the Jordanian delegation.

As everyone came into the meeting, Uri said there were no assigned seats and people should sit wherever they liked. Uri, of course, sat down at the head of the table. Munther told his driver to sit next to Uri. The newspapers carried pictures of the "heads" of the delegations—including the driver.

Munther's story reminded me of the old one about the famous rabbi who changes places with his driver. At the next town, the scholars ask the "driver" a very complicated question.

"Why, that's such an easy question," he replied, pointing to the rabbi, "that even my driver can answer it. Ask him."

There was some serious conversation, too. Munther wanted us to have no more joint meetings in the region or in Europe that involved Jordanians. He didn't want Jordan to be seen as cooperating with Israel and he didn't want Israel to take Jordan for granted, particularly in view of the disastrous *Shin Bet* attempt on the life *Khaled* Mashal, the chief of Hamas's Political Bureau, in Amman less than two months earlier.

Munther also wanted to continue the silence about the project in public, which he feared would be needed for at least a year—pending, I thought, political events in Israel. He mentioned Lenny Hausman's behavior at the Economic Summit in Doha, Qatar, where Munther had put forward his plan for the Red-Dead canal. Lenny had mentioned the water project, and Munther regarded that as a violation of our agreement to be silent.

Apparently, Lenny had also infuriated the Israelis by those remarks, and Meir Ben-Meir had huffily left the room.

I assured Munther that I had had no notion whatsoever that Lenny was going to do that.

I later spoke with Lenny about this. He insisted that he only talked about the possibility of the general method for water management. But, as usual, it was hard to impress on him the need for discretion.

A Last Meeting in Israel, and then Home

After our return to Israel from Amman, we met on Sunday, November 23rd, with the Israeli steering committee. It was a far more interesting meeting than the one we had with the private team. In addition to Uri, Shaul, and the Harvard representatives, it was attended by Dani Katarivas, Yossi Dreizin, Raffi Ben-Venisti, Ram Aviram from the Foreign Ministry, and someone named Micky from the Finance Ministry. No one from the Prime Minister's office could make it, and the budgeting section of the Finance Ministry was also unrepresented.

Uri made a long presentation about the model and then there was a general discussion. Dreizin was adamant that we "didn't have a model" until we had included seasonal variation and inter-season storage. As a result, we planned to try to do that early the next spring without waiting for all the other developments.

When it came to the confidentiality of data and runs, we eventually came to the same agreement as the previous year. The private team would build the

Israeli model with publicly available data, and if Dreizin didn't think it was adequate, he could have a version of his own—but that version would be held confidential, as would any that the government chose to make. Given that agreement about data, I could understand why Uri felt confused about what to do when it came to Katarivas assisting the Palestinians.

We spent a great deal of time on whether internal political aims could be accommodated. The principal example was the reservation of water for agricultural uses in particular districts where it might not be privately profitable (the Negev or the Galilee, for example). Of course, the model permitted exactly that, but it took a while for that to sink in.

Perhaps most important of all was the concern raised by Amiram from the Foreign Ministry that participation in the model could compromise Israel's negotiating position because the model itself might reflect a particular choice of position on controversial matters. One example was model assumptions about the future Palestinian population, a matter of considerable controversy since Palestine expected to have a great many returnees and the Israeli government at the time was determined that no returnees would be permitted. Of course, as in all such issues, we had left the choice of the numbers up to the user, so an Israeli user would not be bound by the Palestinian estimates.

It took me a while to get that across, and I was not sure I fully succeeded. I realized later while it seemed so natural to us because we had gradually moved into a way of thinking that made all politically controversial issues user-adaptable, the "outsiders" hadn't really been exposed to that approach. Of course, anyone who sat down and actually played with the model would find that out right away, but no one on the steering committee had ever done that.

On the whole, though, I was very encouraged by the discussion. The steering committee appeared to be thinking seriously about the prospect that the model might eventually be used. A number of them told me they were on my "side" as regard to the need for a settlement of water issues in the region. More than one also expressed the view that their government was hopelessly divided and did not speak with one voice.

I took that encouragement home as Annette, Dawn, and I returned to the United States that night.

20
Into a New Year

Once home again, I received from Shula—who had stayed behind in the region—some discouraging news. The Palestinians were very concerned about giving underlying data not only to the Israelis but even to Harvard. There were suspicions because of the involvement of Israelis—Shula, Zvi, Ilan, and Uri—and the fact that I am Jewish. I wanted to believe these suspicions came from people outside the project, rather than the Palestinians with whom I had worked with relatively closely.

I told Shula to tell them there was no need at all for me to see the background data. What I needed to look at was the methodology. I didn't care about their underlying population projections, which they were particularly nervous about releasing, but about how they projected per capita demand for water. Of course, they were free to produce as many different population projections as they wanted and never tell me which ones were real (and they did appear willing to let Hans Wesseling look at the underlying data for supply). We arranged for them to feed in their own assumptions about the extent of future returns by refugees and ex-patriots and never have to share them with the world.

All this suspicion meant that the teams would not run the model using each other's data and hence would have some difficulty exploring the values of cooperation. In the short run, at least, that would make no difference, but the rate at which the atmosphere was being poisoned by outside political events was disheartening.

The Dutch, too, were quite concerned. Nelika, in particular, seemed to oppose the project in ways that suggested she really didn't understand much about it. Shula suggested that we hold a lengthy seminar for all the Dutch representatives the next time we were together in the region; it would include not only Nelika, but also the Dutch ambassadors in Tel Aviv and Amman. The former was very supportive; the latter knew very little about the project but apparently also had expressed some negative views.

As this unfolded, there were some other developments. Lenny attended a meeting on refugees in Canada at which Atif Kubursi spoke. Lenny reported back that Atif had cited both the Coase Theorem (see the explanatory note in chapter 1) and me in support of the proposition that ownership and usage cannot be separated—and, further, that ownership must be decided first. I couldn't believe Atif had said that. It was exactly backwards.

Speaking of exactly backwards, a web-based environmental service had picked up a story in the *Wall Street Journal* that quoted me, and ran with the first part of the quote: "It isn't surprising that so many people believe that the next war will be about water ..." In the actual *Journal* article, the full quote appeared, and that sentence ended as follows: "... but that absolutely doesn't have to be the case."[1]

In yet other developments, I was receiving invitations to participate in programs around the world that might extend the project's work. It looked as though there would be a meeting in Oman at which we would attempt to sell the Gulf Water Ministers on building models for their own purposes. Beyond that, I had been asked by Hillel Shuval to participate in a major program by his institution in Geneva on the economics of water and related topics for sustainable growth, and by an English organization to do something similar. There also remained the question of what MIT would do. All of these programs would involve publicizing and extending our methods to other water situations, and while that would have been very desirable, I couldn't do all of them.

Meanwhile, Lenny kept pressing the point that we should be going right to the top and insisting upon the usefulness of the project in negotiations (he did not yet know about the news from the Palestinians). He had also been in touch with Aaron Miller, Dennis Ross's number-two in the State Department, who wanted me to give a presentation to him and water people in early January. Ross and Miller were leading American efforts to promote peace. Miller wanted a presentation from me, one in which he expected to be present in addition to the water people.

Lenny also kept trying to publicize the work, and Shula kept trying to stop him. Lenny pushed very hard for us to expand into the realm of affecting policy, which Shula—who resented Lenny's interference—believed was exactly what she had been doing. For instance, the previous fall, Shula—and then Lenny—had spoken with Zevi Kahanov of the Jewish National Fund (JNF). Lenny had persuaded him to raise funds, in a private capacity (that is, without identifying the JNF), to take the water project into affecting policy. Involving the JNF would have been a signal to the Palestinians that the project really was a Jewish-run operation. Shula opposed this vigorously.

Lenny had also been in touch with the Conference of Presidents of Major American Jewish Organizations, a very right-wing (that is, strongly against a two-state solution) organization I have come to despise. He interested them in a set of presentations about what final status agreements might look like, and lined me up to speak the following February 9—even though I told him that what I could say would have to be extremely limited. The whole thing seemed to me to be a bad idea, even if Lenny said he wanted to do the same for an Arab-American organization.

Then there was a new complication, linked to Lenny's overture to the Conference of Presidents. The extreme right wing, particularly a group known as Americans for a Safe Israel, had subjected to a vicious attack a program the New Israel Fund was scheduled to put on in conjunction with the Smithsonian Institution on the occasion of Israel's 50th anniversary. The NIF was getting a good deal of press coverage, much of it favorable, but the Smithsonian pulled out. I pointed out to Lenny that the Conference of Presidents might very well include people who were out to sink the New Israel Fund. They wouldn't hesitate to do so by sinking the water project and going after me. I did not believe that was a risk we could take.

On top of all this, we were busy preparing for the arrival of people from the Oman desalination center, with whom we would discuss what we could do in terms of offering to produce water models for individual countries. And a small group of programmers arrived from the region to work on AGSM, converting it, among other things, to GAMS with a visual basic interface. Predictably, Ilan Amir was very concerned at our projected departure from a way of doing it with which he was more familiar.

269

By the second week of January, a number of things had happened, but I cannot say that there had been a great deal of movement. Apparently, the Palestinians had sent people to Israel expecting to get data, but instead got a several-hour computer demonstration from Uri. It was totally inappropriate, and when Zvi Eckstein heard about it he was extremely angry—going so far as to declare that he would take over the job of going to Gaza and delivering public domain data. While those meetings were being arranged, the Palestinians were playing games and insisting that before they would attend, they expected to receive a list of the data that would be brought.

Meanwhile, after consulting with Shula, Hans, and others, I had developed a protocol for dealing with sensitive data. It had three components. The first was to distinguish between the sensitive projections data (e.g., future Palestinian population) and actual model-level data. Second, we were not in fact developing a single model, and so it was not necessary that all the parties agree upon a particular set of data or a particular set of projections. Rather, we were developing a set of models (one for each party) into which users could put their own assumptions regarding future events, and then get results. Finally, Delft Hydraulics, and especially Hans Wesseling, would act as a substitute in case parties did not wish to share information with Harvard or me. We believed this would relieve some of the political pressure on the Palestinians.

Growing Troubles at ISEPME

At the same time that we were seeing increasing and very substantial interest in the use of our methods simply for water management, let alone for the resolution of water disputes, problems were continuing to brew at ISEPME and potential conflicts of interests were emerging.

For instance, the people from the Oman desalination center and also representatives from ENI, the Italian energy company, had expressed interest. Apparently, at least the government of Bahrain and another Gulf state were interested in having a model built, and would be willing to pay quite a lot for it. Hillel's institute in Geneva had already chimed in. There was also a group at MIT, centered in the Civil Engineering department, that wanted to have an ongoing water forum that I thought could also include a big part for our sort of modeling. It was difficult to know how to proceed with these things given the restrictions we were under as to what we could say about the project.

Any further development, as far as I was concerned, could not involve me the way I was involved in the water project. In the first place, I didn't want to spend the rest of my life doing nothing but water. I didn't feel I had the energy to do this again, and I couldn't possibly afford to do it *pro bono*. ISEPME couldn't do it, either, in terms of scientific personnel or administration.

Lenny and I talked about this a couple of times, and I suggested the possibility that Charles River Associates, my consulting firm, might take over a large piece of the action. It would, of course, involve carefully working out the contractual arrangements, and, as Ellen reminded me, we would have to steer clear of conflicts of interest.

Sensitivity to potential conflicts of interest had grown exponentially at Harvard following a scandal involving HIID and Russia, and Harvard had become very nervous. During the last week of our most recent trip to the region, it was reported to me that the Harvard Counsel's office was giving consideration to recommending the closure of ISEPME over such issues, although that report appeared to have been considerably exaggerated.

The investigation of ISEPME by the Counsel's office seemed to be marked by relatively little substance and a lot of rumor, so far as I could gather (my reports at the time were coming almost exclusively from Lenny, an incurable optimist and noisy channel). Nearly every junior ISEPME employee had been interviewed, and the Counsel office's investigator seemed to have decided the matter before even getting started. What was most disturbing, of course, was that the allegations clearly originated with Shula either directly or indirectly. Lenny believed that they did so by way of Stacey and Rajai, but there was no question that Shula had filled the air with quite unreasonable accusations of corruption and abusiveness.

When we first returned from the region, I feared the forthcoming Counsel's report would blast Lenny, but as time passed it began to look as though very little was going to happen. Moreover, Lenny claimed to have found three examples of financial irregularities on the part of Stacey Whittle. One of these, indeed, was Shula's admission that Stacey was aware that overtime was ruled out in the first nine months of 1997. This mattered both because Stacey had complained that she hadn't been paid overtime and then received $1,400 or so from the Kennedy School's human resources officer, and because one of Lenny's lawyers told him that a charge of "not paying overtime" had the potential to be the most damaging in the human resources list of charges against him.

I didn't know what to do. Again, Shula was in the thick of this investigation. But while she had been quite reasonable of late, her *Sturm und Drang* fluctuated quite a lot. Much of the time, she was very helpful. Other times, rational discussion was impossible. She obviously regarded the water project as hers, a view that extended *beyond* managing it and directing it for the Institute to the point where she once called me her "partner." Anything that she felt threatened her control of the project or anything else brought a great deal of upset.

Lenny, meanwhile, just wasn't very good at dealing with her. Probably nobody could. He couldn't fire her under the circumstances, and I didn't want to do that. That Lenny was desperate to bring money into the Institute and wanted to use the water project didn't help; neither did the fact that Lenny really resented rather than appreciated the limits placed on publicity.

If ISEPME went belly-up because of the investigation, I was not sure what I wanted do with the future of the water project with respect to Shula. The project itself certainly would not die, but would have to take on a different administrative structure.

I was certain of one thing: we hadn't seen the end of all that yet.

The Ministers Meeting

One of the most interesting events of the early part of 1998 was what was billed as 'the ministers meeting' held in The Hague on March 18 and 19. More accurately, it was a "quasi"-ministers meeting. There had been quite a bit of maneuvering back and forth in advance of the date, with each party keeping an eye on the others to see who would attend. In the end, the meeting happened—but with lower-level participation. None of the actual ministers attended., Even Minister Pronk himself, who was nominally the host by virtue of the meeting taking place in the Netherlands and whose schedule had dictated the dates when earlier in March had been the initial plan, came for only 90 minutes on the second day.

We had done our best to get the meeting postponed, or at least our own participation since the meeting was also going to take up the Porter project. Our best efforts were unsuccessful. In fact, not postponing the meeting was probably a wise move on the part of the Dutch, who may have wished to trade off Israeli reluctance regarding our project against Palestinian and Jordanian reluctance to cooperate on a regional aspect of the Porter project. So, the meeting happened.

Pronk was very annoyed that Porter did not attend.

In the run-up to the meeting, the most important thing that happened occurred in Israel. About two weeks earlier, the Israeli newspaper *Globes*—the Israeli equivalent of the *Wall Street Journal*—ran a lengthy story on our project and on the scheduled ministers meeting. It was obvious it had been based on material supplied by someone within the Finance Ministry. The story was largely accurate insofar as it described our project in general, in the quotes from me of things I had said a couple of years earlier about the value of water, and in reporting that each country was working on its own model. It went on to suggest pretty strongly, however, that we were actually engaged in second-track negotiations on water ("We should live so long!" I thought at the time). That, of course, was not true.

The story came to the attention of Meir Ben-Meir, the Israeli Water Commissioner. Apparently, his head exploded. He took the very strong view with the Finance Ministry that our project was dangerous for Israel because, he said, we were cheating and not producing an academic study of water but rather a study of the value of water. He claimed that we would show that the value of water was low, thus putting enormous pressure on Israel to make water concessions.

Of course, in one sense, Ben-Meir was entirely correct. However, he failed to notice two things. One was that if the value of water to Israel appeared *too* low, it would be because the Israeli user of the model had not put in the appropriate social values. It was not totally obvious that the value would be low if those values were included. The other was that if the value of water really was low even after the insertion of such values, then Israel would actually *gain* by making water concessions. But Meir Ben-Meir, who hated models and was suspicious of academics and Americans (and whom I had still never met) hadn't gone as far as that last conclusion.

Next, Ariel Sharon sent a letter to Minister Ne'eman protesting that the water project ought properly to be under Sharon's purview. I gathered that was going to happen, although the Finance Ministry (largely Dan Katarivas) had fought—thus far successfully—to keep it alive as something to which two successive governments had committed themselves. It looked as though I would have to go to Sharon and try to explain these things. Whether to take advantage of Lenny's connections with friends of Sharon remained to be seen.

There was more to this mess. My own public remarks about the value of water and the nature of the project—quoted in the article and used by Ben-Meir—were well in the past, but they also referred to Lenny's remarks at the Doha Conference the previous November—which had come back to haunt us.

Shaul Arlosoroff, in particular, was furious. He had always believed that disaster would strike in this form. He threatened to write a letter to Lenny— with a copy to Harvard University's president—stating that Lenny was in danger of putting Harvard in the middle of an international incident between the governments of Israel and the Netherlands. First, I, and then Lenny, persuaded him not to do so, as it would seriously damage the project itself. Meanwhile, Ellen and I put the fear of God into Lenny about talking about the project. We managed to cut him off before he could commit me to talking privately to a group of donors to the Jewish National Fund. We hoped we had managed to make him realize that there was something extremely delicate involved with all of this.

This all transpired before we arrived in The Hague. The need to see Sharon weighed on my mind. It had been a year and a half since I'd spoken privately with Dan Katarivas, so I made a point of sitting with him at the official dinner on the last night of the meeting. I also made a point of telling him the following story, for a very specific reason.

My youngest daughter Naomi was in college when the Scuds began to fall on Israel during the Gulf War. It was the first time Israel had been in physical danger that she could remember, and she was quite upset. Some friends asked her why she was so upset and asked her whether she had family in Israel.

"I was going to tell them 'no'," she told me, "and then I thought of Robert Frost's line 'home is where, when you have to go there, they have to take you in.'"[2]

Katarivas seemed to get the message about how my family felt about Israel. Immediately afterwards, he talked about taking me to see Sharon. I told him I would await his instructions.

As for the meeting itself, it was quite successful—although preparing for it had put a halt to work on the project for several weeks. We had to make presentations to the "meeting of experts" on the 19th that turned out to include very few people who were not already members of the project. There were two or three Dutch officials, including, most prominently, Gerben de Jong, who was deputy of the Dutch government's Middle East division and a supporter of the project. Yossi Dreizin was there from Israel; Ben-Meir had finally permitted him to attend at the urging of the Finance Ministry. Also present was Boulos Kefaya from the Jordanian Ministry of Planning, who was very friendly and even helpful.

The presentations were pretty good, or at least two of them were. The best, from our point of view, was the one by Hazim El-Nasir, who was part of the official Jordanian delegation. Most team members had come early for three days of meetings at Delft Hydraulics to practice and prepare presentations, but Hazim had only arrived the night before his presentation. With help from me and from Salem Hamati, he got up to speed very quickly and gave a very strong performance. His talk consisted of some experiments done with WAS 3.2 on the construction of various conveyance facilities in Jordan. More important, Hazim emphasized that use of the model had already begun in Jordanian water planning. He mentioned it was already being included in the Jordanian master water plan. Since a big part of what we had shown was that the model was actually a useful tool, his report was quite important.

Anan Jayoussi, from An-Najah University in Nablus, preceded Hazim. He was attending in lieu of Marwan Haddad, who had become dean and head of the university's Water Institute. Anan discussed the connection of Gaza and the West Bank by a conveyance facility: which way the conveyance system would run depended on how much water in the Mountain Aquifer the Palestinians had and on the Gaza desalination cost. Anan also emphasized that the model had begun to be put into practical use by the Palestinians— specifically, Said Abu Jalallah was using AGSM to make the case to prospective donors for a recycling plant in Gaza.

Shaul Arlosoroff made a third, and not very understandable, presentation about the Israelis experimenting with runs that imposed existing governmental allocations of water on the system and then compared those to runs that optimized. It was a sensible thing to do, but Shaul didn't really tell us much about the results. In fact, his presentation included fewer results than he had shared in the "practice" run in Delft. We did learn that the Israelis were the only ones who had managed to get their new data into the model and were using a primitive version of WAS 3.3, but the version did not yet allow for easy experimentation.

Comments were generally favorable after the presentations. Yossi Dreizin, though, raised objections. He said the model would not be useful for Israel because it could not reserve water for agricultural, or for the environment, and so on. He was, of course, quite wrong, and I was surprised since we had had the same discussion before. He did make explicit that he was "speaking for [his] superiors" and later told Shaul that he had been ordered to say such things by Meir Ben-Meir.

Salem Hamati and Ilan Amir made a presentation on AGSM, and we emphasized how important AGSM could be in its own right in analyzing the effects of changing agricultural prices on crop choice.

When the experts meeting concluded, Gerben de Jong—who had chaired the meeting—prepared a summary and asked me to come to his office separately to discuss it. There he encouraged me to help him make the summary both accurate and favorable. He knew perfectly well what had been happening in Israel. In fact, Minister Pronk had visited Minister Ne'eman earlier that week. De Jong's position was that we must keep the project alive until the international situation turned favorable enough for the model to be used on a regional basis. He was obviously prepared to recommend funding for an additional year for a phase of further model developments as opposed to regionalization. Later discussion with Louise Anten suggested that our proposals should emphasize that the model was beginning to be used and that we had gone to a system of joint programming so that there was cooperation in model development and the teams felt they owned the model.

So, we were still alive. With luck, we would be alive into the following year and beyond.

On another front, I continued to squabble with the Israeli team over what they were doing and over issues of disagreement. I do not mean to suggest that they were always wrong and that I was always right. That was definitely not the case.

A lot of what we had to cope with was a sort of "not-invented-here" syndrome. The most obvious manifestation of this was reflected in Ilan

Amir's refusal to cope with the GAMS version of AGSM or, for that matter, *any* version that enabled the user to put in agricultural prices directly—which was one of the model's major utilities. Zvi Eckstein also reflected this syndrome in his view that one should not have been using Visual Basic, should run the GAMS model only, and that for the Israeli version the programming should be done only in Israel.

Still, there were some good ideas coming out. One had to do with running the system with governmentally imposed water allocations. As Uri pointed out to me, such allocations were how the authorities thought, and the internal political issues had to do with what kind of cost in terms of net benefits those governmental allocations had. This required us to calculate the shadow values not just of continuity constraints for water, but also of the constraints on the governmental allocations themselves when they were put in as equality constraints. (If they had been put in as greater than or equal constraints, there wouldn't have been a problem.) When administrative constraints are imposed as an equality, the shadow value of water at the source is either zero or infinite, depending on whether it is feasible to supply the water in the ways described by the administrative allocations. The shadow values at each other location reflect merely the (system wide) marginal cost of getting water there. The shadow values of the governmental constraints themselves, however, differed from that because they showed the change in net benefits that would occur if those constraints were altered a little. That included not only the change in cost, but also the change in consumer surplus.

Zvi and I had a major exchange concerning capital costs. He convinced me that my position had been wrong all these years, and we needed to deal with capital costs explicitly as part of the model. A way to think about it came up in discussions among Uri, Annette, and me: Think of a multi-year model in which one is optimizing the present value of all net benefits and incurring capital costs when the new structures are built. That is obviously the right way to handle it. In future versions, it could show when to build projects, how big to build them, and what are the optimal projects to build. It was precisely what we ended up doing years later in MYWAS.

In the version at the time, however, if one just did that for a steady state, then it was obvious that the shadow value of water at the end of a new pipeline, say, would include the present value of the capital costs per cubic meter associated with that pipeline. That meant the yearly shadow value

277

would simply include the capital costs per cubic meter. So, we might as well have put them in directly. But that wasn't quite right, not only because putting them in directly would not tell you which projects to build, but also because you wouldn't get the correct size. It was true that you could read an approximation of the size out of the output by seeing how much would flow through the pipe, but that would be correct only if pipeline costs were proportional to pipeline capacity.

I proposed that we proceed by putting in capital costs for new projects and making it a priority for programming to program the multi-year version as I just described so that we could optimize over the correct things.

All of this might have been troublesome, but it was actually a relief to do some substantive work for a change instead of dealing with all the political stuff. As the time marched on to the beginning of June, however, there were new developments on three fronts: the model, politics, and ISEPME.

Model Development and Related Areas

The major push had been to complete WAS 3.3, which was supposed to be our capstone in terms of an annual steady-state model. It was to have corrected data, embody a large number of programming developments (most of them decided the preceding November), be completely "countrified," and to be extremely user-friendly. The last two items were related. Users were supposed to be able to manipulate the assumptions of the model for their countries and then examine the consequences; one country's users would not have access to the data of the others.

WAS 3.3, of course, was the version that we began to report on at the meeting in March. But only the Israelis had made even a partial transition to WAS 3.3, and, at that time, they refused to use the WAS 3.3 Visual Basic interface. It was impossible to say why other than the "not-invented-here" syndrome I mentioned above.

Essentially, all of the programming for WAS 3.3 could have been accomplished at Harvard in relatively short order following the Amman Conference in November 1997. That, however, would have been both impolitic and an inferior method to the one that was used. It would have been inferior because by widening the scope of who was involved, valuable suggestions continued to come in. It would have been impolitic because it would not have involved regional participation, and it was very important that the teams felt they understood what was really going on.

That was our official position. The fact was, however, that the participatory method didn't really work unless the people participating were in the same room. For WAS 3.3, we parceled out the programming tasks to the different teams and asked them to complete them and report—which happened more slowly than had we done things ourselves. Sometimes the slow pace was due to internal politics; sometimes it was due to the difficulties of long-distance and international understanding; and sometimes it was due to misunderstandings, possibly involving different levels of preparation. It was like pulling teeth to accomplish things that way.

We held a modeling-programming workshop for several days at Harvard in mid-May to bring the results together. Participants included Uri Shamir and Shaul Arlosoroff from Israel; Anan Jayoussi and Mahmoud (last name forgotten) from Palestine; Salem Hamati from Jordan; Hans Wesseling from the Netherlands), as well as Annette and me. Of course, Shula Gilad and her staff, Dawn Opstad and Yasmina Mudarres, participated from time to time, as did Louise Anten from the Dutch Foreign Ministry for part of the time.

The workshop not only discussed WAS 3.3 and its program, but we also moved forward to take decisions on model development for WAS 4.0 and, indeed, for WAS 5.0—the set of topics proposed to be studied in the follow-on phase of the contract, if funding was forthcoming.

One step taken at the workshop was to go so far as to permit the user for a particular country to see only screens for that country, except on request. Further, if the request were made, there would be a warning that the data for other countries were not actual ones and the user would have to supply his or her own assumptions.

Further, we decided that the effort in WAS 4.0, to begin immediately, would be devoted largely to producing a seasonal model. There would also be some attempt to generalize the treatment of water quality, in particular to allow mixtures of water from different sources to be studied. We decided to leave open whether AGSM would be incorporated into WAS 4.0, but we were leaning towards giving users the option of doing so as in the prototype program developed a year and a half earlier.

The effort in WAS 5.0, we decided, would focus on building a multi-year model. A principal development in this discussion concerned capital costs, which kept coming up. It made me realize that a multi-year model had more than one part. In particular, it would not be only, or even principally, a matter of modeling the hydrology of how actions in the aquifer in one period affect what happens later on. Rather, a multi-year model ought to involve optimization over time with growing and changing populations. We should be able to provide the capability of deciding not only whether to build a particular piece of infrastructure, but also when to build which piece of infrastructure and how big to build it, taking into account all the other projects that could be built. That we were capable of doing, and it would add immensely to the model's capabilities. It would also deal with any issue of capital costs directly by treating capital costs as expenditures when they occurred. Further, it would contribute to another development we hoped for—the treatment of climatic uncertainty in terms of sequences of unusually wet or dry years.

The level of participation in both model development and the discussion was somewhat uneven at the workshop, and participation and contribution were not totally correlated.

The Israelis tended to dominate the discussion. Uri's comments were, as usual, highly substantive and very useful, but I still had to suggest to him that it was not politic for him to take quite so dominant a role. Shaul, meanwhile, continued to be a problem. As at previous meetings, he simply could not stick to the topic under discussion. Further, he always spoke about the situation in Israel without regard for the fact that this was a project for others as well—others, in fact, whose situations differed from the Israeli one. Since he believed that certain features of the model were simply not relevant to or useful in the Israeli situation, that was quite unhelpful.

The situation was made worse by the fact that Shaul really did not understand how the model worked or how to interpret its output. He seemed incapable of distinguishing between the model output (or input) and the real Israeli situation. Hence, for example, discussion of debugging runs of model output constantly lead to his interjections about how the pipelines assumed were not there, or some similar comment. He did not understand shadow prices (which we had decided to rename "shadow values" to avoid confusion such as his with the prices charged to users).

It was also evident the Shaul had been a somewhat noisy channel, not only in representing the project to Israeli officials but also in interpreting what they said and reporting to us.

Uri and I discussed what to do about Shaul, but answers were very hard to come by. He was very well intentioned and knew a vast amount about the Israeli water system. And he was a friend. But the overall image he conveyed, buttressed by Ilan Amir's refusal to use the new programming for AGSM and Zvi's reluctance to use Visual Basic, was that the Israelis didn't care about the others and always felt that they knew what to do.

Some Political Problems

During this same period between the ministers meeting and early June, we had some relatively minor political problems that concerned Jordan and Palestine, and a growing problem with respect to Israel.

I was not terribly familiar with the problem in Palestine, although I gathered Ms. Nellicka, the Dutch representative in Ramallah, had not helped. So, I requested a meeting with Nabil Sha'ath in the coming July. Shula and I were to have met with him around the time of the Harvard workshop but he had canceled at the last moment.

The planned July trip had another purpose: to meet with Munther Haddadin, who had expressed a certain amount of displeasure toward us that was triggered largely by the appearance of another article about the water project, this time in the *Jerusalem Post*. I had to assure Munther that the article was not the result of my going around and speaking about the project and that, as far as I knew, even Lenny had not done so since the economic summit in Doha the previous fall.

Munther was also disturbed because he felt Jordanian participation was not being given the full recognition it deserved, and that Jordanians were being made to take something of a back seat. That was brought on in part by the Israeli attitude and the behavior of Shaul at the ministers' meeting in March, and perhaps in part because we had had Ilan, Zvi, and Uri visiting Cambridge or Boston for extended periods in recent months. It didn't seem to matter to Munther that Uri and Zvi had not been financed with project money. So, in response, we had told Munther that in the next phase of the project we wanted to have money for extended visits by a Jordanian and a Palestinian. Still, he wasn't happy, and Salem Hamati even told us at the workshop that Munther had instructed him not to give us the programming he had done or any of the experiments. (There was a suspicion that Salem had not actually completed the programming assignments and was using

Munther as cover, but it was no more than a suspicion.)

Finally, we wanted to meet with Munther to urge him to intercede on the project's behalf with Ariel Sharon and Meir Ben-Meir. He had always seemed fairly confident that the Israelis would not cancel an ongoing project if the Jordanians were involved, but I wanted to talk with him about it at greater length.

I wrote a letter to Munther to try to set up an appointment in July. This provoked yet more problems with Shula. The first was that she insisted that the letter be signed by both of us. Further, it had occurred to me that in both the case of Munther and that of Nabil Sha'ath (especially the latter), it might be easier to talk about negotiations were Shula not present. She is an Israeli and the atmosphere was not exactly one of international trust. Admittedly, it was also because Shula was always opposed to talking about negotiations at all. I therefore suggested in my letter to Munther that if he felt more comfortable talking about such matters only with me, he should so indicate, and I gave as the reason that Shula was an Israeli.

Shula expressed considerable disquiet over what I had written to Munther. "We are a team," she insisted. "We must not promote distrust." And so on. She insisted the letter omit the offending clause about not including her and we both sign it. I refused emphatically on both counts.

Shula was aggrieved in the extreme. This came soon after Lenny and I had informed her that, contrary to her belief, she did not make policy for the project. She went so far as to file a complaint with Jean Hood, the Kennedy School's human resources person, to the effect that Lenny was degrading her position as retaliation for her role in the investigation of him.

The letter ended up being sent after Shula had departed for the region, and then I ended up having to cancel my trip on short notice.

Nevertheless, the problems with Shula were not as important as the principal political problem at the time, which seemed surely to lie with Israel. I described earlier how the appearance of a long article in the Israeli newspaper, the *Globes*, had prompted Ariel Sharon and Meir Ben-Meir to make noises about the project. Now the project had been transferred to Sharon's ministry. And then, as I understand it, there was an episode in which Sharon and Munther Haddadin were discussing the question of how much Jordan should pay Israel for extra water delivered from the Sea of Galilee. Sharon proposed that Jordan should pay the cost of desalination, since that was Israel's replacement cost. Munther replied that—while he had no faith in the specific findings of the WAS 3.2 Harvard model since it was based on faulty data—it would have to be wrong by way more than 50 percent before the desalination price was even approached. Apparently, Sharon felt blindsided by that, since he knew nothing about the model. Reportedly, he was very angry.

I also learned that Meir Ben-Meir had given a talk at the Technion in late April or early May in which he excoriated the model and economists generally. His described the model as saying that all agriculture in Israel would be forced to pay the desalination price, revealing that he had not the faintest idea of how the model worked or its results.*

It was obvious from all this that the notion was correct that Sharon and Ben-Meir did not understand the project and were simply reluctant to sponsor it lest it be used in negotiations. They were hearing about it second-hand or, worse yet, through a channel that began with Shaul Arlosoroff who, as mentioned above, really didn't understand it himself.

Based on learning of Sharon's anger and Ben-Meir's speech, we decided the time had come to ask for a meeting directly with Sharon. That "we" did not include Shula, who continued to be absolutely against approaching Sharon for any reason—which she feared would result in his killing the project. Lenny, however, had been very much for approaching Sharon and wished to use various channels he had to get me in to see him. I decided Lenny was right. Uri Shamir concurred, pressing that we should also try to see Meir Ben-Meir.

In retrospect, we had made a mistake over the past year or so. We had confused the policy of not talking in public with the policy of not speaking to high officials such as Sharon when necessary. Those two policies were not the same and, it had become apparent that we would have done better by seeking to meet with Sharon earlier.

Once the decision was made, Shula continued to oppose seeing Sharon. It was that opposition that provoked Lenny's remark to her about not making policy for the project and sent her to human resources. Meanwhile, in the

---

* It was possible some of this was language confusion. At the Harvard workshop in May, Shaul Arlosoroff said he and others in the Israeli water system used the term "marginal cost" to mean the cost of the alternative ("marginal") technology.

weeks that followed, Shula became very upset if I sent a letter or e-mail about the project to anyone in the Middle East without her approval and signature. I did not stand for that.

One day, at the very last moment, we learned that Sharon was going to be in Washington that evening and would be having dinner with his close friend Aryeh Genger, who Lenny knew. Lenny immediately contacted Genger, and I quickly wrote a letter to Sharon asking for a meeting. This letter was faxed to Genger, who promised to deliver it directly to Sharon and to support it. Lenny also wanted the co-chairman of the ISEPME board, who also knew Sharon, to attend any meeting we arranged. Three weeks later, we had still heard nothing in reply.

One effect of all this (prompted, in part by a discussion between Shula and the Dutch Ambassador in Tel Aviv) was that the Dutch suggested we not apply for a second year of funding, since that would require a decision by the parties. Instead, they suggested we ask for an extension of the contract from August 31 to December 31, 1998. They suggested that they could justify this because of a clause in the contract about making up for differences in the money that had been awarded in the contract and the amount we had been told a couple of years earlier, using the decline in the value of the guilder relative to the dollar as the justification. We would have to submit a budget and a proposal for the extension; it seemed pretty clear that it would be approved.

The advantage of doing this within the limits of the existing contract, without requiring the approval of the parties, was to bypass the Israelis.

I mentioned earlier that there were developments on three fronts, and that the third concerned ISEPME. Indeed, that deserves a chapter of its own.

21

A New Home for the Project

After the episode with Shula and the letter to Munther Haddadin, I met with Jean Hood, the Human Resources Officer at the Kennedy School, and complained that Shula had overstepped herself even in reading that letter. I explained that Shula had apparently come to believe that she, not I, made the policy decisions for the water project—which simply was not true. Shula, I told her, regularly wanted to co-sign letters to anyone important, a practice to which I did not assent.

When, in my early Stockholm paper, I had thanked Shula for administrative assistance, she told me that if I wouldn't thank her for anything else, I should not mention her at all. I should have seen then what would eventually transpire.

I suppose it was Ms. Hood's job to smooth things over. She suggested Shula and I sit down with her to discuss these issues. I agreed, but insisted that Lenny not participate. The war between Lenny and Shula would make the conversation fruitless.

As it turned out, the conversation would have been fruitless under any circumstances, as events had overtaken having such a discussion.

On Wednesday, June 3, Lenny met with Joseph Nye, Jr., dean of the Kennedy School, and other officials. He gave me what I must admit is a not totally objective description of what happened. He was presented with the charges made as a result of the report by Diane Lopez, the Harvard Counsel. He had mismanaged ISEPME, the report stated, and was being terminated.

Apparently, the specific charges were unchanged from the previous November. The first was that he had mishandled funds from a particular donor—who was then, presumably, alienated. Lenny replied that that donor supported him, was perfectly happy, and had gone on to make an additional contribution.

The second charge was totally outrageous. A few years earlier, Eytan Sheshinski had asked Lenny and me whether we could wrangle things so his daughter, Yael Venn, could obtain Harvard housing. I felt we owed Eytan a favor, and so at my urging this was done. But the charge was not that Lenny had helped get Yael into Harvard housing, but that he had used Institute and Harvard funds to pay for the housing and even for a hotel for her.

It was a totally false charge; the Sheshinkis *did* pay for the housing. I thought that what probably really happened is that ISEPME, with its poor management, failed to send the bills to Eytan when they were due, but that he did pay them when he finally received them. Of course, had Lenny been a better manager, he would have asked the Dean's Office whether the arrangement was appropriate and would probably have been told it was not.

That, though, was not all. Apparently, Harvard was charging that Eytan "fled Cambridge" to avoid being interviewed about the issue, and as result he had been sent a letter stating that he was no longer welcome at the Kennedy School. It was an absolute outrage, and I called it "scandalous baloney" in a letter I sent to Dean Nye. Eytan, who regularly moved back and forth between Cambridge and Jerusalem, was on a trip for HIID, and his telephone number was well known. He was not interviewed because the counsel's office had not pursued the matter.

Failure to follow up seemed to be standard operating procedure. For instance, I had sent a letter some months earlier to Sheila Burke, the dean of administration, but no attempt was made to interview me during the investigation. In another example, Tamar Miller, the ISEPME associate director, said that someone from the counsel's office had claimed that Anni Karasik was "unavailable," whereupon Tamar asked, "What do you mean by 'unavailable'? She's on the other end of a telephone line!"

From what I could gather, admittedly mostly from Lenny Hausman and his wife Bonnie, and before I had a chance later to read the Counsel's full report, all the charges were of this general nature or related to completely trivial events. Throughout the investigation, the Dean never discussed matters with Lenny or tried to work things out. In fact, Lenny had been asked to keep things quiet. Not even I was supposed to know about it.

The entire process stank. It was obvious that ISEPME's constant deficit did not contribute to Harvard wanting to keep it alive, but something had nevertheless gone very wrong with the investigation. Lenny and Bonnie, of course, believed it had something directly to do with Shula.

I believed they were right in substance, if not in form, and that Shula was wholly responsible. Regardless of whether the investigation was due entirely to Shula, there could be no doubt that the disruptive atmosphere she continually fostered, particularly among lower-level employees, had pervaded the entire affair. On matters of substance, I often thought Shula to be right and Lenny to be wrong, but that was no excuse for what was happening here. As I noted earlier, she and her husband had gone to Lenny and tried to blackmail him, saying that if Shula was not treated appropriately, she would bring charges of corruption against him. Pretty plainly, that's exactly what she had done.

Later, when I finally read the Counsel's full report, I learned that Lenny had been told he would not receive any severance unless he agreed by the end of June not to speak about his dismissal in public. The report also included that Lopez had been urged by Tamar Miller to speak with me, but that upon hearing of the large amount of *pro bono* time I had put in the counsel concluded that my activities were aimed at furthering my academic and professional interest. I took that as an implication that she felt anything I said wouldn't be worth listening to. What an insult! In fact, I was already a very well known and distinguished economist and was publicly working on the project without pay because I regarded it as extremely important—indeed, as my contribution to *Tikkun Olam*, the repair of the world. I still hold that view.

Significantly, the report also said everyone knew that the only project going anywhere was *Shula's* water project! It was clear to whom Lopez was listening.

The copy of the report I received also had a name redacted throughout. Apparently, there had been a witness who talked a lot. The natural conclusion was that "Deep Throat" was Shula herself—despite statements to the contrary in meetings I had later. I was also informed that I did not know the whole story, and that had I seen all the reports I would take a different view of the investigation. But if that were true, then Lenny's counsel, Joel Reck, had not been given all the information—because he had told me all he knew. Further, it seemed Lopez had spent nearly all her time listening to Shula and to people who worked for Shula—two of whom were interns *living with Shula.*

The report also had an appendix in which Lopez described her visit to a former member of the ISEPME board. She asked him whether he thought it was true Lenny had slept with a female assistant. The former board member replied that he was not going to answer personal questions, which Lopez noted and then added, "But I could see from the look in his eyes that he thought it was true." That wasn't even part of the formal charges, and I couldn't help but question what kind of biased investigation was going on!

All in all, Lopez appeared to have swallowed every claim whole. The actual charges centered largely on the fact that Lenny had formed a small consulting group and acted to attract money to it. In fact, those efforts could not have been successful, since over the firm's few years of life, it earned only about $25,000 from consulting. Then there was the charge about the Sheshinski's daughter Yael and her housing costs; the report claimed that ISEPME's willingness to pay Yael's rent was an attempt to attract business from Koor, a large Israeli conglomerate of which Eytan had been the Board Chairman. There was no foundation for this.

Then there was the charge that Lenny had overpaid Bisharah Bahbah, his former Palestinian associate director, who had been involved with the aforementioned consulting group but left to return to Palestine. In fact, Bisharah provided the principal avenue to Arafat and other Palestinians. Indeed, the former head of HIID, Dwight Perkins, said that Bisharah was worth far more than he was paid. But one could not learn that from the Counsel's report.

Dean Joseph Nye joined Lopez in swallowing every claim whole. He explicitly refused to listen to anyone who wasn't a Harvard professor—that is, me—and made Lenny's dismissal a public affair, *contrary to Harvard's usual practice.*

As Alan Altshuler, a Harvard professor who argued on Lenny's behalf, put it, "It may have been necessary to remove Lenny, but why was it necessary to destroy him?"

A New Home for the Project?

When I finally received a formal response from Dean Nye to my letter, it did not respond to the substance of what I had said about the investigation of Lenny. But the dean did emphasize the Kennedy's School's desire to retain the water project and have me continue with it. I was already looking for a new home for the project.

It seemed to me the best place at Harvard would be the Harvard Institute for International Development (HIID), which was part of the Kennedy School, and so I had placed a number of calls, including to Director Jeffrey Sachs. When Jeff returned my call the next morning, he informed me that Dean Nye had spoken to him and asked him to take the water project into HIID. This seemed a complete meeting of minds, and so I was pretty satisfied. Jeff said Theo Panayatou would be the person in charge.

Later that afternoon, Theo and I met to discuss the issues involved. He had concerns about what it would take to manage the project and what would be in it for HIID. We dealt with those matters, and then I raised the personnel issue, making clear that if Shula Gilad was to be involved I would not, under any circumstances, continue. Theo assured me HIID did not want Shula— not only because Theo obviously regarded Shula as a difficult problem, but also because HIID already had a staff and would not gain from the project if it had to take on new staff. I made a good-faith effort to preserve the jobs of Dawn Opstad and Yasmina Mudarres, but Theo seemed adamant. We discussed the prospects of doing with a much smaller staff, including keeping Anni Karasik on part time to provide both contacts in the region and political advice.

We ended our meeting with an agreement that Theo would speak with Sheila Burke about project finances, after which we would meet again. This was to take place in the week of June 15, since Theo was going to China and it turned out I would be off to Pittsburgh because of the birth of my sixth grandchild, Eve Joelle Zikmund-Fisher.

Meanwhile, I let Shula know the project would be moving to HIID, with details to be worked out.

While I was in Pittsburgh, there was yet another incident involving Shula. I wanted to write to Nabil Sha'ath and ask him to see me to discuss the use of the model in the context of negotiations, among other things. I knew that was a subject Shula did not wish me to discuss with him, despite that she really had no idea of what I had in mind—which was nothing beyond using the model to examine the consequences of different agreements to the

Palestinians, a suggestion we had already made to Nabil Sharif and others. At Ellen's strong suggestion, I decided the time had come to make sure of my ability to reassert control over the project's communications and contacts.

In view of the uncertainty of the ISEPME situation, I sent the letter to Sha'ath from MIT and asked that he respond directly to that office—which upset Shula very much. When she asked if at least she could co-sign the letter and I said no, she told the Kennedy School administration that I was sending out a letter on MIT stationery about a Harvard project. This prompted calls from Jean Hood and Sheila Burke. When I finally spoke to Burke, I told her this was more of Shula's continuing resolve to monitor my communications, which I would not permit. She agreed it was improper.

Also during the week that I was in Pittsburgh, I spoke several times with Anni Karasik. She informed me that she was planning to accept a position with the Peres Peace Institute. Its director, Uri Savir, had been the chief Israeli negotiator before the change in government. We explored arrangements under which she could both work on the water project and do that—an arrangement that looked as if it might work, given that she was not going to resign fully from ISEPME. Moreover, it seemed clear to me that informing the Peres Institute about the water project was something that might be very useful. It might prove to be a funding source, should the Dutch decide they couldn't continue, or perhaps the Peres Institute might be interested in funding the general project on management of international waters that Uri Shamir, Howard Raiffa, and I had been discussing.

Anni informed me that many people at the Peres Institute knew about the water project, but they all knew conflicting things. So, I arranged to see Uri Savir on July 3.

Over the next week or so, I went back and forth on the question of Shula who had, after all, contributed a great deal to the project. Lenny, of course, urged me not to keep her, but I couldn't really pay attention to that. After all, Lenny considered her to be his personal devil. But both Ellen and Anni concurred, pointing out how extremely manipulative she was. I also knew how inefficient she was. While disposing of her services seemed difficult at the time, I believed that not to do so would, as Anni put it, set me up for much bigger difficulties in the future.

Upon returning to Cambridge, I had a more-difficult-than-expected meeting with Theo and Richard Pagett, also of HIID. It turned on the question of whether it would be to HIID's advantage to take the project at that time. One issue involved Shula: if we didn't keep her on, she would complain, and regardless of what the truth really was, HIID really didn't want more bad publicity in the newspapers after its scandal a year or so earlier about its projects in Russia. I said I would take the hit for not retaining Shula, and Theo and Richard agreed they would talk to Sheila Burke the next day, and make it clear to her that I would not stay if Shula did and that they considered me essential to the project.

It came as no surprise that this provoked a call from Sheila Burke's office to set up a meeting. It took place on June 22 with her, Jean Hood, and and Robert Uliano of the Harvard Counsel's office.

The meeting began propitiously. Burke began with the invoice to Harvard from the MIT Economics Department for reimbursement for my released time of eighteen months earlier. Payment had been unbelievably tangled up in bureaucracy, but the Kennedy School was obligated to pay as part of the Dutch budget. I had been forced to keep pressing, through Shula, to get it taken care of, even threatening to stop work if it was not. Burke said she didn't understand why it hadn't already been paid and that she would order the check be cut immediately. That set the tone for the meeting. It made clear that the Kennedy School wanted to be helpful to me and not aggravate me unnecessarily. It also sent a message that I was dealing with grownups— people in authority who could actually get things done and who were experienced in administration.

For the first hour, we discussed the question of what had to be done in the Dutch contract and the possibilities for extension into later years. I brought up the issue of Anni Karasik and encountered no difficulty. Dean Burke reiterated the Kennedy School's strong wish to keep the project and me.

"You know," I said as politely as I could, those are the same thing." There was general agreement.

Dean Burke also informed me that they would like the project to remain with ISEPME, which was not to be dismantled, at least until the end of the current Dutch contract in December. Theo did not feel able to take it over yet, but would become more active beginning in October, after which the project would be transferred to the new Center for International Development (CID)—which was to be a joint effort of HIID and the

Kennedy School.

Eventually, we came to the Shula question. I made it clear I would not continue if Shula did. Sheila Burke explained that one reason for the proposal that the project stay at ISEPME during the current contract and then move was that, when CID was ready to take it, the new organization could decide on its own personnel, and that would provide a natural changeover. But Bob Uliano pointed out that the issue concerned Harvard's university-wide personnel code and was not just a matter of the Kennedy School or institutes within it. Hence, the change of home within Harvard would not solve the problem, and we would be facing in December the same problem we were already facing.

So, I insisted the problem be solved before I would proceed. I explained the various frictions that had arisen with Shula's aggrandizement of her role and what I considered to be her great inefficiency, which was a particular waster of my time.

Ordinarily, I was told, a situation like this would lead to intensive counseling between me and Shula, followed by mediation—something I refused to do. I pointed out that it was bound to be ineffective, would waste a very large amount of my time, and that I wasn't going to put up with the emotional nonsense that had characterized previous relations in ISEPME during the Lenny-Shula wars. I went so far as to observe that we were meeting on my 40th wedding anniversary and that my wife had assured me that if Shula and I continued, there wouldn't be a 41st.

There was general sympathy for my position. I sensed pretty strongly that they knew Shula was very, very difficult; indeed, one of them used "pain in the ass" to describe her. But we were still in search of a solution.

Burke reiterated Harvard's very strong wish to retain the project and me. I said I was happy to have the project remain at Harvard, but I would move it if necessary. I omitted any mention that at I had, at Ellen's suggestion, already explored whether a threat to move the project to MIT would be a credible one (I had spoken with Paul Joskow, then the head of my department there, as well as Maureen Maguire, the administrative officer) and it seemed it would have been.

The meeting adjourned with agreement that we would all think about how we might manage the Shula issue before reconvening a week later.

That evening, when I discussed the meeting with Ellen, she came up with an excellent idea. Were Shula's job to disappear, rather than her being replaced, the personnel code would presumably permit it. It made a lot of sense: what Shula considered her job *was* going to disappear as I asserted sole control of policy and contacts with people who count. I needed efficient logistic and financial administration, not someone to usurp my role as chair.

I called Bob Uliano the next day and suggested it as a possible solution. He said he had to think about it, but that it might very well work.

Meanwhile there had been some other developments. Even before I had met with the deans, Shula had shown me a passage in the Dutch contract that required us to discuss political issues with the various Dutch ambassadors. It applied to the projected meeting with Sharon, she said. Why she hadn't pointed this out earlier was unclear, but in any case, she now claimed that the letter from Gerben De Jong asking us to delay applying for a second phase effectively said not to go near Sharon. That was not my reading, but I agreed to speak with Dutch representatives, beginning with Louise Anten.

At first, the call with Louise was going to include Shula, but then I decided there was no reason it should. The Dutch had the benefit of what I was sure were extensive conversations with Shula on the subject. So, I told Shula I would speak with Louise alone—which made her very upset.

As it turned out, Louise Anten's advice was not to see Sharon, and she was quite persuasive. I had told her that if Israel was going to drop the project, I would hate to have it done under circumstances in which the people making the decision obviously did not understand it. Louise agreed, but pointed out that there was no indication whatsoever that Israel was planning to drop the project. She also argued, and I thought she was right, that regardless of whether Sharon understood the project, the more he *did* understand it the less likely he would want it to continue. Someone who was using the water issue as an excuse to avoid giving up land would not be pleased with a project that makes water an issue of secondary importance.

I agreed I would not go to see Sharon. Lenny thought my decision was all due to Shula's manipulations, and it put him in an awkward position, since he had to go back to Sharon's friend Genger and make excuses. I told him to blame it on me and to say that personal reasons made it so I could not meet with Sharon at that time.

Meanwhile Eytan Sheshinski had arrived in town. We discussed his situation at the Kennedy School. He had received a letter from the dean that suggested he should avoid coming there to teach.

I told him I felt someone had slandered him, and that some legal advice might be in order from a "hardball" lawyer. I suggested our mutual friend Robert Rifkind, who was not only the president of the American Jewish Committee, but was a partner at Cravath, Swaine & Moore.

Later, Rifkind and I had a long talk about the matter. I was heartened by his involvement, and hoped that something would come of it, at least as regarded the Sheshinski affair.

That was where things stood. We were involved in trying to plan a trip to the Middle East in early July. It was hampered in part by Shula's insistence that one had to meet with everyone, and that she, I, and Hans Wesseling had to see everyone *together*. That would make for a schedule that kept me in the region for longer than I wanted to stay, and I resisted.

When Burke, Hood, Uliano, and I reconvened on June 29, one of the things they raised was to assure me that Shula was not, in fact the principal complainant in the investigation. I took this to mean that they were leaning towards Ellen's proposed solution to the Shula problem, because even if what they said wasn't true, it was something they would have told me because they needed to protect her as a "whistleblower" if they were going to move her out of her job. I still thought that she had at least generated a great deal of poison that was then reflected in the Counsel's report.

The remainder of the meeting was extremely satisfactory. We agreed the project would move to CID, probably at the end of December. Theo Panayotou would become much more active beginning in October. There would be an appropriate transition of personnel at the end of July, as CID would wish to name its own personnel. The dean's office would take care of all the arrangements. Shula would be needed for other activities at the Kennedy School and ISEPME, which would be kept going. In fact, there was a search for a new director, and suggestions were welcome. The list at the time included people who might not accept; Dennis Ross was given as an example. I suggested Yair Hirschfeld.

When the short meeting ended, I commented to Ellen that my only regret was that I had rather enjoyed behaving like an 800-pound gorilla and now would have to give that up.

V. July 1998 to September 2000

22

We Return to the Region

At the beginning of July, Anni Karasik arranged a meeting with Uri Savir at the Peres Center for Peace in Israel. Savir, the former chief negotiator for Israel in the peace process, was the center's director. Anni and I also joined Avi Gil, Savir's assistant, Professor Samuel Pohoryles, director of the Center's agricultural department, and Yitzhak Abt, another person from that department.

I was delighted to discover that Pohoryles had already read about the project (in a 1995 *Haaretz* piece) and at the time had written a letter to then Foreign Minister Peres urging him to do something about it. That was a good start to a meeting that became more positive as it proceeded. Scheduled for an hour, it lasted two—during which I gave a thorough presentation of the model.

No commitments were made, but it seemed clear we were going to develop a very beneficial relationship for both sides. The Peres Center had close contacts with the Jordanians, Palestinians, and Dutch, was interested in agriculture, and was quite serious about preparing the people who would have to negotiate on such issues when those negotiations became possible.

Savir saw great potential value in the project and was obviously very interested in moving it forward. He said they would work up a proposal for how we might proceed and that we should meet again either in Israel or in Boston a few weeks later.

The meeting could not have gone better. It had been a while since I had had the experience of sitting down with a group of people who had taken a serious look at the project and underwent a sort of conversion.

I was also scheduled to meet with Shimon Peres himself that day. Anni told me that Savir had at first thought it would not be a good idea for me to do so: Peres, he worried, might become overenthusiastic, and Savir wanted the chance to think about things. But by the end of the meeting, there was no doubt that he wanted me to meet with Peres as soon as possible. Apparently, the Peres Center was involved in a program to consider the post-peace economics of Israel, Jordan, and Palestine—something into which the water project would fit very nicely.

"Wow!" I thought. After all the problems, political and otherwise, the meeting and its initial outcome seemed like a big thing.

On Friday, July 10, Hans Wesseling, Shula Gilad, and I met with Ambassador Como van Hellenberg Hubar of the Netherlands at the embassy in Tel Aviv. It was my second time meeting with him, and he could not have been more helpful. He spent three-and-and-a-half hours with us discussing the future of the project and having us demonstrate the software.

Hellenberg Hubar made it clear that his government wished the project to continue, preferably in its form at the time—but they were prepared to be flexible. Should the Israelis pull out, for instance, the ambassador believed the Dutch government would be interested in going forward under a different arrangement, possibly one in which at least the Israeli piece became private. He suggested that the European Union wanted to become involved in final status issues and that it might be that the Dutch would consider bringing the project to the European Union and continue to finance it as the Dutch contribution to that effort. He said he would be in touch with his government about all of this.

We next met with the Israeli private team for a couple of hours. It was a largely technical discussion. I observed some friction within the team. Zvika Eckstein and Shaul Arlosoroff did not see eye-to-eye on how to allocate costs of conveyance systems to different users. Zvika wanted to do this according to economic principles; Shaul believed the way that it was currently done was according to appropriate principles and so forth. Zvika was probably right, but without more information, I was in no position to decide. The friction stemmed from the fact that Zvika had apparently done very little work and Shaul felt he was only dropping in, looking at what had been done by others, and then criticizing. As usual, though, Shaul thought that various kinds of model development were not worth doing. He didn't think much of the seasonal issue and, as his discussion with Zvika suggested, didn't think refinements would make much of a difference. I was not sure that he really knew what he was talking about.

The next day, Hans, Shula, and I went to Gaza. We arrived at the Erez checkpoint a little early. Al-Azhar University had sent a car for us, but the driver didn't see all of us, and thought Shula was a soldier out of uniform. So, he left to find a phone—which meant we couldn't see him. Finally, we called the university and they tracked him down. We ended up arriving nearly two hours late at the university, informed the private team that we had to go immediately to the Ministry of Planning (MOPIC) for our main meeting, and rescheduled for later.

The meeting at MOPIC was led by by Ali Sha'ath and was attended by most of the people there with whom we had worked. Fortunately, Khairy El-Jamal was also there.

Ali Sha'ath began with the obligatory political speech, in which he emphasized that the Palestinian government would not agree to be bound by the results of the model, nor to negotiate according to those results. They wanted to avoid an "imbalance" with respect to the Israelis. If the Israeli government was going to withdraw, he said, so would the Palestinians. This position was mitigated by Khairy's comments regarding how useful the model actually was.

It developed that the real position was that the Palestinians would like to see the project continue, but if Israel was to withdraw, the Palestinians would require that the project continue only as a private enterprise with no explicit government involvement. I could not believe, however, that Ali Sha'ath would wish to let this out of his control in such a way.

Khairy informed us that the Palestinians were in discussions with the Israelis about constructing a recycling plant in Gaza that would sell retreated water to the Israelis for agricultural use in the Negev. It would be in exchange for fresh water delivered from Israel. I pointed out that such an endeavor had been marked more than two-and-a-half years earlier as a really good undertaking in terms of the model. I don't know whether they had gotten the idea from us, but it was good news in any event.

As usual, we parted in a pretty friendly way, and we then walked over to the University where we were late for a meeting with Riyad El-Khoudary, the president, who was also a member of the Palestinian National Authority executive committee. I had last seen him in Casablanca in 1994. I was glad to see that his attitude towards the model appeared to have changed, perhaps because his university was getting some money. But, of course, he was as pessimistic about the current situation as everyone else.

After meeting briefly with the Gazan private team, we returned to Israel. I had dinner that night with Ruth Sheshinski and Rachel and Alon Liel. Six months earlier, Rachel Liel had become the Director of Shatil, the New Israel Fund's major subsidiary that provides technical and organizational assistance to what I have long called "self-help groups"—a particular kind of small non-governmental organization (NGO) that does home-grown advocacy and service. Rachel was doing a wonderful job. I had become the president of the New Israel Fund and had seen her during our three-day meeting in Jerusalem that had ended a couple of days before. I had first met her months earlier, when she first took the job, but it had taken me several months to realize she was married to Alon, who had been the director-general of Beilin's Ministry during the happy days when we had enjoyed Israeli government support for the water project.

We were all very glad to see each other, and the Liels and I became fast friends.

Sunday, July 12, began with my trip to Tel Aviv to see Shimon Peres. A Sunday morning was the only time he could meet, which required that I ask Shula to remake our appointments in Jordan. She wanted to know whom I was seeing, but I thought it best not to tell her—I didn't want her to argue about the role of Anni Karasik. It had also occurred to me that Shula might ask Rafi, the taxi driver, where he was taking me, but I thought it would be going a bit too far to ask him not to say.

When I got into the taxi, I told him to take me to 8 Shaul Hamelech Street.

"Oh, you are going to see Shimon Peres," Rafi said. In fact, it was an enormous office building—and not someplace that would be obvious as holding an office for Shimon Peres. But Rafi knew everything.

I confess that I ended up asking him not to tell Shula.

The meeting with Peres also included Pohoryles and a Dutchman named Pieter (whose last name I cannot recall) who had attended the presentation at the Peres Center. We had run into him the morning before at the Erez crossing.

The presentation to Peres didn't go as well as my earlier one. He didn't seem to want to sit still to understand the entire project, and I worried that he might think it was nothing more than a project to create a recycling plant in Gaza. He did say he might speak with Sharon, but I urged him not to do so without coordinating with the Dutch ambassador. He agreed. He also asked me to leave him material to read.

As I was to discover the next day, things had gone better than I had thought.

After the meeting with Peres, it was off to Jordan, with Shula and Hans.

To Jordan and the West Bank

Once in Amman, we first went to Hazim El-Naser's office at the Water Ministry, where Salem Hamati joined us. There was some brief discussion about Jordanian experiments, and it turned out that real experiments with actual data had to be approved by Hazim, who did not yet have a working version of the model or a usable computer. Shula believed that Salem was not doing any work and was resisting giving things to Hazim. I didn't believe it.

Hazim mentioned that Uri Savir had been there a day or so before and wanted to discuss our project. Shula didn't hear him say that, and I asked him not to talk about it except with me. As events were to prove, this was pointless. Munther Haddadin mentioned it himself in Shula's hearing later that day. Then I had no choice but to talk quite openly about my meetings at the Peres Center and with Shimon Peres.

Afterwards, Shula predictably said, "The Peres Center: isn't that where Anni Karasik is taking a job?" Shula also complained to me. "I thought you were my friend," she said.

The main point of our meeting with Munther, though, was something different. He informed us that he had met more than once with Meir Ben-Meir concerning our project, and had told him that Jordan considered the water project very valuable both for its own domestic purposes and possibly as leading to cooperation in trade and water. He asked Ben-Meir about Israel's attitude regarding the project's continuation. Ben-Meir had promised to think about it.

Munther had phoned Ben-Meir the day of our meeting. "Professor Fisher and his team are here now. What should I tell them?" he asked.

301

Ben-Meir replied that Israel would not oppose the continuation of the project so long as it proceeded on a country-by-country basis. It was all we could ask for at the time, and I expressed my gratitude to Munther in no uncertain terms. (After we left his office, I said to Hazim that I needed to hug someone, and I couldn't very well hug the Minister. So, I hugged Hazim.)

Of course, there would be a price to pay, but it seemed a very reasonable one. Munther wanted to be sure that Jordan had parity with the other teams in recognition and activities. We spoke in terms of bringing a Jordanian to Harvard for an extended period in the second phase. Munther also wanted to have Harvard's name taken off the project, since it had become a very cooperative project involving five different countries. I agreed we would certainly do that.

Later, when Shula, Hans, and I discussed new project names, Shula made it difficult. She insisted that we should ask everyone on the project how they felt about changing the name and what name we should use. She also suggested that changing the project name might lead to contracting difficulties—which I told her was absurd. In any case, the best we were able to come up with was "Middle East Water Optimization Project," or MEWOP. "Middle East Water Allocation System" (MEWAS) would preserve the WAS model name, but "allocation" tends to mean assignment of ownership rights, and that was not what the project was about. I did try to work hard on finding a name whose acronym would be something like RAINDROP, but the best I was able to come up with was WATER—as in "What About Trying Economic Rationality?"

That evening, Hazim, Salem, and their wives joined us at dinner—breaking early because of the World Cup finals. The next morning, we returned to Hazim's office and attempted to run experiments. It failed, in part because we found a small bug but mostly because Hazim's computer wasn't yet ready. I think we succeeded in fixing the bug; Salem promised to fix the other problem.

The next morning, we left Amman and traveled across the bridge and to Ramallah. There we met Michel, the assistant at the Dutch Representative's office, and reported on our most recent meetings. We had also hoped to meet with the West Bank private team of Anan Jayyoussi and Amar Jarrar, who had worked quite hard on the draft Palestinian reports on WAS 3.3 and AGSM, but it turned out that would have to happen with Hans and Shula the next day, after my departure for the United States. It became urgent that we meet instead with Ambassador Hubar again.

We called his secretary at home on Sunday, and she arranged for us to see him in Tel Aviv the next afternoon.

The Dutch Ambassador

On Monday afternoon, July 13, the Dutch ambassador spent more than ninety minutes with us reviewing the events of the previous few days. He began, however, by saying that Pieter had called him to arrange a meeting between Hubar and Shimon Peres for later in the week.

Clearly, there had been no point in trying to keep any of my meetings from Shula. And the meeting with Peres must have gone rather better than I had thought. When I remarked later to Anni that the people at the Peres Center certainly wanted to hit the ground running, she didn't know the idiom.

We discussed at some length the possible involvement of the Peres Center, including that we might be training future negotiators. Ambassador Hubar seemed much more optimistic than he had three days earlier, and predicted some kind of change in the Israeli government in a few months. Our main topic of conversation, however, concerned our talks with the Palestinians—especially the news concerning Israel we brought from Munther Haddadin. Ambassador Hubar, who had already sent messages to The Hague after our first meeting, now proposed to send a message suggesting they dispatch someone, probably Louise Anten, on a fact-finding mission to the three parties to ask them how they would feel about a second phase of the project on a country-by-country basis. Of course, we believed they would all say yes.

The ambassador suggested as well that someone from our team might accompany her, and I said it I wished to be the one (Of course, I didn't want it to be Shula.) Hans was raised as a possibility, which I thought was perfectly sensible. He could correct misconceptions about the nature of the model and the project, and would also be seen as a representative (of sorts) of the Dutch government—which I surmised would probably be good.

We felt pretty good about all that as the meeting ended.

The trip, however, did not end on an upbeat note. Just before I was to go to the airport, we were having dinner at a restaurant in Bat-Yam, overlooking the sea. We received a call from Dawn, Yasmina, and Annette, who said they had received a fax from Gerben De Jong stating that Minister Pronk had refused our request for funds for the extension to December. It was causing some panic back home. Someone had also told Lenny about the fax, who had called Anni about it. And by the time I got home, Anni had phoned Ellen, who had spent a sleepless night.

Yasmina and Dawn were particularly concerned, and rightly so, about their jobs. They went to see Jean Hood, who told them the Kennedy School might be able to pick up the slack for a time.

For our part, we phoned Louise. She was disappointed, but expressed the view that it was a matter of internal reorganization in the Dutch government. There had been elections a month or so earlier, and the new government was still in formation. Pronk didn't want to commit funds and bind his successor, who was not yet named. In fact, Shula had gotten the Harvard Office of Sponsored Research to agree to the same arrangement as last year—namely, that payment could be received only in the next Dutch fiscal year. But that would have involved funds under the control of Pronk's successor.

There seemed to me to be no cause for alarm. The successor would probably be named soon and the situation would be clarified. Meanwhile, Louise urged us to apply for Phase II as quickly as possible.

Back Home and Making Project Progress

When I returned to the United States and addressed the financial matters, I confirmed that this wasn't a big problem. Yasmina and I took a preliminary look at the project finances, which suggested that we could run at a low level for several months without using up our allocated funds and still pay the teams for their deliverables when they produced them. We were trying to get a more refined estimate.

The whole thing with finances reminded me of a time in 1984 when the New Israel Fund had just acquired a formal board. I was a member of the development committee, and I was called to New York because the Fund was going to run out of cash *the very next day* to meet payroll. The idea was that we would somehow raise the needed money in a day!

Before departing, I sought some advice from Ellen, who had served as the chairperson of the Town of Concord Finance Committee. She introduced me to the idea of a cash flow chart that showed the town's cash position and how it would change as monies were expected to come in and go out.

Ellen suggested I ask the people at the fund to show me *their* cash flow chart. "What is a cash flow chart?" was the response. Then they prepared one. It showed that while there really was a crisis, it wasn't going to hit for another month.

They told me I was a genius, that no one understood these deep things the way I did, and that I must therefore become the treasurer of the New Israel Fund. I responded that they really wanted my wife, but that didn't work out, and I became the treasurer for twelve years—until they finally persuaded me to become president (I failed in my attempt to lose an uncontested election).

Something similar appeared to have happened with the Dutch funding for the water project.

A good deal else happened over the next few months. All three teams were slated to finish reports on experiments with WAS 3.3 and AGSM. As it turned out, the Israeli and Palestinian WAS 3.3 reports were received only in draft form, and there was no Jordanian WAS 3.3 report at all—although I was assured it was in the mail.

The Israeli draft was in fairly good shape. It got into even better shape when Uri Shamir could work on it, made difficult by his being at Harvard through August. When the Israelis ran the model, they found a number of anomalies. I made a number of suggestions that they continued to act on.

The Israelis claimed it was impossible to use the model adequately if you had to add an additional pipeline to an existing one. They were determined to charge for capital costs in the model, and so that "impossibility" was because they wanted to spread the capital costs of the new pipeline over users of both, which obviously would give the wrong result. I pointed out that if they would simply treat capital costs as a lump sum that had to be recovered, they could assess the cost and benefits of an additional pipeline very easily. But the problem may have been symptomatic. There appeared to be a battle between Shaul and Zvika over the allocation of costs in the carrier. According to Zvika, Shaul insisted on doing it in the way Mekorot, Israel's national water company, had always done it: allocate much of the costs to the first section of the line. Zvika, the economist, kept trying to explain that doing so did not provide an adequate answer regarding how the costs were actually incurred. The costs involved here were the variable costs, but the problem was

305

symptomatic of how difficult it is to make engineers—and especially Shaul—
break from the view that the way they have always done cost allocations must
be economically correct. (All these issues regarding capital costs were
ultimately resolved in MYWAS, the multiyear version of WAS, simply by
treating capital cost expenditures as cash outflows as they occur.)

The Palestinian draft report was not in as good shape as the Israeli one.
As I understood it, the draft had been prepared entirely by Anan Jayoussi and
was nothing more than the same experiments he had done the preceeding
March in The Hague—based on the assumption that Palestine owned vast
amounts of water. I sent it back, urging him to do experiments with other
amounts of water. He had focused on an interesting question—which way
one would want to run water in a line from Hebron to Gaza—but it was
interesting only as a function of the amount of water they would actually get.
Like the Israelis, the Palestinians had yet to make use of the many features of
WAS 3.3 not already in WAS 3.2.

At the end of August, Annette and Hans had gone to the region to assist
with the reports. When the Palestinians ran the model for 1995, it appeared
that water was not scarce. A close look showed that the Palestinians had
misunderstood what they were to do, and had taken demand in that year to
be the amount of water *actually* consumed rather than the amount that *would
have been consumed had water supplies been available*. It was interesting substantively,
and somewhat disheartening in terms of the progress being made.
Substantively it showed, as did some even worse problems discovered with
Jordanian experiments on the same trip, that one couldn't be really sure that
one had the data right until one used them to exercise the model. The model
was so intertwined and complicated that it could turn up results we didn't
even know were there. In the case of the Palestinians, this showed up in the
shadow values—we no longer called them "shadow prices" to avoid
confusion with actual prices—of water in the West Bank in 1995.

In terms of the educational progress being made, the problem was far
more serious. We were convinced that most of the people on the Palestinian
private team were deadwood who didn't do anything at all and didn't bother
to run the model. Anan Jayoussi was not the only exception, but he was the
outstanding one. In contemplating the next phase of the project, we were
considering how to break out of that problem.

The situation seemed even worse in Jordan, where the failure to produce an experimental report appeared to have been deliberate, although we weren't sure who had made it happen that way. Hazim El-Naser, the Jordanian team member who seemed most able to run experiments, had been unable to get a working model of WAS 3.3 up on his computer because mysterious events kept intervening. One view was that it was deliberate sabotage by Salem Hamati, the team coordinator, but why Salem would want to do such a thing was not clear. Another view was that Munther Haddadin did not want to expose experiments made with actual Jordanian data. Apparently, he had asked Shula, "You don't suppose the other people are giving you their data, do you?"

Obviously, we would have to deal with the Jordanian situation as we went forward. It was unclear whether that was going to be easier given Munther's resignation in August as Water Minister after a scandal involving algae polluting the Amman water supply—for which he was not responsible directly. Hanni Mulki, former head of the Royal Scientific Society, succeeded him and—as was reliably reported—had restated Jordan's commitment to the project. Meanwhile, Munther became leader of the Jordanian private team and, I presumed, would devote more direct attention to the project. And, of course, he remained the real power in Jordanian water, so far as I could tell.

Unlike the model report, Jordanian agricultural report showed signs of real progress. In the first version, the team just ran a lot of scenarios for the Jordan Valley and reported the results so the numbers could speak for themselves. I pointed out to them that they had what appeared to be an anomaly—specifically, they had some cases in which the shadow value of water in the wet season was higher than in the dry. In the revision, they went back again and calculated the water requirements for growing the exact cropping pattern that was grown in the Jordan Valley in 1995; they compared that with the actual water available; and then they still found that the deficit was greater in the wet season than in the dry.

I thought this was quite an intelligent thing for them to have done, and it revealed that they were thinking along the lines that demand as well as supply both determine shadow value and scarcity. Indeed, since some crops require water in both seasons, it was perfectly possible to get their result, since a little more water in the wet season might enable higher-value crops to be grown and a little more water in the dry might not, depending on the different configurations of water requirements. I was glad to congratulate them on this, although—as Annette pointed out—the fact that they found the cropping pattern in 1995 could not be grown with the water available suggested there was something wrong with the calibration of the model. (They had also commented on the discrepancies between the optimal pattern and the pattern grown but had not come up with this answer.)

As all this unfolded with the reports, things were continuing to become complicated in other areas of the project.

Complications

It wasn't the only thing that transpired while I had been away in the Middle East. There had also been some developments at Harvard.

I had learned while in Israel that the Kennedy School had changed its mind about where the project should be housed. It had decided to try to keep it at the Institute, which was no longer to be called by its acronym ISEPME (although I shall continue to call it that in this book). The restructured ISEPME would have Robert Blackwill, a former diplomat, as acting head; Richard Falkenrath, then a Harvard assistant professor of public policy, would be his deputy. Blackwill saw the water project as the only solid thing the Institute was doing, and had asked that we remain with ISEPME. The dean had agreed.

Meanwhile, Sheila Burke assured me Blackwill understood the "staffing problem" perfectly and that the earlier commitments remained in force.

I let Annette know there was absolutely no question: she would remain employed. No matter what might happen at Harvard, I intended to retain Annette myself and would do so through FMF Inc., my professional corporation, if I had to. I didn't know whether it would be sensible to retain Yasmina and Dawn completely on the project, because I would like to have a more experienced financial officer, but I let them know I thought there would be no problem and I hoped to ensure that they remained employed.

Upon my return from Israel, I met briefly with Blackwill to discuss the project's future. He made clear to me that he was heir to the commitment to get rid of Shula and asked me to wait until around the beginning of November. By that time, he wanted to have a report on Phase I and a Phase II proposal already in Dutch hands. Indeed, those documents were to be submitted on October 1. He also wanted me to accompany him to The Hague at the end of October. In the meanwhile, he would find a suitable job into which to transfer Shula.

From that time in early July until October, I never saw or spoke with Blackwill again. My dealings were all with Richard Falkenrath. They were, on the whole, quite satisfactory, although it became obvious that, not surprisingly, the two men's agenda was first and foremost to support the Kennedy School, second to preserve the Institute, and to support the water project third. While we didn't share agendas, there was usually little conflict. Nevertheless, it was irritating to have to clear with them any statement made to the outside world.

Whatever problems we had, though, were overcome by the pleasure (not always entirely worthy) I took in seeing what happened when people determined to be serious and take charge collided with the unprofessional style of ISEPME in general and of Shula in particular. This played out first upon Shula's return in late July, when we began to try figuring out just how much money we had left in the Dutch budget and to decide what would happen for the rest of Phase I. It took a full month to determine how much money we had actually spent and get the accounting correct.

Falkenrath ran the first meeting we had on the subject. He thought in a linear, business-like way: How much money is there? When are reports due? Who will do what? This collided head-on with Shula's usual scatter-shot way of bringing up anything she wanted to discuss, whether appropriate or not. What began as a determination to figure out what we had to do next and how we were going to report to the Dutch became tangled with a theme that Shula could not possibly abandon: that "we're all in this together" and that this was going to be a cooperative project." (Even someone as rational as Annette had her good sense swept away for a few minutes by Shula bringing up something so irrelevant.) Shula proved completely incapable of sticking to the point and concentrating on the fact that, given that the Dutch were not funding the extension, we could not proceed simply as we had always done.

Falkenrath told me later it was the most remarkable meeting he'd ever been in. I had been pretty quiet throughout.

Shula was quite unhappy that there was going to be someone in control in a way Lenny never was. As the summer wore on, things became increasingly embarrassing as she and Yasmina could not figure out how much money we actually had. Finally, after a month had gone by, Shula produced a memo about how difficult it was to produce a final report in only a few days time—but her real issue was that she was not in charge.

Meanwhile, Falkenrath asked me not to make things more difficult for them with Shula, bearing in mind their commitment to moving her out. I did not take that to mean I should simply sit still for whatever came along.

Shula began to offer suggestions—not cleared with me first—for how to save money. She suggested, for instance, that I might be "replaced" on missions scheduled for that fall. While it may have been that she had become confused about me being "in the budget" (in fact, my work was *pro bono* and I was only reimbursed for expenses)—and hence that money could thus be saved—that's not what the emails she sent out to everyone said. I emailed her back publicly—by this time, we were under instructions to let everyone see all the exchanges—and pointed out that this had not been discussed with me, that I did not consent to it, and that it was certainly not her place to suggest in any way that I should be "replaced." She replied that I had misunderstood her, but I responded again in much the same form.

I might not have repeated that I could not be "replaced" had another incident not been brought to my attention. While I had been in the Middle East, Tamar Miller had informed me of a staff meeting at which Yasmina said, out of the blue, that I thought I directed the project but that it had been the intellectual work of others before me. Presumably, she meant Fishelson and Eckstein. Since Yasmina could not possibly have had independent information on the subject, it was clear where she got it.

In one meeting I did attend, she actually said that were I to engage in a project on negotiations outside the water project I would be taking intellectual property. I asked forcefully what that meant. "Whose intellectual property? Shula's?"

Yasmina backed down on that issue quickly. What she didn't know was that sometime earlier I had taken legal advice and obtained contracts from the relevant people indicating that I held the intellectual property. But she also said that the project was at the Kennedy School only because the school had "policy people"—by which she evidently meant Shula.

It was all quite remarkable. Yasmina and I got along on the whole pretty well, although her attitude over time had made pretty clear that what mattered for me as project head definitely did not come first in her book. The meeting and others simply stiffened my resolve that I was doing the right thing to replace Shula, despite her considerable contributions.

When August rolled around, Shula resigned before being replaced—although it was not to take effect immediately—and then left on an extended vacation. I heard about it some five days later from Lenny, who had also heard about it indirectly. She said it was because she wanted to finish her education in social work, but I believe the real reasons were rooted in her discovery that she was not going to be in control and that I was not going to be easy about it.

Later, Falkenrath described Shula to me as "the most high-maintenance individual" he had ever met. It was a pretty good phrase—and nothing like what I said when I got the news. I came outside from our vacation home in Marion, Massachusetts and said to Ellen, "This is a day of independence for all the Munchkins and their descendants!"

When I mentioned that to Theresa, my fabulous secretary and assistant, she told me I was "bad."

Shula's was not the only resignation. Dawn Opstead left for vacation in early September without mentioning that she had found another job. Others not involved in the water project, including Tamar Miller, left ISEPME. As time went on, Blackwill and Falkenrath began to realize that the Institute had become an empty shell and that perhaps the only competent person, Anni Karasik, was now leaving for a job at the Peres Center—despite entreaties to her to stay.

Work on the report for Phase I went on, however, and in early August, I produced a draft proposal for Phase II, bearing squarely in mind Gerben De Jong's discussion with me in The Hague the preceding March. He had pointed out that we all knew the political situation would not permit us to move ahead in terms of regional cooperation and had urged me to produce a proposal for technical model development so the project could be kept alive until, as we all hoped, the political situation would change. Hence my draft proposal concentrated on technical developments.

Uri Shamir offered extensive comments and I then revised it. I took his suggestion to have the model operate in a decentralized way in terms of the technical development with cooperation inside the region between scholars. I added to that (using it as a reason) that the time had come to remove the secrecy and allow publishing and speaking. Scholars working on aspects of the model had to be able to publish their results. I pointed out that Hazim El-Naser had already asked me whether the Jordanians could publish their experimental results. I was very glad to hear he wanted to do that. (I would have been even gladder if I had ever seen the results.)

The Phase II proposal began with a discussion of where we were: WAS 3, the annual steady-state version, which was essentially complete but had to be moved out into the field, experimented with, and actually used. There then came WAS 4, the seasonal version. On that, we had a programming workshop in late July and were some distance towards having a useable program. Certainly, we were farther along than when Annette, Ilan, and I had been dealing with it two years earlier.

Aside from various questions, quite important, involving uncertainty and long-run issues and AGSM, the proposal focused principally on WAS 5, planned to be a multi-year model (and that later became MYWAS). While an important aspect of it was likely to be aquifer management, I had come to realize that one didn't need to do that to have a very useful multi-year model. One could, in fact, model demand over a large number of years and then think about infrastructure planning in terms of maximizing the present value of benefits. That would enable the timing of infrastructure and the order in which different pieces were to be built to be taken into account in the optimization. It would also permit an absolutely correct treatment of capital costs, ending what otherwise seemed an interminable effort to make engineers understand about sunk costs and the fact that charging for capital costs didn't necessarily mean charging for it in the price of the water. With a multi-year model done correctly, capital charges for new projects would show up in places and years in which they were appropriate.

I got my technical draft in, as promised, before August 15. Shula had been charged with preparing the part about how the project would actually work, along with a draft budget, but she never did anything.

Still, I was quite excited about the prospect of doing what the proposal laid out. Other aspects of the proposal for WAS 5 were also technically exciting. They included aquifer and year-to-year water resource management generally, as well as the treatment of hydrological uncertainty. We would be able to produce a model of unprecedented power as a management tool.

While Uri and I and some others could certainly have done that, the questions remained whether the people working on the model generally had the capacity to do. As Hans Wesseling pointed out after his and Annette's mission to the region in late August, the Palestinians and Jordanians didn't seem to have yet absorbed WAS 3. Hans asked whether we shouldn't be concentrating on that rather than on moving ahead with WAS 5. He asked for some indication from the regional teams that they wanted to move ahead, and he got agreement from the Jordanians and especially from the Israelis, who said they would really like a multi-year model.

The response was, of course, symptomatic of what typically happened in the project. The Israelis were technically more sophisticated than the others (although the problems of dealing with the Israeli engineers, who thought they knew everything, were immense). The Israeli water system had been using optimizing models of different kinds for a while, although nothing that could do what we were doing in terms of demand.

Yossi Dreizin, head of the water commissioner's planning office, was always telling us that we needed to incorporate various features into our model to make it sufficiently sophisticated to be useful. Sometimes he didn't remember that he'd wanted us to do certain things. Other times, I thought he gave us difficult and pointless tasks just to keep us from getting ahead with the work, since he was no friend of dealing with Arabs.

The Palestinians, in contrast, were certainly not so advanced (and the private team was not advanced at all), and both they and the Jordanians would plainly have found WAS 3 a useful tool. So, of course, would the Israelis had we been able to persuade them to give it a try.

In any event, Wesseling's advice that we should concentrate on getting WAS 3 into use ended up being taken quite seriously by his government, which caused me to have to make a serious revision to the proposal.

Minister Pronk had moved to another job in August—from Minister of Development Cooperation to Minister for Housing, Spatial Planning, and the Environment. A new development cooperation minister from his party, Eveline Herfkens, was appointed; she was a largely unknown quantity when it came to the water project. The change compelled me to step outside the narrow confines of what was happening and write a letter to Pronk, dated August 24, 1998, to express how I really felt about the importance of what he had done. I phrased it in terms that meant the most to me, trying my best to explain them to him. I reproduce it below in its entirety.*

---

* I should point out that despite having resigned, Shula still held some sway at ISEPME at the time this letter was written. Just as Shula had insisted with previous

As you take up your new responsibilities, I want to express myself concerning your old ones.

I am often asked why the government of the Netherlands is supporting the Harvard Middle East Water Project. My standard answer is because they are nice. But I am very much aware that niceness is an inadequate characterization. In this activity and the wider initiative, as well as in others, you have led your government in what Jewish tradition calls *tikkun olam*, the repair of the world. There is no higher calling, private or public.

I do not know whether the Water Project will succeed in its aim to remove water as a cause of contention in the Middle East (and elsewhere). The overall political situation is certainly gloomy. But I am determined to press on to the extent of my abilities, and there are others also dedicated.

However it turns out, I am proud to have been associated with you.

The fact there was a new minister compelled our friends in the Dutch Ministry to rethink what they wanted us to do. We thought about it, too.

Rethinking the Proposal to the Dutch

Plans went forward for discussions with the Dutch. In early September, Richard Falkenrath informed me of his plan to visit The Hague in the middle of that month to introduce himself to the various Dutch authorities—including the new minister (although later the Dutch Foreign Service said meeting her was not necessary)—and to begin negotiations over the level of Dutch funding. Falkenrath's view was if the Dutch were thinking only in terms of $1 million per year, as Shula had reported, it might not be enough for the Kennedy School to carry the project's administrative structure. I told him I thought the administrative structure was too heavy in any case, but we agreed there was no point in the Kennedy School subsidizing the project.

Falkenrath was contemplating taking Shula with him on the trip; I told him I objected very strongly to him not taking me. There was very little merit in taking Shula, I said, and some downside given that we would have to discuss with the Dutch the issue of Shula's successor and Shula was certain

---

letters, Yasmina insisted that everyone on the project should sign the letter to Pronk. I pointed out it was not a camp reunion, and that we weren't thanking him for having arranged some dinner we all attended. In fact, I was the only one he had ever spoken to personally. The letter went out over my signature alone.

to be difficult. Falkenrath said he felt obligated to take her given his agreement at the time of her resignation in August, when she had threatened to quit instantly. Apparently, he had promised her he would treat her in the way she thought she ought to be treated. Indeed, when he had *hinted* to her that he might not have her travel to the Netherlands with him, she had behaved in such a way that it became clear to him that she would "go bananas" if she didn't make the trip. That was from a conversation that took place on an international phone call! He didn't think the project could afford to take both of us.

I informed Falkenrath that if he took Shula and not me, *I* would go bananas. He immediately said I would go—which is what happened. Annette Huber came, too, taking advantage of my ability to get two-for-one tickets in business class thanks to the American Express platinum card.[*]

Shula's response to not being tapped to make the trip was, essentially, to drop out of the project altogether. Then the question of her successor became a matter of some urgency. In part, it involved a discussion of the role of the Peres Center, since the obvious person to succeed Shula as the arranger and administrator in charge of the region was Anni Karasik. Indeed, Anni had made a very favorable impression on both Blackwill and Falkenrath, who asked her to stay on—but she decided to take the job at the Peres Center.

While the Center would certainly have made her available to work on the project part time, we wondered whether that was politically safe. We weren't concerned about the Palestinians and Jordanians; despite being Israeli, the Peres Center had good relations with all the participant countries. Rather, it was whether an association with the Peres Center would poison the project in the eyes of Israel's Likud government. Blackwill and Falkenrath were justifiably nervous about the possibility. I had no independent information on the subject, but Shula—quite predictably—had written from Israel in July that "the Peres visit is known and is viewed unfavorably here." Of course, she didn't mention by whom other than herself. Yasmina, who couldn't

---

[*] Laying out money for the trip meant that the project now owed me more than $10,000 in reimbursable expenses. It was typical: Yasmina was always too busy to bother putting in my expense reports in a timely manner. She submitted my expenses for the July trip in late August, and I didn't see a reimbursement until early November.

possibly have known without being told (by Shula), blurted out in one of our meetings that the Peres Center was known to be out only for publicity.

I counted neither of these statements as informative. We decided to put the matter squarely to the Dutch and invite their opinion on the subject.

The possible involvement of the Peres Center, however, stretched beyond direct involvement in what I shall now to refer to as the "Dutch Project." Following our meeting, Uri Savir expressed increasing interest in the water project. In my discussions with Anni, several possibilities came up, including sponsorship of a program on how to get a model like ours into negotiation. Uri Shamir had written a proposal for that the previous spring, but we decided it was too dangerous to include in the proposal for Phase II. Another possibility was to train people who might become negotiators should circumstances change, and also using the model to effect cooperation in other areas of the Middle East.

Most important from the Peres Center's point of view, though, was that the Center was very anxious to propose the use of the model in connection with its Jordan Valley project on agriculture. It was a project they were running with the knowledge and consent of the Israeli government, together with the Jordanian government—including Munther Haddadin. Application of the model to that relatively small but important area would lead to a practical and cooperative use.

All this was backdrop to the meeting with the Dutch, scheduled for Thursday, September 17 in The Hague.

23

A Visit to the Dutch

We met with the Dutch in The Hague on Thursday, September 17, in the afternoon. Falkenrath went a day early and had an informal dinner with Gerben De Jong and Louise Anten. I could not leave the United States until the previous evening, since I was first supposed to be deposed in and later had to prepare my testimony for *U.S. v. Microsoft* (a major antitrust case involving software, in which I worked for the Department of Justice). I went right from that to the plane, where I met Annette, and right from the plane to the meeting with Falkenrath.

When Annette and I met with Falkenrath, he told us his dinner with De Jong and Anten had been very successful. They came across as strong supporters of the project who wanted to push it to their minister. The one negative aspect was that he had raised the Peres Center question and was told quite definitely that it was not acceptable, because the Dutch had a government-to-government relationship and could not directly involve the Israeli opposition. I wished he hadn't brought that up.

Falkenrath, Annette, and I then went to a small lunch with Louise Anten and then to the larger meeting, which was chaired by Marcel Kurpershoek, De Jong's boss and the Director of the North Africa and Middle East Department of the Dutch Foreign Ministry. He had been mainly responsible for the Porter Project and, judging from my conversation with him the preceding March when I had sat next to him at a dinner, knew relatively little about the water project (at least then). Hans Wesseling was also there.

Kurpershoek took a tough line that I thought was quite reasonable. In his view, since the prevailing political situation made cooperation impossible, we had to provide serious reasons for why the Dutch should go on funding the project. They would have to sell it to their minister, he said. The other Dutch representatives echoed these sentiments. They sympathized with the fact that we all wanted the same thing, but were not overly impressed with Falkenrath's plea that this was a particularly important time to move our project into regional cooperation—perhaps because he had no suggestions for how to overcome the political difficulties involved.

The emphasis quickly came down to this proposition: if, as we claimed, we had built into WAS 3 and AGSM tools that could be used, the project should push hard to see that they *were* used. That should be the principal aim of the Phase II proposal, with model development—or anything else, for that matter--secondary.

The Dutch also made clear that they were not prepared to make a commitment of longer than a year—and even preferred nine months. The implication was also that they were not prepared to fund at an extremely high level.

So, we discussed how to make all this happen. One suggestion was that we offer the model as a way of resolving internal debate, or at least of providing a common language for internal debate between agricultural interests and budgetary interests within a country. Zvi Eckstein later worked on this suggestion.

I mentioned that the Israeli team had suggested the possibility of applying the model within a district to study the water problems there, and, more important, that the joint Peres Center–Jordanian project on the Jordan Valley would be an excellent "laboratory" for trying out cooperative practical use. That suggestion, of course, renewed the Peres Center question. We agreed that the project could not directly involve the Peres Center as a full participant, but that if the Jordanian project was happening with the full knowledge and consent of the Israeli government, we could involve water modeling if the government consented. That became an important part of the Phase II proposal.

The subject of Shula's replacement also came up. At one point, Louise Anten suggested to Hans that Delft Hydraulics might find a person to take over her role. Hans was quite reluctant.

Overall, the three American participants were not particularly happy about the meeting. On the way home, Falkenrath and I reopened our discussion about whether the Kennedy School would wish to continue. The first round of that discussion had been about the likely size of funding. Now he questioned continuing because of the Dutch conditions.

Annette and I felt somewhat differently. We were both in this, of course, for the cooperative part of the project. If the strictures of the Dutch project were going to prevent us from dealing with cooperation, then it might be a good idea to get outside those strictures. We had also obtained permission from the Dutch to approach the Peres Center for the funding of such an effort. Hence, what was involved had become the continuation of the countrified model-building project under Dutch auspices. We felt it would be too bad were that to cease, and particularly if it were to cease abruptly, but that it would not be the end of the water project.

The Kennedy School Pulls Out

On my return home, I resumed work on the Phase II proposal along these lines, and also continued drafting the Phase I report, with considerable assistance from Annette.

We needed to impress on the Dutch the many things we had done to maintain a cooperative project and the importance of continuing the project in the face of considerable adversity. I took a piece Shula had faxed from Israel (and that Annette had edited a bit), removed it from the Phase I report, and included it in a cover letter to the Dutch. It made a very good case for all the obstacles we had overcome and the cooperation we had managed to engender. From how it read, one would have thought cooperation was the project's main point.

Monday, September 21 was *Rosh Hashanah*. On Tuesday, I spent the day preparing for my *Microsoft* deposition the next day. It wasn't until Thursday that I was able to get back to the water project. I finished the next draft of the Phase II proposal and e-mailed it to Louise Anten and Hans Wesseling. Louise was to leave for Cairo on Saturday, and I wanted her to be able to read the proposal on the trip and, if possible, give me her comments.

Falkenrath, Annette, Yasmina, and I met that Friday, and we agreed that I would phone Louise in Cairo over the weekend. I was unable to reach her when I tried on Saturday morning, and then discovered an email from

319

Falkenrath asking me not to contact Louise before speaking with him. He said he had yet to get the budget straight and run the question of the project through the Kennedy School. The October 1 deadline was fast approaching, and we had agreed we would send the proposal and report no later than Tuesday or Wednesday to catch Gerben De Jong before he left for the Middle East. But I was unable to reach Falkenrath directly and kept being put off by messages that said that he had to meet with others as before.

When I awoke on Tuesday, September 29, I felt pretty sure the Kennedy School was going to pull out of the project.

Late that morning, Falkenrath finally phoned me and conveyed that Dean Nye had taken such a decision for several reasons. The Dutch would be told that the Kennedy School did not feel a professional interest in a project that would have to be restricted to avoid international cooperation. ISEPME was said to have left the School with obligations amounting to well over $1 million, and it did not appear that the water project would pay for itself in terms of the administrative time involved, since they had also discovered that there was no competent person available to do the administrative work. (Of course, some costs are sunk, so the amount of the debt had nothing to do with whether it was worth going forward. Further, they were spending money on people who were not now working and found that they didn't have any other projects that needed an administrator.)

There was more. The dean regarded it as an anomaly that the two people doing substantive work on the water project, Annette and me, were not Kennedy School employees and that the principle investigator was an MIT professor. Of course, he had known that all along.

Finally, I suspected an unstated reason. I had not endeared myself to the dean by continuing to consider Lenny Hausman's dismissal an outrage.

At first, this decision made me furious. I pointed out that I could have taken the project away from the Kennedy School in the spring but had refrained from doing so because the School had made all sorts of statements about commitment to the project. Promises had been broken. Moreover, I said, Harvard was pulling out at a terrible moment: the proposal for Phase II was due in two days; I was due to leave on a four-week vacation in Africa during which I would essentially be out of touch; and the Dutch budgeting cycle was such that they would have to decide on Phase II during November. If Harvard had not waited so long to pull out, I would have had time to find another administrative home. Worst of all, Harvard was pulling out abruptly and hence quite possibly giving the signal that there was something wrong with the project.

During the negotiations in June, I had ascertained that both the National Bureau of Economic Research and MIT would have taken the project, but now there was not time to make such arrangements.

For me, this could not have come at any worse time. The preceding afternoon, I had felt perfectly dreadful. I didn't know just what a nervous breakdown was like, but that thought did occur to me. And that was before this news.

Clearly, the promises from Harvard had been good for only four months. The Kennedy School was pulling out because its expectation that the Dutch would continue to fund the project at the same rate as when it was funding Shula's bloated bureaucracy was wrong—on top of which Harvard's own actions had put an end to grants to ISEPME from people who had been dealing with Lenny.

I also learned that Dean Nye, whom I had never met, had no intention of discussion these things with me in person.

I asked Falkenrath whether the Kennedy School still had an interest in a wider water project, and he reiterated the idea of an executive program in the management of natural resources. He said he hoped that I and water would to be a significant part of it. I replied that it seemed doubtful I would be feeling kindly toward the Kennedy School, and he said he quite understood.

He asked me not to talk about the decision until I met with him and Blackwill. The next day was *Yom Kippur*, the holiest day in the Jewish calendar, which began at sundown in only a few hours. I knew I had to consult a higher authority, and so we set our meeting for the morning of Thursday, October 1—the same day I was to leave for Africa.

Yom Kippur Sets the Tone

For *Yom Kippur*, the Jewish "Day of Atonement," the most important item on my agenda was prayer. Beyond the water project, I had a lot on my mind. My son-in-law, Brian, would almost certainly need a bone-marrow transplant soon. My close friend of 40 years, Zvi Griliches, had been diagnosed with pancreatic cancer—an incurable and painful disease. But I also had much to be thankful for, especially the births of more grandchildren. On the previous

Yom Kippur, I had prayed in particular for Eve, the sixth, who was very much a "high-tech" product and still uncertain at the time. Now there was the seventh, Herschel Bateman Fisher Humphreys, whose parents had a proven record of producing adorable children, but who hadn't even been conceived a year earlier.

Heaven knows I had much to atone for, too. I prayed for the ability to pray properly. I prayed for the guidance and strength to continue the water project, an enterprise I still believe to be God's work.

By the end of the *Kol Nidre*, the evening service that introduces *Yom Kippur*, I had a sign of sorts. On a previous *Yom Kippur* a few years earlier, I had prayed hard for the strength to carry the project forward. At the end of that day, a man next to me turned said, "You really prayed." That had pleased me, but was also quite startling; after all, one doesn't often say such things to strangers. Now, at the conclusion of the *Kol Nidre*, the woman next to me said that I *davened* (Yiddish for "prayed") so well that it was a pleasure to be next to me. I hoped so.

The next day I was able to say, and I hope meant, *hineni*— "here I am." It is what one says in answers to God's call.

By the end of the day, I felt certain that all would be well.

My certainty was quickly born out. At the Thursday morning meeting with Falkenrith and Blackwill, we discussed Harvard's letter to the Dutch minister, which would be sent that afternoon, expressing the Kennedy School's willingness to go on with a broader project of cooperation but its unwillingness to continue in the present form. Money was not to be mentioned. Falkenrath would draft the letter and I would offer comments before it was sent. After that, I would be free to email Louise Anten and Hans Wesseling to express my own willingness to go forward and to offer suggestions for an alternative administrative home. The hope was that getting that email out quickly would get me a response the next day in Frankfurt, where Ellen and I would stop overnight en route to Africa.

Blackwill, Falkenrath and I also agreed that Harvard would complete Phase I, I would to be completely free to use the software and do anything I wanted with the project; and that I would have complete access to the project's files and ISEPME's database of names and addresses.

I returned home and called Annette to let her know where matters stood. I also spoke later with Anni Karasik. As the Dutch had agreed I could, I had already sent the Peres Center the proposal for a project on negotiations and included the Jordan Valley Project in the proposal for Phase II. Further, as an independent academic, I was free to keep anyone I wanted informed of my work—and I certainly planned to take up Uri Savir's offer to train those who might (again) be Israel's negotiators. But I could not invite the Peres Center directly into the Dutch Project.

Anni told me of a meeting in Jordan earlier that week that included Savir, Crown Prince Hassan, Jawad Anani, who had become a royal counselor, and Hani Mulqi, the new Jordanian Water Minister. Savir had brought up water and was told that Jordan was very much committed to the use of our project. We agreed that should the Dutch fail to fund Phase II, I would turn to the Peres Center.

That afternoon, I sent my email to Louise and Hans. I was extremely anxious to get the email out because I would be unreachable after October 3, on safari in Namibia. I knew Louise was returning from Cairo that night, and hoped she would get the email the next morning. I gave her and Hans the phone number of the hotel in Frankfurt where we would stay, as well as that of the hotel in Windhoek, Namibia, where we would be for one night before heading out on safari.

In my note, I suggested several possible alternatives, first of which was Delft Hydraulics, Hans' own firm.

In Frankfurt on October 2, we returned to our hotel room after a late lunch to find a message from Hans asking me to call him at home in Utrecht. I tried again and again, but the line was always busy. Then the phone in our hotel room rang. It was Louise Anten. They had been talking.

The news was very, very good. Louise could not predict what her Minister would do, but thought it very unlikely that he would agree to Blackwill's suggestion of a wider project at this time. That made the question of a new administrative home an important one, and Louise urged me to go with Delft Hydraulics, assuring me that Hans wanted very much to take the Project. She also said there was plenty of time, because the Minister was to be in Washington for ten days.

Louise also asked whether she and Hans could work on the proposal along the lines Hans had suggested in an email earlier that week, which I thought was a very good idea. Hans's suggestions were largely organizational, designed to let readers of the proposal understand the "deliverables" more clearly and making the write-up more accessible to funders.

It didn't escape my attention that if Hans and Louise worked on the proposal, the chances of selling Phase II to the Minister would certainly be higher. So, it was agreed.

When I spoke later with Hans, he said the reluctance about Delft Hydraulics taking over administrative duties he had expressed two weeks earlier had been political. I took this to mean he did not wish to be seen to compete with Harvard. The rest of our conversation focused largely on administration for the rest of Phase I, during which Harvard would remain in charge and Annette would be the central person in my absence. We talked about the possibility of Delft Hydraulics using Marten-Pieter Schinkel, a good Dutch economist who had spent a year at MIT and had edited the fourth volume of my collected works.

Next, I spoke with Annette and asked her to inform the regional teams and ministers of the change (Blackwill had written to them the day before). I also left a message for Anni.

That was where things stood as I headed to Africa. There was, of course, no guarantee the Dutch would fund Phase II, but there would be a smooth transition regardless, and I was confident the project would survive.

Freedom!

And I was free! Free from the Lenny-Shula wars! Free from Shula's emotional storms and machinations! Free from having the project viewed and used as something to save ISEPME—first by Lenny and then by the Kennedy School—which was understandable but wasn't my priority! Free from the total lack of professional administration that characterized ISEPME! Free from having to clear everything, as I had been forced to do those last few months, with Falkenrath and Blackwill--who did a fine job but naturally had their own agenda!

A long chapter was ending. The Harvard Middle East Water project was over. But the Middle East Water Project—under some new name yet to be determined—was very much alive.

Anyone who has read this far will know that I considered the project to be my personal contribution to *tikkun olam*, the repair of the world. Two sayings had been running through my mind:

Rabbi Tarfon says: "The day is short, and the task is great, and the workers are sluggish, and the wages are high, and the Master of the house is pressing." (*Pirkei Avot* ["Ethics of the Fathers"] 2:5)

... But Rabbi Tarfon also says: "You are not obligated to complete the work, but neither are you free to refrain from it." (*Pirkei Avot* ["Ethics of the Fathers"] 2:21)

I certainly had not refrained. But though the work was not finished, my obligation was far from over. Besides, when I could leave the external and, especially, the internal politics aside and actually work on the economics and the model, it was lots of fun.

As it turned out, the shift to Delft Hydraulics turned out to be one of the best things that ever happened to the project. But that did not make me any happier about the shabby way I had been treated by the Harvard Kennedy School. Some years later, I got a chance to say something publicly about that—but so you understand what I did, I must first tell a story concerning Sir Francis Crick, the great biologist and Nobel Prize winner with James Watson for their discovery of the double-helix chemical construction of DNA.

Cambridge University is beautiful, dating from the Middle Ages and with marvelous architecture—of which the jewel is King's College Chapel. In 1960, a new college within the university, Churchill College, received its Royal Charter. It was to be a science-oriented college, and was named in honor of Sir Winston Churchill. Crick accepted a fellowship on the condition that no chapel would be built in the college.

Then, in 1963, a benefactor of Churchill College offered the money for a chapel. The majority of college fellows voted to accept the offer, arguing that some members of the college would "appreciate" a place of worship. Crick countered with his view that many more might "appreciate" the amenities of a harem, to which he offered a financial contribution.

Of course, the offer was refused. Crick resigned his fellowship.

Crick subsequently received a letter from King's College offering him a fellowship there since it had come to the college's attention that he was without a college association at Cambridge. Not one to miss a historic opportunity, Crick replied that he would be honored to become a fellow of King's, and stated the assumption that, of course, the college would be tearing down its own chapel.

A few years after the expulsion of the water project from Harvard, while Ellen and I were in London, I received an email from someone at the Kennedy School whom I did not know. It stated that the school had received a serious grant from the Kuwait Institute for Scientific Research. Apparently, the grant was for a study of the economies of the Emirates, and the study was to be led by the "Kuwait Professor." Someone had suggested me for that position, and the emailer wished to know if I was interested.

I knew immediately that I did not want such a position: I was happy at MIT. Further, I was not interested in supervising the work—except as it might involve water. So, I was certainly going to refuse. But, of course, there was the issue of my past treatment by the Kennedy School, which raised the question of *how* to reply.

That's when the story of Crick and the chapel came sharply to mind. Here was an opportunity to express myself, even though I did not have the wonderful canvas on which Crick had written. I drove Ellen crazy for two days trying out different replies.

At last, I settled on a reply that opened by stating that I was flattered and honored by the inquiry. I then went on to explain, in only moderate detail, that my previous go-round with the Kennedy School had resulted in very shabby treatment. I followed that with these words: "Given that experience, I regret to tell you that I am likely to be interested in an appointment at the Kennedy School only when I have received reliable information that it has begun to snow in an area popularly supposed to have a climate even hotter than that of Kuwait."

I really enjoyed that, and contemplated sending a copy to Dean Nye. Ellen said it would be "bush league," and so I refrained.

I never received a reply.

24

Keeping the Project Alive

When I returned home from Africa at the end of October, I was quite annoyed to discover that very little had happened at Harvard. Even the Phase I report I had written, which was to have been sent to the Dutch on October 1, had not been sent. But, as it turned out, that didn't matter.

The Dutch had already told me that they knew our aim was for the project to be used for cooperation, but since the parties weren't yet ready for that, the Dutch wanted to keep the project alive until they were. Conditional on the agreement of the parties, the new Dutch Minister had been persuaded by her staff to approve 1 million guilders (approximately $500,000) for an "interim phase" of the project that would last nine months and be concentrated on perfecting WAS 3.3 and getting it into use.

A mission to the Middle East was scheduled for early December. The various Dutch embassies and representatives officially requested appointments with appropriate government officials in the region. Louise Anten, Hans Wesseling, Annette Huber, and I would go.

Things took a refreshingly easy turn.

In late October, the Israelis and the Palestinians had signed the Wye River Memorandum, aimed at resuming implementation of 1995's Oslo II Accord. Netanyahu and Arafat had both signed it at the White House, and there was some sense of relaxation in the overall atmosphere. We even felt it in the project. When Hans Wesseling called Ali Sha'ath about a meeting, Sha'ath was enthusiastic and offered to set one up with himself, Nabil Sharif, and Nabil Sha'ath. What a contrast from the preceding summer, when Sha'ath had not been exactly forthcoming, and had said the Palestinians would not negotiate on the basis of this project, that they really did not want to be part of a project in which they could be seen to be cooperating with Israelis, and yet, that were the Israeli government to withdraw, they would also no longer wish to be part of the project. The Palestinians wanted "parity"—whatever that meant. That meeting was to take place on December 3, although Nabil Sha'ath probably would not be there.

Khairy El-Jamal also showed considerable enthusiasm for a meeting.

I had a sense that they were looking at the possibility of final status negotiations actually beginning fairly soon, and that they needed to be as prepared as possible on the water issue.

Meanwhile, back at Harvard, I had been attempting to arrange the appropriate transition. Phase I was to expire two days after my return, and no one had applied for an extension. There was a meeting scheduled for the very afternoon I arrived from Africa; no one had bothered to find out when, exactly, I was returning. I attended, jet lag and all. We agreed to apply for a time extension, and that in a couple of weeks, we would put in another application detailing how we were going to finish Phase I.

I wrote a one-page description of what we were going to do. It was a royal pain: everyone—from the contracting officers to Yasmina—felt entitled to edit it, sometimes changing the substance of what it said. A few comments were appropriate given that Harvard was committing to do something, but all of the others were total wheel spinning by a bureaucratic system wound around its own axle.

The Dutch approved the brief extension.

I was glad that phase was coming to a close. To help move it along, I pursued the question of the transition of files, permissions regarding software, and contacts. I asked the attorney for Charles River Associates (CRA), my consulting firm, to call Bob Iuliano of the Harvard counsel's office, who had been working with the Institute and with the various deans in their negotiations with me. He said he thought the agreement signed a couple of years earlier was sufficient to give me all the rights I wanted, but he couldn't find a copy. He believed that Harvard had a policy of retaining the files a project produces, but didn't know whether the Kennedy School would wish to pursue that policy. I pushed to get an answer.

I thought the best outcome would be for CRA to take the files and permit Harvard to copy them if desired. Second-best would be for Harvard to retain the files but allow us to copy them when we wanted. But it was looking more and more likely that no one would take responsibility, which meant that soon enough, no one would be able to find anything—especially since ISEPME would cease to exist the coming November 15. I supposed we'd know the outcome soon enough.

Meanwhile, in the run-up to our mission to the region in early December, I met with Aaron Miller, Dennis Ross's number two in the State Department. A year earlier, Lenny Hausman had suggested to Miller that we meet. Lenny told me Miller was interested, and I prepared a PowerPoint presentation on the project to show him, which I later used with Uri Savir at the Peres Center. But Miller ended up being too busy to get together.

Now, with the apparent beginnings of a new atmosphere after the Wye River agreement, I sought that meeting myself. First, though, I asked Louise Anten what she thought. She consulted Gerben De Jong, who said I should go ahead. So, I worked through Debra DeLee of Americans for Peace Now, who had good contacts with Miller and Ross. She had been, among other things, a chair of the Democratic National Committee. She had sought me out to work with Americans for Peace Now, and we had become friendly.

DeLee phoned Miller on my behalf; he said I should call him. I did, and waited about a week-and-a-half for a call back. We agreed to get together on November 23.

The meeting began slowly, but ended splendidly. Miller had read the "Liquid Assets" report a year earlier, but didn't really know much about the project. Water, he said, was one of many subjects he had to be concerned with; early in the meeting, his level of interest was polite but not very great. By the end, though, he was registering considerable enthusiasm. He plainly believed we really were onto something, and, while I was sure the State Department would be reluctant to push the project heavily, Miller said that he would like to see something happen. He had grasped the proposition that monetizing the water problem might make it highly solvable.

We discussed the where the parties might stand. He said he would be very happy if I could produce some movement, particularly in convincing the Israelis. He told me the time had come for me to see Sharon, who was then the Israeli Foreign Minister, and was, according to Miller, preparing himself for final status negotiations in all areas.

I told him we had arranged to see Ben-Meir first, on the advice of the Dutch Embassy. He thought about it, and then said it was probably the right thing to do. He suggested, though, that it would be important for me to see Sharon within the next month or so.

The appointment with Ben-Meir had already been set up. I knew Ben-Meir had talked to Sharon about it. We would see what developed. Meanwhile, Miller and I would both be in Jerusalem the next week—he in connection with the visit of President Clinton—and he asked to see me.

During our meeting, Lenny Hausman's name came up. Miller remarked that one of the ways he knew about me was through Lenny. "Lenny has a good name here," Miller said. Lenny was glad to hear that.

Meanwhile, preparations for the current mission went on through phone calls and emails between Annette, Hans, Louise, myself, and a young man named Jan Verkade, whom Hans had retained as a project manager. This process was a pure pleasure. Verkade took charge of the arrangements and did a terrific job, with the assistance of the various Dutch Embassies and representatives. I was beginning to realize that we would accomplish a great deal without Shula or the Shula equivalent. Moreover, for the first time we were not traveling with an additional entourage of one or more of Shula's assistants, and Verkade did not come on the mission.

The days when we had to rush madly back and forth cajoling people to meet with us seemed over. There was a certain loss of frantic behavior and an increased sense that we could do all this in a sensible, rational way.

Back In the Region

Annette and I arrived in the region on Sunday, November 29, and went to Haifa the next morning to meet with Uri Shamir, two of his students, and Ilan Amir. Our discussion focused on scientific plans for going forward. That afternoon, in Tel Aviv, I gave a talk on innovation and monopoly leveraging, with particular reference to the Microsoft case. The audience had been organized by Zvi Eckstein and included the partners in his consulting firm as well as David Tadmor, the head of the Israeli equivalent of the U.S. Department of Justice's Antitrust Division.

Meanwhile, a curious problem had arisen. Anni Karasik had warned me that doubts had been growing at the Peres Center concerning our project and that the meeting scheduled for December 1 would be difficult. On the night of our arrival, she had called to say that the three people involved—Yitzhak Apt, Shmuel Poraheles, and Shmuel Kantor—had decided it would be a bad idea even to have the meeting because the results would be so negative both for the project and the Peres Center itself.

That made very little sense to both Anni and me. It suggested that the people involved had simply talked themselves into a position they did not want disturbed. I had met Apt and Poraheles on my previous visit to the center and they, particularly Poraheles, had been enthusiastic. I had met Kantor in 1994 when he served as translator during my discussion with the previous water commissioner, Gideon Tzur. He also came to my lecture at Fishelson's *shloshim* memorial service in January 1996. I had never regarded him as particularly friendly to the project.

I did what I always do in such an emergency situation: I called Uri Shamir. He told me all three of those might have had some ongoing connection with the water system, and particularly with Meir Ben-Meir, the water commissioner. Kantor, in particular, was an advisor to Ben-Meir, as he had been to water commissioners from time immemorial. Uri said he knew Kantor quite well and would be perfectly happy to call him. He did so, and then reported back to me that Kantor thought there would be a meeting and that he himself was neutral. Uri's report contradicted Anni's probably correct report that Kantor was the most negative of the three.

In the end, the meeting did happen. A group of us—Anni, Annette Huber, Uri, Hans Wesseling, Johanna Van Vliet of the Dutch Embassy, and myself—met with the three men. It was a blazing success, largely thanks to Uri. (I later realized I made a big mistake by forgetting to invite Shaul Arlosoroff, who was quite upset.)

We learned that, thanks to my presentation in August, which had emphasized the grand picture, we now had to persuade them that the model could actually be used in relatively small situations such as the Jordan Valley and as an optimal management tool. Further, they were convinced of the need to find new sources of water. We had to point out that the model could elucidate that need by showing what new sources of water would do and under what circumstances they would be worth having at different costs. We had to emphasize further that the model did not prevent a subsidization of agriculture or other policies, nor did it suggest that more water would not be needed.

We also discussed the political question of where the model stood with the Israeli authorities and whether it would be adopted. In view of later events, there may have been something they knew that we did not.

The meeting was a blazing success. The Peres Center people would think about the use of the model. Kantor wanted to know how this stood with Nabil Sharif, the Palestinian Water Commissioner, with whom he spoke with some frequency. In view of our meeting with Sharif two days later, I believed their discussion would have a good outcome.

Unfortunately, a greater problem soon arose, one not connected with the Peres Center. On the following day, December 2, we met at the Dutch Embassy with Ambassador Hubar and Ms. Van Vliet, who the day before had warned us we were in for a difficult week. Now we found out why. The ambassador, who was obviously quite sorry to have to deliver his message, told us that he had met with Meir Ben-Meir and had been informed by him that Israel did not want our project to continue. He gave as the reason that the model would make transparent the cost of subsidies to agriculture.

Several things stood out to me. First, it was quite true that the model would make transparent such costs. By saying so, Ben-Meir had for the first time revealed the possibility that he might actually understand something about the model. I had no doubt that Ben-Meir regarded such transparency as dangerous to his own personal predilections regarding water policy, but it was a truly remarkable statement to make in public—and one that could not have possibly been agreed to by the rest of the Israeli government. Rather, it was an outright statement that Meir Ben-Meir wished to suppress the model because it would lead to a rational discussion of water policy and, moreover, to outcomes that opponents of his water policy in the government, such as at the Finance Ministry, could use. Indeed, I considered it a statement that he could not afford to have publicized, which is just what would happen if Israel were to withdraw from the project.

Ben-Meir, in making that statement, was also apparently withdrawing from the commitment he had made to the Jordanians more than once that Israel would not oppose the continuation of the project on a country-by-country basis.

The Dutch Ambassador told us that, in the light of this, the meeting that was to have taken place with Meir Ben-Meir and Dan Katarivas on

332

Wednesday, December 9, would now be a "government-to-government" meeting at which Annette and I would not be present. What I called "our side" would instead be represented by Hubar, Louise Anten, and Hans Wesseling (as technical advisor). The Ambassador felt, correctly, that he had to take very seriously this "official" no—which he described as "the flattest 'no' that I have ever had in my life." It did not mean, however, that he could not explore the extent to which that "no" would hold up.

We discussed strategy, with the Ambassador encouraging us to do everything we could think of to deal with the situation. Then and over the next few days, several ideas arose. One was the possibility of encouraging the Finance Ministry, or perhaps the Ministry of Foreign Affairs, representatives of both slated to attend the upcoming meeting, to oppose Ben-Meir's position. The Ambassador made clear that he could not possibly call Dan Katarivas or anyone else, but it was also equally clear that he wasn't telling me not to make the calls. He also said he would begin the meeting by asking whether he had understood Ben-Meir's position correctly, allowing others to intervene if they so chose.

Uri ended up speaking with Katarivas, who predictably said he wouldn't stick his neck out. Uri learned that the Dutch would, however, be presented with a group of other possible projects about water so that good relations would be maintained between the two governments.

I pointed out to the Ambassador that Jordanian pressure might be secured and that we would be exploring that shortly. I also asked him about Sharon. I had met one of Zvi Eckstein's consulting partners, Kalman Gayer, who had been an advisor to Rabin and had been talking to Sharon a good deal. He had offered to try getting me in to see Sharon, and now the Ambassador said quite definitely that we should pursue that possibility.

That evening, I discussed seeing Sharon with three leaders of the Israeli team, separately. Predictably, Zvi was anxious to do it. He also told me Gayer did not want his relationship with Sharon publicly known. Uri did not feel he knew enough to express an opinion. Shaul was vigorously opposed, basically because Sharon didn't like mediators and that such an attempt might predispose him to kill the project. That view, which Shaul had held for months, no longer seemed particularly persuasive if the project was about to be killed anyway. I told the Israelis that this had to be an internal Israeli decision.

Finally, of course, there was the question of what would happen to the project if Israel withdrew and how that would appear to Ben-Meir. In the first place, it was likely, although not certain, that the Dutch would choose to proceed, particularly with the Palestinians. Were that to happen, the Israeli private team would certainly continue to operate, although we would have to find them other funds. Moreover, the project's model, WAS 3.3, was then in existence, especially for Israel, and it wasn't going away. It would become known; the Israeli team would certainly publish. Ben-Meir's opposition and the reasons given for it would also become public. Since all Israel was being asked to agree to for the interim phase was to examine the model and the extent to which it was useful, it couldn't be a good political idea for Ben-Meir to kill it in the way he was contemplating. In one way or another, I thought, we should be able to put that to him—politely.

I was not moved to give up faith. As I told Hans and Annette after the meeting with the Dutch Ambassador, the project "died" every three months, but was still alive. Moreover, we were in an area of the world where resurrection has a long history.

That night, we had dinner at a terrific southwest/Mexican restaurant with Anni Karasik, and her husband, Lester Thurow of the MIT Sloan School, as our guests. I asked Anni for any help she could provide. Also, faith or not, I did considerable damage to two margaritas.

Gaza

The next day, December 3, we drove to Gaza, where Michel Rentenaar, the assistant to the Dutch representative in Ramallah, joined us. I knew Michel as someone who could not conceal his bias in favor of the Palestinians. Moreover, he had never been enthusiastic about the water project. On this occasion, though, he was.

It was my first visit to Gaza when we were actually on time. At the Ministry, we met first with Ali Sha'ath, Nabil Sharif, and Khairy El-Jamal. It went tremendously well, largely thanks to Khairy El-Jamal who, as Hans put it, was the Uri Shamir of this team. Khairy did a wonderful job explaining the contributions the model could make to Palestinian planning and how it did not prevent him from making any negotiating claim he chose. (I told Khairy later that it was unusual for me to find someone who could explain things better than I could.*)

The Palestinians very much wanted to proceed and do so on as large a scale as possible by introducing the model into their water planning. They wanted to use it in connection with the strategic planning they were then setting up. Further, they wanted to take it to a regional basis.

Nabil Sharif emphasized that his negotiating position would not be based on the model but would be based on ethics. He was going to demand that water consumption per capita, including consumption for agriculture and industry, should be the same for Israel and Palestine. By that, he meant water ownership per capita.

Immediately, Khairy told him that was a wonderful principle, and the model would certainly not negate it. But what would happen, he asked, when Palestinian economic circumstances were such that the demand would not be there for such water consumption?

"Oh, in that case," answered Sharif, "I will gladly sell to the Israelis." In the case of excess demand, he would gladly buy from the Israelis.

Khairy emphasized such trades were exactly what the project was ultimately about. There wasn't any doubt that they wanted to explore that and wanted it badly.

I mentioned it wasn't going to be easy, particularly because one of the questions would be how many people were to be involved, thus bringing into the picture forecasts of how many Palestinians would return from the diaspora. Indeed, since claims about population would matter, I suggested to Sharif that he ought to popularize the slogan "Go to bed for Palestine." He told me they were already doing just fine in that department, thank you, and probably didn't need any further encouragement.

We then headed upstairs and met briefly with Nabil Sha'ath. That meeting was also very revealing. Sha'ath had just returned from a conference of donor countries in Washington at which, as he put it, he torpedoed a meeting on water the Israelis wished to have. In doing so, he rejected the

---

* I did not fail to compliment Uri, too. After the Peres Center representatives (except for Anni) had left the room, I told him I wanted to say something publicly. It was that Eytan Sheshinski and I had for many years done favors for each other and that I had already told Eytan that the biggest favor he ever did me was to introduce me to Uri. Uri thought about that for a minute and said, "That sounds like a compliment." Exactly.

suggestion from Ben-Meir (who he referred to as "Ben-Meir Ben-Meir Ben-Meir Ben-Meir"—to which I added "Abu Meir") that the way to solve the water problem was to build giant desalination facilities: one in Hadera on Israel's northern Mediterranean coast, and one near Ashkelon on the border with Gaza. The latter would supply the Palestinians. The "Harvard/Fisher model," Sha'ath had said in rejecting Ben-Meir's suggestion, showed that desalination was a terribly inefficient idea and would simply result in delivering very high-cost water to the Palestinians.

Sha'ath also said me that he had mentioned the model in a similar way on a previous occasion. I began to see why Ben-Meir might not like me very much, but I was delighted to see the use of the model increasing (in this case by the Palestinians, and previously by the Jordanians), even in preliminary ways in negotiations.

Reiterating his complete support for the project, Sha'ath said he believed the Israelis would end up using it, too, and that we would go on to the regional cooperation phase. From his mouth to God's ear!

At lunch, Ali Sha'ath expressed how very, very happy he was to have learned of Harvard's disappearance from the project and the emergence of Delft Hydraulics. He said he had never regarded Harvard as neutral and that he thought the new arrangement would be much better.

Frankly, I was not exactly sure of what his non-neutrality referred to. I did know he regarded it as an imposition that Shula was always telling him whom he had to send to meetings and that a year earlier, at the meeting in Amman, he felt the Palestinians had been asked to present their data when the Israelis were not being asked to present theirs and had, in fact, done practically no work up to that time. He may also not have liked that Shula was an Israeli citizen. (Perhaps he didn't like that I'm a Jew, but I don't think so.) I understood the annoyance factor. For all my complaints about Shula, however, I thought the accusation was unfounded—at least at the conscious level. At an unconscious level, there may have been something in it. At the end of her tenure on the project, Shula had become increasingly annoyed at the behavior, as she saw it, of both the Palestinians and the Jordanians in fulfilling their obligations. Perhaps she would not have been so annoyed had they been Israelis. I had seen her quite annoyed at Ilan Amir in her dealings with him, for example.

The day was not all joy, however. Also at lunch, Ali Sha'ath mentioned that the Public Management Unit (PMU), the governmental part of the Palestinians participation, had never received the $180,000 it had expected. No one at lunch knew what had happened to the money, and I promised to look into it. When we asked Louise Anten about it the next day, she said the money was gone. When budgets had shrunk because of the guilder devaluation, Shula had apparently asked to convert some money to the central team's budget. No contract arrangement had been arranged with the PMU; doing so would mean the PMU would have to deal directly with the Dutch. I remembered that from the previous June, but I had no idea that the money would simply disappear. Worse yet, I had no idea Ali Sha'ath would never be informed about any of that. We decided Harvard (definitely not me) would have to send a letter to Ali Sha'ath to explain and apologize. It was not going to help matters.

We were back in Israel the next day, December 4. Louise Anten arrived on the KLM flight that comes in at 1:00 a.m. She knew the project was in trouble and what Ben-Meir had said, and I thought her visit risked a certain amount of political capital. I still think that was at least partly true, although Louise always denies it. Needless to say, I was very glad to see her.

We spent much of the day meeting with the Israeli private team. Things went quite smoothly, with a good deal of our time being devoted to discussing the political problem. All the team members were committed to continuing with the project under any circumstance, even if it meant without funding. I did not believe it would come to that, since there were sources of funds other than the Dutch Government. Louise promised to speak with her Minister.

The Israeli team's commitment could also have been important, I thought, to indicate to official Israelis that the project would not be going away no matter what they did. As one of the Israelis said, "He can't do that! This train has left the station. We will publish regardless of Ben-Meir."

On to Jordan

That evening, Louise, Annette, Hans, and I flew to Jordan for a series of meetings that the Embassy had set up for the following two days—meetings the Dutch Ambassador Bernard Tangelder would attend, since ours was a relatively official mission.

The Ambassador felt, however, that it would be inappropriate for the Embassy to arrange a meeting with Jawad Anani, because he was not then in the government. Relevant Jordanian government officials, the Ambassador said, would not welcome Jawad's intervention, and he was no longer important. I did secure permission, though, to arrange my own meeting with Jawad, which I set up for December 6.

On the morning of December 5, we met first with our old friend Hazim El-Naser at the Jordanian Water Ministry, who reaffirmed Jordan's commitment to the project and its desire to participate in the interim phase as described to him. He talked about integrating the project's software into Jordan's strategic plan in the coming months, which he regarded as a big job but one very well worth doing. He made it plain that the Jordanians continued to regard the project as quite important, although largely for their own internal purposes.

We then all went to a meeting with the Minister, Hani Mulki, scheduled to last about 15 minutes or so. I could not have dreamed what happened.

I had met Mulki previously, when he was president of the Royal Scientific Society (RSS), which, just before the Dutch Initiative, was supposed to become our Jordanian partner in the project. I hadn't seen him for three years. Our relationship had always been perfectly fine, but I did not recall him being extremely enthusiastic about the project—not that he was particularly negative, though. Certainly, as President of the RSS, he had lent us his support.

At this meeting, however, he was fantastic. I suggested to Hazim before the meeting that while I couldn't ask Mulki myself to do so, it would be very desirable if he could call Ben-Meir or someone appropriate in Israel. Then Mulki himself suggested that Jordan should ask for a tripartite meeting of the people concerned with water and with the project, and particularly the Water Ministers—including Ben-Meir. He believed he could absolutely convince the Israelis that they should participate and that the project would be good for everyone, including for regional cooperation—and in this case, probably between Jordan and Israel in resolving the problems in carrying out their treaty obligations. He wanted that to happen very quickly. He suggested further that some mild pressure from the United States might be helpful. Specifically, he had in mind that it would help if President Clinton, during his trip to Gaza scheduled for the following week, would say that cooperation would be welcome in various areas and mention water. Mulki proposed to speak about this with the U.S. Ambassador to Jordan.

Mulki also mentioned that there was to be a major water conference in Amman, probably the following March. It appeared to be a joint effort of the Center for Middle East Peace and Economic Cooperation, Green Cross International, and the World Bank.* The first two were the organizations that I had put in touch with each other nearly a year earlier and had not heard from since. Mulki wanted our project to be featured and discussed.

In short, the meeting was terrific. Mulki remarked how he had been convinced by me years ago, and again and again reiterated Jordan's complete support for continuing the project. Since Jordanian Ministers do not do these things without consultation, it was a fair assumption (backed up by things we already knew) that the matter had been discussed up and down the line. I expected the attitude to continue in our meetings the next day. Certainly, even if the Israelis were not convinced, there would now be a major case for the Dutch to continue in some form.

I should add that a feature of Mulki's idea regarding the tripartite meeting was that I would participate and explain things. That would have to be very carefully prepared, particularly because Ben-Meir might be the Israeli representative and it would be our first meeting. That was not certain, however. Apart from the question of the level at which such a meeting would be held, rumors in Israel and a positive statement by Nabil Sharif did indicate that Ben-Meir would soon be out—which I supposed could be a good thing for the project, or not. One possibility was that the Israeli Water Ministry would be returned to the Agriculture Ministry, where Minister Rafael Eitan was about as right-wing as it was possible to be.

Nevertheless, for the moment, I was a very happy and grateful camper. As we left the meeting, I asked Louise what she thought.

"Can't you see I'm dancing?" she replied.

It was good to feel that a higher authority supported the project, and I don't mean ministries.

The next day, we met with Jawad Anani. As always, he was welcoming

---

* The Center for Middle East Peace and Economic Cooperation was co-founded in 1989 by Wayne Owens, at the time a Democratic member of the U.S. House of Representatives from Utah, and S. Daniel Abraham, then the chairman of SlimFast Foods. Green Cross International, founded in 1993 by former Soviet Union President Mikhail Gorbachev, is a non-governmental organization that focuses on environmental issues and their interconnectedness with security and poverty.

and helpful. The principal outcome of our meeting was that he volunteered that day to talk directly with the Crown Prince, after coordination with Mulki, and urge him to call Sharon or someone else appropriate.

Our remaining official meetings in Jordan were at the Planning Ministry with Boulos Kefaya and the new Minister, Nabil Ammari. They both reaffirmed Jordan's support for the project, although Kefaya engaged us in a discussion of what Jordan would get out of it—a reasonable thing to do. Of course, we already knew we would get such support since the Planning Ministry was only formally the ministry involved. The real action was at the Water Ministry.

In the afternoon, we met with the private team and with Munther Haddadin, who joined us near the end. It was an amicable get-together in which we settled details for the wind-up of Phase I and discussed the interim phase of the project. The Jordanians seemed quite accepting of the idea that the interim phase would not involve a direct, specific contract with them but rather that they might be hired as private consultants as the Ministry so chose. Munther agreed that the Ministry should have the say, but urged that the Ministry not sign the contracts, since that would require a competitive procurement process.

We told Munther about the meeting with Mulki and Anani, and he said he thought we were bringing up much too heavy artillery. He thought he could handle Ben-Meir himself, reminding us that Ben-Meir had twice promised him that this blockage would not occur. I responded that he was perfectly free to consult with Mulki and Anani and agree with them on a coordinated strategy.

We had the feeling during the meeting that Munther was rather subdued. Hans even said "downcast." Munther had been having a very difficult time coming to grips with what Hazim said was his quite unfair dismissal and the possible legal problems that followed. Certainly, he could have felt out of the loop, disappointed that he no longer was a locus of power in the project. That seemed to me to be perfectly understandable.

News from Ben-Meir

We returned to Israel late that evening. For a few hours the next morning, we met again with Ambassador Hubar. The Israeli Government had decided that rather than a single meeting with many attendees, the Ambassador,

Louise, and Hans would have a series of separate meetings with individuals. The first was to be with Meir Ben-Meir, that very evening, to discuss desalination. The next day, they would meet first with Dan Katarivas, others in between, and then end with Shmuel Kantor to discuss joint Israeli-Palestinian projects. We surmised that the Kantor meeting would be a welcome way to avoid direct confrontation over the model and present the Dutch with alternative projects they might support.

Annette and I then spent the afternoon in Ramallah and Nablus, meeting with Palestinian officials and with part of the Palestinian private team. These were quite good meetings that reaffirmed the Palestinian commitment to the project. Hans and Louise went to Tel Aviv for the meeting with Ben-Meir.

Getting to the West Bank had been an adventure. Raffi Cohen, our taxi driver and organizer extraordinaire, sent an Arab taxi driver to take us. He obviously did this in view of the fact that there had been riots in both towns the day before—the unrest in response to Israel meeting its Wye Agreement promise to release prisoners by freeing common criminals rather then political detainees. The driver, however, had an Israeli taxi with Israeli plates.

As soon as we crossed out of Jerusalem into the West Bank, the driver took down the sign on his windshield that identified him as serving a Jerusalem hotel and took out two strings of beads, hanging them over his rearview mirror. It is a common custom among Arabs, especially taxi drivers. He then immediately stopped and bought an Arabic newspaper, which he prominently displayed on the front windshield while driving, and read while waiting for us.

The day passed without any untoward incidents. When we returned in the afternoon, the driver removed the beads and the Arabic newspaper and replaced the sign for the Jerusalem hotel as soon as we approached the Israeli checkpoint. It reminded me of an old movie, *Hotel Sahara*, in which Peter Ustinov plays the proprietor of a hotel in the northern Sahara during World War II. The region is first conquered by the Italians, and then by the British, and then by the Germans, then by the British again, and so on. As each wave of conquerors comes, Ustinov hurries to take down, for example, the picture of Mussolini and replace with a big picture of Churchill or whoever needs to be displayed next.

When Annette and I returned to our hotel in Jerusalem, I had a message from Anni Karasik asking me to phone her, which I did. She said Shmuel

Kantor was anxious to learn the results of our meetings in Gaza and Amman. I assumed the reason was a good one. If there were a flat-out "no" to the project coming, why would he care about those meetings?

But the best news was yet to come. While we were waiting for Louise and Hans to return, Annette and I walked over Mount Zion and down to the Western Wall. I prayed as hard as I could. When I finished, I recounted to Annette a Hasidic saying: "Pray as if everything depended on God; act as if everything depended on you."

"Well, we've done both," said Annette. "Can we go back to the hotel now?"

We walked back slowly through part of the Old City. On the way, I called Ellen to tell her where things stood and also to assure her that I was okay despite the violent demonstrations in Ramallah and Nablus the day before. I also told her that, in a water meeting in Ramallah, someone had suggested I should be referred to as "Abu Maya," which in Arabic translates as "father of waters." I had told him it was quite flattering but that in the Native American Algonquian language it would translate into "Mississippi River." I thought I had better not let my grandchildren hear about it or I would have a new name.

We had settled on Hans Wesseling's suggestion of "Abu Model."

Hans and Louise returned from their meeting with Ben-Meir at about 8:00 p.m. It had gone fantastically well. Ben-Meir invited them and the Ambassador to talk about the model and seemed extremely open-minded. The "no" that the Ambassador had characterized as the most definite one he had ever heard in his career had been rescinded. Instead, Ben-Meir wished to convene a two-day meeting some weeks later at which his experts would meet with us to study and evaluate the model. (Apparently, he did not want our Israeli team to come.)

Ben-Meir was clear about internal policy arguments with respect to the model. He had also made a number of statements about the model that were plainly not true, but that didn't matter. Evidently, it was to be given a serious hearing. Moreover, Ben-Meir stated that paying enough to desalinate 100 million cubic meters a year could solve the problem of water in Israel. That perception, if extended in dealing with the Palestinians, was what the project started out to be about.

The report back from the meeting was amazing. Ben-Meir had been quite friendly, interested, and polite. Something—we will probably never know exactly what—had plainly happened to effect this change. What I did know was that the change in Ben-Meir's attitude seemed to have occurred shortly after we left the Western Wall. I suppose the Almighty acts through human agencies in these matters.

I was reminded of a story I had told earlier that day. In the darkest days of World War II, a famous rabbi is supposed to have met the Polish General Sikorsky. The rabbi told Sikorsky that there were two ways the Allies could win the war.

"What are those?" asked the general.

"By a miracle, or naturally," replied the rabbi.

Sikorsky asked how they could win "naturally, and the rabbi replied, "Oh, that would take a miracle."

"And how could they win it by a miracle?" asked the general.

"Naturally!" the rabbi responded, with a broad smile.

Well, we appeared to have won naturally through a miracle.

Louise expressed the opinion that while the nature of the interim phase of the project might depend on the outcome of the two-day discussion that we were going to have with the water commissioner's experts, it was obvious there *would be* an interim phase. And it looked as though we might go on beyond that. I assumed that discussion would include Yossi Dreizin, and he had already said privately that the model was the best tool Israel had.

The rain was very late that year. I had been thinking of the story of Elijah. Earlier in the day, I recalled—I think from the *Talmud*—the saying that Elijah's greatest deed was not defeating the priests of Baal and bringing the rain but that when he did so the people praised God. Hans said he assumed that meant that if Frank Fisher solved the water problem, people should praise the Lord. I said I would be happy to accept that outcome.

It occurred to me that it was just possible that our meeting at the Peres Center the week before had convinced Shmuel Kantor to talk to Ben-Meir and put to him the proposition that the model was, in fact, a highly useful tool and that Israel did not want to be in the know-nothing position of rejecting it sight unseen, particularly if the neighbors were going to embrace it. But again, I supposed we would never know.

Late that night, Louise remembered that Ben-Meir had said he was speaking under conditions of extreme confidentiality. That probably applied to the time when he talked about his opposition to the model having to do with internal reasons and the threat to the position of the water commissioner that he felt it might represent. But we decided we had better keep everything secret. Unfortunately, by that time, I had already spoken with Zvi Eckstein and Yona Shamir, and Annette had spoken with Yossi Yakhin. The next morning, I tried to reach all of them, but when I called the Shamirs at 7:30 a.m., Uri had already left home to walk to his office at the Technion. Yona reached him and he phoned me a half-hour later.

It was too late. Uri had already talked with Yossi Dreizin. We agreed that even though Dreizin was to meet with Ben-Meir tomorrow, it would be worse to call Dreizen back. Uri recounted that Dreizen took the view that the change in attitude must have come from above, namely from Sharon. He said Sharon was interesting himself more and more in water, presumably as part of the responsibilities of the Foreign Minister in preparation for the negotiations. If so, and that apparently extended to our project, it had to be good news for us in affecting the national situation.

Whether any of this was thanks to the Jordanians, we did not know. Sharon had been in New York ever since we were in Jordan, and of course we didn't know when his change of heart took place. But it was no small thing to have changed the way two countries thought about water. Now we had a chance to convince the third.

A Few More Things, and Home

I arrived home early on the morning of December 10. Before Annette and I left for the airport the night before, though, Louise and Hans returned to the hotel after their day of meetings. They reported that Dan Katarivas, with whom they met first, was very supportive—as expected. He told them he had done considerable work on our behalf. Most of their other meetings did not involve discussion of the model at all. But their last meeting with Shmuel Kantor did, and it went extremely well.

Shmuel had begun with some skepticism but warmed toward the subject as the day went on. He emphasized that we needed to present examples really relevant to Israel and to the Israeli master plan. The Dutch delegation replied that that was a really good reason for why Uri Shamir and the other members of our Israeli team should be included (a proposition for which Ben-Meir had shown some dislike.) Shmuel agreed it would be extremely important to include Uri and said he planned to call him.

Louise Anten was floating on air over the results of the meetings. She had obviously put herself very much on the line for the mission and the project.

It was decided that all the people we saw officially in Palestine and Jordan were to be informed by the respective Embassy or representative office. But it was left to me to inform Jawad Anani (Ambassador Tangelder did not want to see him officially) and Munther Haddadin, which we agreed I could do in writing or email with a simple statement— "the problem we discussed has disappeared"—and a request for them to call. I sent Munther an email and reached Jawad on the phone as soon as I got to Cambridge.

When I told Jawad the problem had disappeared, he said he had spoken to the Crown Prince, and I believed I understood that he had taken some action. I thanked Jawad and he congratulated me.

"It's raining today," he added. There had been a major drought in the region, and everyone we had seen had remarked about it.

All in all, the trip was a great success and an emotional experience for each of us. When I returned home, I sent an email to Louise, Hans, and Annette that adapted Shakespeare's *Henry V* as follows:

> We few, we gallant few, we band of brothers (and sisters),
> Gentlemen at Harvard now abed,
> Shall think themselves accursed they were not here,
> To fight with us upon Saint Crispin's Day.

25

The Meeting with Ben-Meir and Its Aftermath

During December and January, work went on to close down the Harvard operation. It was made extremely tedious by the Kennedy School's attitude with respect to permitting the transfer or even the copying of the project's files. Harvard appeared to be frightened that we would find or take something that bore on Lenny Hausman's dismissal. Or perhaps it was a case of incredible bureaucracy. In any event, despite repeated queries beginning in November and despite letters and calls from CRA's legal counsel, it proved nearly impossible to get even a reply. Finally, Jean Hood agreed to let Annette mark and copy the academic files.

Meanwhile, the project's fame appeared to be spreading. After some pressure, the Jordanian interest in having me speak at the big water conference in Amman sponsored by the Center for Middle East Peace and Economic Cooperation and Green Cross International resulted in an invitation. Apparently, the organizers from the Center thought the Project had been "fired" and gone out of existence. But the death and mourning period for King Hussein ended up postponing that conference indefinitely. Not so a conference on regional cooperation in the Mediterranean area to be held in Arles, France, in May. Rafi Bar-El, then at Ben Gurion University (one of the sponsoring institutions), invited me to be one of the speakers at a plenary session. Also, I was invited to be the keynote speaker at a conference sponsored by the Sea Grant program in the United States relating to the combination of economics and "hard science" in land and water management. And inquiries about the Project kept coming in.

It was all quite remarkable given the ban on publication we had been observing. But with that ending, we agreed it was time to begin a book.

Meanwhile, the promised workshop with Water Commissioner Meir Ben-Meir took place. Hans, Annette, and I traveled to Tel Aviv, met with the Israeli private team on February 7, and then with Ben-Meir on February 8. The Dutch Embassy made the arrangements for the meeting with Ben-Meir and others. It was notable that Ben-Meir had insisted we not bring any members of the Israeli private team.

The meeting went extremely well. We could tell how the wind was blowing by watching and listening to Shmuel Kantor from the Water Commission; the Israeli private team had told us he was there to see how his boss Ben-Meir, the Water Commissioner, was leaning.

The weather vane was pointing more and more in a favorable direction. At the very start of the meeting, Kantor looked at the schematic diagram of the Israeli conveyance system put together by our team and pointed out what he thought was an error—which may not have been there and was, in any case, totally trivial. Ben-Meir stopped him cold, saying we were not there to talk about data, but about concepts. By the end of the meeting, Kantor was trying to formulate the terms of cooperation.

We all worked hard to bring that about. As suggested by the private team, I began by saying that the members of the Israeli Water Commission were the true experts on the Israeli water system, and that we were there to see whether we could provide a useful tool. We wanted that tool to meet their needs. I emphasized that, unlike what was often heard from economists, we were not recommending free markets and that such a recommendation would not work.

As planned, we began with an illustration of a drought scenario for 2010. In it, desalination looked like a serious remedy and one could see how the model produced criteria with which to judge such a proposition. Of course, we had been forewarned that Ben-Meir strongly favored desalination to guard against the unreliability of the natural system—that is, the possibility of droughts. For certain levels of desalination costs (about $0.60 per cubic meter), the Israeli model suggested he was right.

Ben-Meir, at least at this meeting, was a far more reasonable person than everyone who spoke of him had led us to believe. He had strong views about

347

economists, which I did my best to dispel (at least as they applied to me). He also had strong views about the necessity of preserving green spaces in central Israel and agriculture in the Negev, both for environmental purposes and to prevent Arabs from taking the land. He was explicit about that. Whether water policy was the appropriate instrument to achieve such aims or whether the Water Commissioner was the appropriate official to decide that Israel would embrace those aims was not discussed. Rather, the discussion turned on whether the project's tools could be employed to assist in the efficient application of such policies and an evaluation of their costs.

The Water Commissioner was also seriously interested in issues such as aquifer and year-to-year management that we would only be able to address in the context of the multi-year model, which was yet to be built.

Frankly, we got on extremely well. I was agreeable and even charming without rolling over and playing dead. I "spoke truth to power."

In the course of the meeting, as the atmosphere warmed, we traded jokes and moved to a first-name basis. At appropriate moments, I made sure Ben-Meir knew I was Jewish, was unpaid for my work on the project, and that I cared personally what happened in the Region.

I also mentioned, in Hebrew, that I spoke a little Hebrew. In the afternoon, when we explained AGSM, Ben-Meir—who grasped the explanation much faster than his colleagues—asked me whether it was okay if he explained it to them in Hebrew. When I agreed, he said that my Hebrew would doubtless allow me to follow along.

As soon as he began, I called out loudly, "*Lo nachon!*" ("It isn't right!"). Ben-Meir stared at me and laughed. Two days later, I could not believe I had done that. Ellen said I must have been manic, and she was right.

The meeting ended with a promise from Ben-Meir to cooperate with the interim phase, provided he could find someone to assign to it. (Uri said that this was not an excuse; Ben-Meir is very strapped for personnel.) If we were lucky, he would ask Yossi Dreizin to take care of it, and Yossi would consult Uri, who was an advisor on modeling to the Planning Department.

Of course, as Shaul thought, the favorable outcome of the meeting may have been pre-ordained either because of the intervention of the Crown Prince of Jordan or because the Israelis wished to maintain good relations with the Dutch government and so Ben-Meir had been instructed to cooperate. The statement in the meeting by Johanna Van Vliet that the Dutch

348

had decided to go forward with the other parties regardless of Israeli participation may have had an effect, too.

I like to flatter myself into thinking that our presentation—my portion in particular—had something to do with it. Near the end of the meeting, Ben-Meir asked me whether, were I unemployed, he could hire me "for this." I wasn't exactly sure what he meant, but I was pretty sure it was a favorable question.

Ben-Meir did not want to discuss regionalization yet, and we assured all the Israelis that participation in the interim phase did not commit them to that. Such a decision would come at the end of the interim phase (approximately the end of 1999). The forecasts were that, by then, there would certainly be a different Water Commissioner and—depending on the outcome of the Israeli elections in May—quite possibly a different Israeli government with a different attitude.

The Project's Survival

It appeared the project would survive. When I think back to how it managed to do so through the political vicissitudes of the two-year period up to the spring of 1999, I think of the story of the thief who was sentenced to death by one or another King Louis of France. The thief won a year's reprieve on the condition that he would teach the king's horse to talk.

"How could you make such a promise?" asked a friend of the thief.

"Easily," replied the thief. "In a year, anything could happen. King Louis might die, I might die, the horse might die. Or the horse might talk."

Nevertheless, it took time for the horse to perform. There followed a long period of waiting. The Dutch sent out the proposal for the interim phase on March 20, but replies were slow in coming. Meanwhile, I started to write a book on the project.

During this period, Uri heard from Meir Ben-Meir's people that Ben-Meir would participate, but only if there was money to pay the salary of whomever he put on the project. I pointed out that this was not a matter of whether the Water Commissioner's office had the money but rather of whether the Israeli government would put up the funds. I said I thought it was a matter for diplomatic negotiation.

Louise Anten read my reply and emailed that she couldn't agree with me more. If the Israeli government would not put up the funds, she added,

349

perhaps our attitude should be *"tant pis pour eux"* ("too bad for them"). I couldn't resist adding an extra letter and replying that the (im)proper phrase was *"tant pis pour eaux"* ("too bad for water"). Louise really enjoyed that.

Meanwhile, on a different front, CRA and Delft Hydraulics agreed in principle to go into the business of building WAS models for other countries. Hans, Annette, and I spoke with Uri and Ilan about this. All of us, especially Uri, wanted to make sure this didn't interfere with the existing project and we were determined not to let that happen. Khalid Kanoo, a former member of Lenny Hausman's ISPEME board, agreed to take up our commercial work with the Water Minister in his home country of Bahrain. More concretely, MIT had become involved with the government of Singapore concerning water planning and pricing, and I was asked to participate. It was made clear that such participation would be on behalf of CRA–Delft, and it looked as though Hans and I would travel to Singapore in the summer to assist in writing a proposal for a large study (although, like many other prospects, this came to naught).

There were two pieces of bad news during this time. Louise Anten informed me that the Dutch representative in Ramallah had only gotten around to forwarding the proposal for the interim phase to Ali Sha'ath the first week of April, despite that it had been sent from The Hague on March 20. More important, Louise told me that she would be reassigned inside the Dutch Foreign Ministry beginning that summer, as would her boss, Gerben De Jong. It was a great loss. Louise had been a tower of strength, imbued with zeal for the project. I continued to believe that she had taken a serious career risk the previous December when she persisted with the mission even after Meir Ben-Meir's refusal to Ambassador Hubar.

I sent Louise a note expressing, rather inadequately, my deep regret that she would no longer be working with us.

There was good news, too—indeed, very good. Uri sent an email informing me that two days earlier, Meir Ben-Meir had sent a letter to Ambassador Hubar stating that Israel would participate in the interim phase and believes the project to be an important potential contribution to Israel and the region. Further, Yossi Dreizin would be in charge. Contact would be directly with the Water Commission, not through the Ministry of Foreign Affairs (which would be kept informed).

I couldn't help note that this Israeli position was taken before the coming election, while Netanyahu was still the prime minister.

I replied, with copies to Louise, Hans, and Annette, stating that it may be that "the stone which the builders rejected has become the chief cornerstone." I assumed they would know how the quotation ends.[*]

The Arles Conference

In late May, I attended a conference in Arles, France, on "Forging Regional Cooperation in the Mediterranean Basin." The conference was mainly about Israeli-Palestinian relations. It had a large attendance, including a number of liberal Israelis and two board members from the New Israel Fund (Simone Suskind of Belgium and Henriette Kahan Dalev from Ben Gurion University). Naomi Chazan, a former board member and at the time a Knesset member from the Meretz party, was also present.

The big conference party dinner was held at an oxymoronic "fixed" Romany camp with two sorts of entertainment: men riding horses and throwing oranges at the audience (one hit Rafi Bar-El); and the other singers and (definitely) female dancers showing a considerable amount of flesh. I was standing there with my friends Simone, Henriette, and Naomi, who were and remain very active feminists. I inquired as to whether, considering my company, I was permitted to watch. They caucused and replied that I should consider the performance as a cultural and artistic one, and that I was therefore permitted to enjoy it intellectually.

I spent the better part of an evening with Palestinian delegation head Ali Sha'ath. It was a very friendly meeting. He told me that Nabil Sha'ath was more committed than ever to the support of the project. He also, however, asked my help in recovering the money from Harvard that was to have gone to the Palestinians. (Indeed, I may have brought up the subject before he could.) I told him I would certainly support him but, to put it mildly, I had no influence with Harvard. I do not think he understood that.

In any event, it was plain that the Palestinians had been waiting for the Ehud Barak government to be formed and to move on the peace process before agreeing to the interim phase. The same appeared to be true of the Jordanians: the Water Ministry, of which Hazim El-Naser had become

---

[*] "The stone which the builders rejected / Has become the chief cornerstone. / This was the Lord's doing; / It is marvelous in our eyes." (*Psalms* 118:22–23)

351

Director General, signed off on the interim phase in late March. (Of course, with the new Israeli government, the interim phase might have blended quickly into discussion of regional cooperation.)

The next day, I sat down at lunch next to Princess Wijdan Ali of Jordan, who was well known to my feminist friends and with whom I had an animated and very pleasant conversation. She was vice president and dean of the Jordan Institute of Diplomacy, and was kind enough to invite me to speak at the institute when I was next in Amman.

On the final day of the Arles conference, I gave a paper on the project. It was supposed to be a big plenary session at which Munther Haddadin, Nabil Sharif, and Meir Ben-Meir would also speak—but they all canceled. I may have aroused some interest nevertheless.

After my talk, the conference went immediately into its closing session, which featured a large number of formal talks. The room was extremely hot, despite the air conditioning. I was seated behind the Princess and Yitzhak Navon, the former president of Israel.

The first hour of that session was devoted to a lengthy talk by the French Minister of Justice that had little to do with the subject of the conference. When she was done, and it became apparent that the session was far from over, I let out a groan. The Princess turned around and said the session couldn't end because she had not yet given her paper. She was not listed on the program, so that came as a surprise.

I asked her how long it was. "Ten pages," she said.

"Single-spaced or double-spaced," I asked. She said single.

When I asked the font size, she said, "There you are in luck; it is 12 point. My eyes aren't good enough for anything smaller."

When the time came, the session adjourned with no paper from the Princess.

"You know," I told her, "you're trickier than most princesses."*

The Beilin Meeting and Back to Cambridge

In June 1999, there were more developments. First, the Israeli team met with the Yossi Dreizin's planning office at the Water Commission to discuss how

---

* I received a kind and sympathetic email from Princess Wijdan Ali immediately following the 9/11 attacks in 2001. I thanked her. I wish I had kept her address.

to proceed. The meeting was described as friendly, but everyone was awaiting the appointment of a new Commissioner.

Another development concerned my meeting with Yossi Beilin. Immediately on hearing the welcome news about the election—in which Ehud Barak, who ran on a strong platform of peace negotiations and withdrawal from Lebanon, defeated Netanyahu in a landslide—I sent an email to Alon Liel, who had been Beilin's Director-General and had been enormously helpful to me in the past. I thought the time had come to remind Alon that the project was very much alive. I wanted his advice regarding whom to speak with in the new government, even though that government was still in formation.

I had sent the email from Seattle, where our son-in-law, Brian Zikmund-Fisher, was receiving a lifesaving bone-marrow transplant. I received no reply. I tried again when we returned to Cambridge. Only in mid-June, when I received my materials for the New Israel Fund Board meeting (Alon's wife Rachel had become head of one of the fund's primary subsidiaries), did I realize that I had been sending the emails to the wrong address. So, I tried again—but again received no reply.

On June 25, Ellen and I left for Israel to attend the New Israel Fund's board meeting, at which I would retire as president. On Sunday, June 27, there was an opening reception at the Zionist Confederation House in Jerusalem, and one of the first persons I saw was Rachel Liel. It turned out the emails hadn't mattered.

Rachel told me that Alon wanted to set up a meeting for me with Yossi Beilin, who was then waiting to find out what role, if any, he would have in the new government (the general belief at the time was that he would become Minister of Finance). I pointed out that such a meeting was the most important thing I could do on my trip, and that if Beilin wanted to see me, I would be available at his convenience.

After considerable checking by Rachel, an appointment was set for Beilin's office at the Knesset for the coming Tuesday, late in the afternoon.

It was a wonderful meeting. Alon Liel attended, as did Daniel Levy, Beilin's young assistant who later became important in the peace movement. Beilin remembered the project quite well. He was delighted and congratulatory that it had survived the three years of the Netanyahu government and that the Dutch were still interested. In his view, the project was very important and was ready "just in time" both for negotiations and for the analysis of desalination options.

A couple of times, Beilin noted how fortunate it was that I had never seen Sharon—who, he was quite sure, would have killed the project.

Our discussion was totally open and frank. After three years, it was a great relief not to have to soft-pedal the international cooperation aspects of the project. Beilin asked for a document describing what had been done, and I agreed to send him the Phase I report when I returned home.

When we left, Alon assured me that even if Shlomo Ben-Ami, the other rumored candidate, got the Finance Ministry, Beilin would probably still take the project.

The next day, Rachel Liel told me she understood it had been a good meeting. Ellen told her I had come back "glowing."

Unfortunately, things did not work out so well. To everyone's surprise, neither Beilin nor Ben-Ami was chosen as Minister of Finance. Instead, the post went to Avraham Shochat of the One Israel party, about whom I knew very little. Beilin became Minister of Justice.

I never received the promised note from Levy, Beilin's assistant, requesting the report and providing the appropriate email address to which it should be sent. Nevertheless, I sent the report through Rachel and Alon.

Back home in Cambridge, I had lunch with Gilead Sher, a top advisor to Barak. Annette joined us. Sher, who was in Cambridge for two weeks, was apparently advising the prime minister on peace negotiations and was in the process of preparing a memo on water. Despite knowing nothing of our project when the meeting began, he appeared to catch on very quickly, and told us he expected to change his memo to feature our work. He was interested in how to proceed in terms of using the work for negotiations and promised to get back to us in about two weeks. I gave Sher the Phase I report and the paper I had delivered at Arles in May.

Both Annette and I thought this was a remarkable meeting with a smart, cooperative guy, and we were quite excited about it. It was also remarkable to think that the meeting had come about through Alan Altshuler, who knew about the project because of his involvement with me in defending Lenny Hausman after Lenny was fired. I supposed that the connection would have been made some other way, but it was still a strange chain of events.

In early July, at the suggestion of Louise Anten, the Netherlands Ministry of Foreign Affairs commissioned Delft Hydraulics to undertake a study of Israeli-Palestinian water issues using WAS. It was the first use of the new WAS 3.3 to examine regional issues. I took it as a positive sign.

The study was to begin in mid-July and finish a month later. Hans Wesseling arranged for Shaul Arlosoroff to consult on Israeli data issues and Anan Jayyousi on Palestinian ones. Annette Huber was hired to deal with programming and modeling issues. Given the small $25,000 budget, it was felt that even my expenses could not be handled, so we agreed I would participate by email, principally in the interpretation of results.

Of course, it did not turn out quite that way. As the time neared for Annette to go to the Netherlands on August 8, and she and I discussed various issues, I became more and more uncomfortable about not being there in person. I expressed this in an email to Louise, telling her I felt like the father whose children were about to drive the family car.

Louise replied with a pretty persuasive paraphrase of her 15-year-old son: "We have a license. You gave us the keys. If you are so concerned, why aren't you here?"

Given the high value of the opportunity costs of my time, which I donated anyway, it seemed silly not to go just because I would have to pay my own airfare. So, I made reservations just in case, still assuming I would end up not going. My daughter Naomi remarked, "Couldn't stay away, huh?"

The Netherlands in August

Of course, Naomi was right, and I ended up going to meet with the Dutch. The immediate reason for my "change of mind" was a call from Annette and Hans on August 9. They had just come from a meeting at the Ministry of Foreign Affairs at which they learned that the Dutch Foreign Minister was to go to the region on August 21. They hoped this would push the interim phase forward; the Palestinians had still not signed. The Foreign Minister wanted to have a memo from us in his pocket that set forth the results of the new project using WAS 3.3 to examine regional issues. Hans and Annette thought I should come, at least to write the memo.

I arrived in Delft on Friday, August 13 and returned home on August 19. It was a terrific experience. Hans, Annette, Anan, and I all made substantial contributions, as did Shaul, although he was not present in person and his contributions had come earlier in the form of resolving various data issues.

My presence was particularly fortunate because there were several results that had been obtained that could not have been right in view of economic theory, and being there to point those out led to various corrections. (Bugs had not been discovered previously because it was the first time WAS 3.3 was being run as a regional model.)

All of us, perhaps no one more than Anan, really enjoyed ourselves. To paraphrase Anan's attitude, the individual country runs of the model were finger exercises. Now we saw the real power of the method. The results were remarkable. I reproduce here the conclusions part of the memo we gave to the Dutch Foreign Minister (with minor changes of tenses).

Economic Feasibility of Water Scenarios in the Middle East
Conclusions

Cooperation is a "win-win" policy that will be worth $50–150 million *per year* by 2010. It is far more valuable than are changes in the ownership of the water itself. All parties gain from such cooperation.

The water in dispute between the parties is not sufficiently valuable as to hold up a peace agreement. Water can be traded off for other concessions, and the trade-offs need not be large.

For a stable peace agreement, water should be divided in a flexible manner so that changing situations (for example, droughts) do not strain the agreements reached. Cooperation in water as envisaged in this project provides such a flexible method and does so efficiently. In fact, the gains from cooperation increase with increasing drought.

By 2010, the gains from cooperation will be on the order of $50–150 million *per year*. Starting from the ownership allocations of 1995, the Palestinians would gain $123 million and the Israelis (who would own most of the water) $13 million *per year*. If, on the other hand, one starts with an ownership allocation according to (expected) 2010 population, the Palestinians would gain $45 million *per year* and the Israelis $36 million *per year*. Starting with an allocation in which the Palestinians owned an amount of fresh water equal to 100 cubic meters per capita per year, the Palestinians would gain $46 million *per year* and the Israelis $2 million *per year*.

With cooperation, the value of the entire Mountain Aquifer would be less than $100 million per year in 2010.

Desalination on the Mediterranean coast would not be needed in normal years. With cooperation in water and the construction of infrastructure (recycling plants and conveyance systems, largely for the

Palestinians), there would only be a need for additional sources of water in 2010 in years of drought.

These conclusions would not be altered if the number of Palestinian "returnees" were greatly to exceed current forecasts and total 2–4 million persons.

The need for desalination would crucially depend on cooperation in water. Without such cooperation and with the 1995 ownership allocations, the Palestinians would find desalination at Gaza an attractive option by 2010. This would even be true if the Palestinians achieve their goal of equal water ownership per capita (unless they are able to construct a major pipeline across central Israel). In that case, the Israelis would probably require desalination plants as well.

The construction of recycling plants in the West Bank, particularly in Gaza, would be beneficial (some $ 160 million per year) regardless of water ownership or co-operation.

Just before leaving for Delft, I had received an email from Gilead Sher, who had been up to his neck as Israel's chief negotiator in the revised Wye River agreement—signed soon thereafter. He asked me to call Yossi Kucik, Barak's Director-General, who would be in charge of non-security negotiations. He also told me he would be involved with water, as would Alon Liel. I took this as very good news. Alon's involvement could only be because of the project.

Since I was about to leave for the Netherlands, I put off calling Kucik until I could consult with the Dutch Foreign Ministry. With their agreement, I put in calls to Kucik on August 18 and again on August 19, when I returned home. I was told he would call back, which he did not do. So, I emailed both Alon and Sher, neither of whom replied. I assumed this was because of the attention focused on the Wye River amended agreement, but that things should begin to move.

Probably for similar reasons, things appeared to be moving with the Palestinians. Ali Sha'ath traveled to The Hague in connection with a visit by Arafat. He met with people from the Foreign Ministry, together with Hans. While Ali was still pressing for assistance with the money from Harvard, he stated that he expected to approve the interim phase within a week.

The budget for the interim phase provided funds for a full-time local expert. Hans strongly suggested Anan—who was, by the way, a member of the technical team for the negotiations. We all hoped that would happen.

Ali Sha'ath asked that the interim phase be shortened to five months. Since that would mean it would end in February, and that month had been set for the start of serious Final Stage negotiations, we took it as a strong suggestion that the Palestinians wished to use the model in negotiations.

While we were together in the Netherlands, I discussed with Hans, Annette, and Anan my discomfort at continuing to give papers in which I was listed as the sole author, but also that I couldn't possibly list everyone involved in the project. Hans made the very sensible suggestion that I should list as co-authors those who contributed most directly to the results being discussed in a given paper. I did that with my paper for the Sea Grant conference in Chicago that was coming up shortly, listing Hans, Annette, Anan, Shaul, Uri, and Zvi; I apologized to the Jordanians for leaving them off, explaining that the results mainly concerned Israeli-Palestinian issues.

This trip to the Netherlands also marked the departure of Louise Anten from direct connection with the project. She was succeeded by Leslie D'Huy. I was reminded of the story of when Thomas Jefferson succeeded Benjamin Franklin as U.S. Ambassador to France; Louis XVI is supposed to have said to him (presumably in French), "Ah, you replace Dr. Franklin."

"No, your Majesty, I succeed Dr. Franklin," was Jefferson's quick reply. "No one can replace him."

Readers can only have a partial idea of my feelings on the occasion of Louise's "departure." An email from Louise reminded me that those feelings were shared. I began thanking her in papers as one "who shares the vision"— but that still seems inadequate.

Stalled

The water project was pretty much stalled for the next six weeks, but there were a few developments. In early October, Annette and I went to Princeton University and spoke to Hal Feiveson's class at the Woodrow Wilson School. Eytan Sheshinski had suggested to him that he contact us concerning the class he was giving on water in the Middle East, and he had come up to Cambridge that summer. We gave him the Oz version of the model as a teaching tool and looked forward to hearing about the results.

As for developments in the region, we heard nothing from Ali Sha'ath until mid-October. At that point, the Dutch Foreign Ministry received a letter approving the interim phase, but on the condition that the Palestinian government receive $150,000 as settlement of the debt they believed—I think correctly—was owed to them by Harvard for Phase I. Hans told me Gerben de Jong's successor said he would find the money.

The remaining snag was the concern of the Dutch lest Harvard assert rights over the software. The fact that Harvard and we could only find an unsigned agreement was not reassuring. Harvard's pettifogging and obstructionist lawyer Bob Iuliano asserted the agreement *was* signed, which turned out to be the case. Of course, no one could find the signed copy, and I no longer had access to ISEPME's files. Peter Rosenblum, CRA's lawyer, complimented me on the substance of the agreement, and we told Harvard we were acting as though it was in force. The agreement gave licenses to the Dutch and the three governments in the region, and permitted either me or Harvard to do essentially whatever we wanted with the model, provided we gave the other party 90 days' notice. We gave that notice immediately.

As for Jordan during this "stalled" period, the only thing that came up concerned authorship of papers. I had been invited to be a keynote speaker at a session of a giant conference scheduled for the Netherlands in March 2000: the Second World Water Forum. I planned to give much the same paper as at the Sea Grant conference in Chicago, but with more detailed results. I thought it unfair to leave the Jordanians off the title page this time, so I sent the paper to Hazim El-Naser and asked him to suggest which Jordanians should be listed, stating that, of course, I planned to list him. After an exchange of emails, he made it clear that he would find it embarrassing to appear on this title page. The reason was pretty clear: anti-Israel feelings in Jordan, about which there had been newspaper stories—even suggesting death threats for Jordanians involved in cooperation. So, the Jordanians stayed in a footnote.

Regarding the Israelis, Kucik never returned my calls, and Gilead Sher left the government (for technical reasons, although he was still important). Alon Liel told me at the end of September that the project would be under the new Regional Cooperation Ministry, led by Shimon Peres, at which Avi Gil, formerly of the Peres Center, would be the Director-General. Apparently, Alon himself wrote in the "Fisher Project" as part of the agreement establishing the Ministry.

The Ministry itself had taken a long time to establish—fine, I thought, so long as it would be an active one. However, I had no word from anyone there or from Alon for a month. When Rachel Liel was in Cambridge in early October and asked me how the project was going, I mentioned that. She promised to galvanize her husband when she got home.

Meanwhile, when Uri was in Cambridge in early October, he told me that a principal reason nothing had happened was that the Barak government had yet to put its water team in place. Meir Ben-Meir was still Water Commissioner—with the authorities possibly waiting until the end of the drought to replace him. Further, Uri said, there might end up being different people in charge of domestic issues and in charge of negotiations.

When I asked Uri who the new Water Commissioner was likely to be, he said Noah Kinarti was often mentioned, especially for negotiations. That was a role he had played in the previous Labor government. Uri was close to Kinarti, but hastened to add that this did not mean Kinarti would accept the model.

Yossi Dreizen had also been mentioned. When I asked Uri whether he himself had been mentioned, he said he really hoped not. He then went on to tell me all the reasons he would rather be a technical advisor but added that, if asked, he would have to take the job. I pointed out that he wasn't really answering my question, and he finally admitted that his name was also being mentioned. When he left, he told me to wait a week for news—which I did, and then emailed him, asking what was up. I had the feeling that it was a very real possibility he would be the new Water Commissioner. But that did not happen.

Uri's visit to Boston had been specifically for my 65th Birthday Conference and celebration and for Annette's wedding, on October 9 and 10, respectively. The birthday conference and dinner, put on in my honor by the MIT Economics Department, were wonderful. I had asked the Department to invite Uri and was extremely pleased when he agreed to come. Much was said at the conference and dinner about the water project. In particular, Uri gave a paper on it, discussed by Hossein Askari, a student of mine 30 years earlier.

Dennis Carlton, another former student, announced the donation of a Chair to the Department, with me the first occupant.* Dennis and his wife

---

* Dennis said he was doing this in part because of the Jewish tradition that one who comes into money acquires an obligation to use it for good. Lexecon, his consulting

Jane also wanted MIT to give a course in Jewish Studies or Middle East Peace. I later twice participated in teaching a course on the economics of Israel and its neighbors in association with Ephraim Kleiman of the Hebrew University, who taught the lion's share.

This donation had several consequences for me. First, I became the Jane Berkowitz Carlton and Dennis William Carlton Professor of Economics. Second, it probably delayed my full retirement from MIT (where I was already on half time). Third, the Carlton Chair carried with it $20,000 per year in research funds, which I thought might help to keep Annette on the project. Such help was needed. Annette had earned her doctorate and needed to find gainful employment. Her preference was employment on the water project, or at least to find something that would permit her to join the water project when there was something for her.

We celebrated Annette's marriage to William Lee the next day. When they left on their honeymoon, there wasn't much prospect of employing her on the project. We investigated whether CRA could hire her on other things. And there was, at least with the prospect of the interim phase and the money from the Carlton Chair.

Meanwhile, we still had some prospects on the commercial front, but none of them were really alive as yet. Lenny Hausman wanted us to submit a proposal to Kuwait, and I had to write one up that could be generalized for use elsewhere. He had a Dutch contact who was said to be willing to take such a proposal around Europe. He also wanted me to write op-ed pieces; I was less convinced that doing so was a good idea.

---

firm, had won an immense damage verdict for slander against the law firm of Millberg, Weiss, and Dennis received a large share. Later, I found myself sitting at the same table with Mel Weiss, with whom I had earlier worked. I told him that I was effectively the Millberg, Weiss Professor of Economics at MIT. When I explained why I had said that, he responded that he was glad the money was being put to good use.

26

Nothing Is Ever Easy

After considerable delay, Prime Minister Barak appointed Noah Kinarti as
his chief water negotiator with the Palestinians. Eli Suissa, the Minister of
Infrastructure from the ultra-Orthodox party Shas, saw it as an attempt by
Barak to control water directly in violation of the agreement for a coalition
government, and forbade anyone under him—including the Water
Commissioner's office—to speak with Kinarti. So, Meir Ben-Meir resigned
as Water Commissioner, but stayed on for three months.

Kinarti had been Rabin's chief water negotiator. He was a *kibbutznik* and
highly disposed towards agriculture. During the earlier negotiations, it had
been pretty clear that he saw his job as retaining as much water for Israel as
possible, regardless of the effects on the Palestinians, and a lasting peace
arrangement. He was also said to be opposed to models and modeling.

"That dinosaur!" Eytan Sheshinski had exclaimed to me upon hearing of
his appointment.

Uri Shamir was appointed as Kinarti's technical advisor. This would have
seemed to be a good development. Indeed, Uri urged him to use the WAS
model, but Kinarti refused even to discuss the matter. Alon Liel told me
Kinarti had once tried to get Peres to drop the project. So far as I knew,
Kinarti had never bothered to find out what the project or its tools actually
did.

Despite all this, I believe Uri made the proper decision, given his new
position, that his first loyalty had to be to Kinarti. He remained friendly and

sympathetic to the project, but felt he could no longer take an active role. It was a great loss.

Work on the Model

Despite what was happening with water *politically* in Israel, we still had the project. Modeling and thinking about the model was fun. Annette and I began work on the intertemporal version, which was to become MYWAS. We got around the problem of large fixed costs and integer programming by having the model choose capacity additions and costing those with a formula including a term *(1 − exp (−□ {□ capacity})),* where □ is very large. We also dealt with the extraction rate for fossil aquifers—those that are not replenished by inflows.

On a different front, we discovered a minor bug in the existing model, although not one that affected most existing results. When a piece of infrastructure is capacity constrained, the shadow value of that constraint is included in the price paid by purchasers of the facility's output (either private users or the government when it sets prices and, as it were, resells the water). Such higher prices reduce buyer surplus. But the money paid goes somewhere. When the facility was entirely within a district, we treated the higher price as paid to the water producers. But when the facility was an inter-district pipeline, we had the buyers paying a high price at the receiving end and the producers getting a low one at the exporting end. That would be correct when there is no capacity constraint, because then the difference corresponds to the operating costs of conveyance. When pipeline capacity is constrained, however, the part of the payments involving the pipeline's shadow value does not correspond to any direct cost to the system. (This could be seen by observing that such payments were not subtracted from the objective function.) Hence, we needed to account for the receipt of such payments.

It should be noted that this does not matter to the optimal solution; it matters to the calculation of social welfare. The problem arises only when there is constrained inter-district pipeline capacity. That affected some Jordanian runs and possibly the Israeli experiments with an additional pipeline to Jerusalem and the bringing of water from the Litani River. But I did not believe that the qualitative conclusions would be affected.

We adopted the convention that the payments are received by the exporting district, which was therefore assumed to "own" the pipeline. That convention, which could be overridden at the user's option, mattered only for international pipelines. The experience showed the usefulness of being sure that the social welfare calculations were such that the sum of all social welfare was the objective function.

I found this problem while reflecting on an issue that had arisen the previous summer when we were doing regional runs at The Hague. At one point, we appeared to have found that permitting trade led to one country losing rather than gaining. That had occurred because Annette and I had, without reflection, permitted trade runs to involve simply sharing the profits from any common pool resource. It turned out that was *not* the same thing as buying water.

To see this, suppose that Israel and Palestine have equal shares in a particular common pool. Suppose that, without trade, Israel pumps its share but Palestine, with much deeper wells, does not. Suppose the same thing happens with trade. Then Israel would acquire no water from Palestine, but sharing the profits would mean Israel would pay Palestine half the producer surplus from the common pool in question. In effect, it would mean that Palestine would share the shadow value of the Israeli wells that stems from the fact that they are relatively low cost *and* constrained in lifting capacity, since otherwise the Israelis would pump more under trade. Reflecting on the shadow value of infrastructure led to the discovery of the bug.

It was a reminder of a recurring moral: When there is something you haven't completely understood and absorbed in a model as complicated as ours, it will come back and bite you.

Throughout the Christmas vacation and into January 2000, Annette and I worked to prepare new runs of the model. The experience showed again that one never knows about a complex model until every bit of it has been exercised. The constrained capacity mistake caused us to realize that the sum of social welfare, including subtraction of government cost, over the three countries, ought to equal the value of the objective function. So, we had begun printing both the sum and the value of the objective function as part of the output as a check on our post-processing accounting. When we did that, we found an exact match most of the time. But when we put in fixed-price policies towards agriculture in Israel, there was a discrepancy of $4 million per year, with the value of the objective function lower than the welfare sum. That was too big to be rounding error.

Finding the answer to the problem took ten days, or rather ten sleepless nights, while Annette was with her family in Pittsburgh and we communicated by email. I was reminded Robert Solow's remark. A colleague in the MIT Economics Department, he was asked how one finds a theory thesis topic. "You don't have a theory topic," he said. "It has you."

Repeated reviews of the program failed to turn up an answer. Moreover, two phenomena eventually suggested that the problem could not be a programming bug. The accounting relationships were such that any mistake in the calculation of the reimbursement paid to the government by consumers would cancel out of the final sum. (I actually experimented with that by adding an arbitrary figure, and it was so.) Indeed, the accounting was so straightforward that pure programming errors were very unlikely to produce the problem. That was consistent with the fact that the problem did not appear when fixed-price policies were used in Jordan or Palestine.

Experimentation showed that nearly the entire problem was coming from the Israeli district of Bequat Kinnarot—the district in which the Israeli National Carrier began and nearly all Sea of Galilee water was extracted. When no fixed-price policies were imposed for that district, the problem dropped to the level of rounding error.

After brooding (but not sleeping) on those facts for several days, it became clear to me that there must have been something in the data for Bequat Kinnarot that was creating the problem. Moreover, that something had to involve a cost that was properly accounted for in consumer prices when they were not fixed, but was not being accounted for in the social welfare sum when they were. Such a phenomenon had been found a long time ago in the accounting for environmental costs and recycling profits; this had to be something else.

I decided the only thing that could be doing that would be non-zero intra-district conveyance or leakage costs—costs incurred when water was used within a district. Since the interface showed that the Israeli team had assumed zero intra-district leakage, and since the interface could not be used to read intra-district conveyance costs, I emailed Annette that the answer had to be the latter. That seemed particularly likely, since the Israeli team had assumed no extraction costs from the Kinneret.

Annette's response— "I think you've got it"—prompted me to reply with an email from Eliza Doolittle to Henry Higgins asking her to check that other districts such as Hereford or Hampshire did not have the same problem. Her mother was much amused.

As it turned out, the answer proved to be correct. There was a $0.09/cubic meter intra-district conveyance cost in Bequat Kinnarot and none elsewhere. While this affected all water consumed in that district, agriculture was the only consumer large enough for the problem to show in the results beyond rounding error.

I began to feel much better about the entire enterprise.

Pressing Political Events

Meanwhile, political events were pressing. Serious Israeli-Syrian talks were to be held in Shepherdstown, West Virginia, with water on the Golan said to be a major stumbling block. Noah Kinarti would lead the Israeli water team, which would include Uri.

Our model runs showed results of immense importance for the negotiations. We ran the model for 2010 removing 125 MCM per year from the Jordan River—roughly equivalent to the flow of the Banias Springs on the Golan. The result of this for a year of normal water flow was a loss of $4 million per year to Israel when the then-current fixed-price policies were assumed still to be in place, and $10 million per year without such policies. In effect, the model showed that the loss of the Banias water could be made up by extraction from a relatively high-cost source in the Jerusalem Mountains plus a considerable increase in the amount of treated wastewater sent from near Tel Aviv to the Negev for use in agriculture. Since conveyance costs from the Sea of Galilee would be saved, the net costs were only $4 million per year. Without fixed price policies, there was a deadweight loss as consumers cut back on water consumption in response to (small) price increases.

In the event of a 30 percent drought, the harm from the loss of the Banias was less trivial—about $40 million per year. Further, the desalination ceiling meant such a loss could never exceed about $75 million per year.

Obviously, our result was big news. Even the larger sums were too small to be worth holding up negotiations. But Israeli water negotiations were in the hands of Noah Kinarti, who had never been willing to listen to any serious presentation of the model and would not even talk about it.

The time had finally come to do what Lenny Hausman had been pressing me to do for years: go public. I began to meet with or speak with Lenny several times a day to strategize. We came up with a three-pronged campaign: alert higher-up negotiators; speak with government officials in Israel and the United States; and get press coverage.

I asked Alon Liel to set up appointments for me with various higher-ups in the government. There was a strong feeling that working through the Finance Ministry might be the best plan, and so I put Minister Shochat and his Director-General Avi Ben-Bassat—someone Zvi Eckstein knew and thought highly of—on the list. At Eytan's suggestion, I also added Haim Oron, the Minister of Agriculture from the Meretz party, whom Eytan regarded highly.

By all reports, the people around Prime Minister Barak who really counted were all ex-generals. One of these, Danni Yatom, was the Prime Minister's right-hand man; Lenny urged me to see him. Another ex-general, Uri Sagie, was to be in charge of the Syrian negotiations under Barak.

Fortunately, Baruch Levy, himself an ex-general as well as a former member of the ISEPME Board and a long-time supporter of the project, knew both Yatom and Sagie personally. After speaking with Lenny, Baruch agreed to hand-deliver to the two men a one-page letter I prepared for them on the day before their departure for the negotiations in West Virginia.

Baruch reported back that the position that water was not worth much had penetrated the Israeli team, although one of the ex-generals remarked that "Kinarti will be difficult" in the negotiations.

Another Trip to the Region

The work Annette and I had been doing on the model was in large part to prepare for the Fishelson Lecture on January 11. It had been four years since I had spoken on water at Fishelson's *shloshim* observance; now I was to give the fourth annual lecture in his memory. The trip was also a trip for Hans, Annette, and me to try to get the Interim Phase started; we would be exploring cooperation in the Interim Phase with Yossi Dreizin, as well as with the Agriculture and Finance Ministries. For that, I enlisted the aid of Haim Ben-Shahar, also known as "Habash," whom I had known for years and who was both very well disposed and very well connected.

I was determined to use my presence in Israel to go as public as I could

367

regarding the water project, and I resolved to make considerable noise.

The Fishelson Lecture was well attended and caused a stir—at least among the audience. A reporter (I think from *Haaretz*) attended, but no story appeared. A day or so later, however, I was interviewed by a reporter for *Ma'ariv*, and that story did appear and drew attention. It was picked up in a major Lebanese newspaper. Its emphasis was the proposition that the water on the Golan (or in the Kinneret) was not worth a great deal.

The *Ma'ariv* reporter had interviewed me at the suggestion of Alon Liel, with whom we met a few days later and who, as usual, gave good advice. We agreed it would not be a good idea for me to keep reiterating the point about water on the Golan being worth very little, since a good deal of the message for the Israeli negotiators was that the Syrians, not believing this, might be willing to give concessions for water. To keep pressing the point would be to inform both sides, and the Syrians were not my clients—whatever the hope for the future. (In fact, the Syrian–Israeli negotiations ended up breaking down over other issues, principally Syrian unwillingness to offer anything at all. But that was a couple of weeks later.)

As a result of Alon's advice, I emphasized the same point about the value of Golan water to Eta Prinz-Gibson. Eta, a freelancer for the *Jerusalem Post*, was an old friend from my early days in Boston's Friends of Peace Now in the 1980s. She had come to the Fishelson Lecture, interviewed me at great length both then and afterwards, and went on to write a very big article about the project and me that appeared a few weeks after the trip.

Eta was also quite friendly with Gilead Sher, and persuaded me to call him. Sher and I then met late one evening in Jerusalem. He was still very much involved and very close to Barak, although he did not want this generally known. He took a memo from me to Barak (with no apparent result) and suggested strongly that I get a legal advisor to prepare a draft of what treaty articles with the Syrians would look like. Sher was very enthusiastic when I suggested Robert Mnookin of the Harvard Law School, a well-known mediator who was on the board of the New Israel Fund. After the trip, Mnookin and I did draft the treaty articles, which I sent to Sher— and never received a reply.

We also had meetings at the Agriculture and Finance Ministries and at the Water Commissioner's office. We were joined in those by Zvi Eckstein and Shaul Arlosoroff. Shaul also joined us when we met with Shimon Peres.

The meeting at Agriculture, arranged by Habash, was a great success. We met with Haim Oron, the Minister of Agriculture, and his staff. Oron was known as "Jumes." Eytan Sheshinski had recommended Jumes to me as very intelligent—a liberal man, but, of course, committed to the traditional protection of Israeli agriculture as an almost Biblical imperative.

I went to the meeting to placate any opposition from Agriculture and reassure the Ministry the model could be used. I was pleasantly surprised by the outcome. Both in that first meeting and in a follow-up with the staff a week later, at which Ilan Amir joined us, we found the officials friendly, cooperative, and willing to listen. We concentrated on AGSM, and left with the impression that the Ministry could use the project in evaluating policies.

We met the same day with Yossi Dreizen at the Water Commissioner's office. While still a bit formal, it was a much friendlier meeting than previous ones. Yossi was very interested in building a multi-year model. He wanted to study the effects of different rainfall patterns and, not surprisingly, given that drought, inter-year storage, and aquifer management were much on his mind. He wanted, however, the model kept as undetailed as possible.

The following week, we met at the Finance Ministry with Director-General Avi Ben-Bassat and others, all economists who spoke our language. Afterwards, they contacted Shaul and asked him to have us use the model to evaluate desalination versus purchases from Turkey at the same ports.

From the figures we were given the choice was obvious. Turkish water was assumed to be sold with prices on a declining step function, with desalination costs being the same as prices on the middle step. In normal years, neither option was efficient but, for drought years, some extra water was required, particularly since price policies were specified, although the agricultural policies were modified from the existing ones in some of the runs. That implied water quantities were fixed by demands, given the specified prices.

The required extra water quantities fell in the middle step of the Turkish step function. That made the costs of the two options practically identical when extra water was needed. Since the Turkish option would be interruptible with the costs largely variable, and the desalination option would have heavy capital costs, the Turkish option was obviously to be preferred. By July, Israel—which had been taking bids on desalination plants—appeared to be moving to the Turkish option. (Some years later, largely thanks to the efforts of Alon Liel, the two countries came very close to an agreement that Israel proposed to sign because it wanted to please Turkey, which had set up the facilities for the transfer. But, as it turned out, the costs were found to be too high.)

We had two other meetings in Israel that January. One was with Haim Ramon, Minister without Portfolio. Gideon Sher had suggested we see him; he had read the *Ma'ariv* article and wanted to see us. He had already realized that water is just a matter of money. In part, at least, his conviction as to water and money was probably due to various proposals (including that of Wayne Owens of the Center for Middle East Peace and Economic Cooperation) to build desalination plants. It was less clear to me that he understood that the water in dispute was worth considerably less than the desalination price (let alone less than that price minus the costs of the naturally occurring fresh water), and I did not believe that he understood that cooperation in water would benefit everyone. But we had only a very short time with him, and he seemed to have an open mind.

Finally, there was a very good meeting with Shimon Peres at the Knesset building. Hans had left for home; Annette, Shaul, and Alon Liel joined me. Avi Gil, who had become the Director-General of the Peres Ministry, also attended. It was my third time with Peres, but the first time that he appeared to be really listening and engaged. The meeting ended with Peres and Gil following us out into the hall and asking me to produce a sample scenario showing how trade would work. Peres promised to take it and the cooperation proposal to his colleagues.

When that sample was sent a couple of weeks later and Peres did share it, it lit a spark. I do not know all that happened, but Shaul informed me that Meir Ben-Meir (then still water commissioner) went ballistic, felt betrayed, and ordered his people to have nothing to do with us.

Later, in March, Uri told me that going to Peres had been a great mistake. But I was not engaged in the project simply to assist or cooperate with the Israeli water establishment. Some effort had to be given to the cause of international cooperation and the use of the project in negotiations. If the water people like Ben-Meir and, especially, Kinarti were blocking it, then we needed to speak with those who had a broader view. Peres, then the Minister for Regional Cooperation, and considered then the probable next president of Israel, was high on that list, even though his ministry was regarded as having little power and people surmised it would probably disappear when he became president.[*]

Significantly, in the same conversation in which Uri insisted speaking to Peres had been a big mistake, he also informed me that the view that the disputed water with the Palestinians was of little value had penetrated the upper levels of the Israeli government. Of course, Uri was still limited in what he could say because of his position under Kinarti.

While in Israel, we also had a farewell meeting with Ambassador Hubar and his assistant, Johanna Van Vliet. Hubar was going to become the Dutch Ambassador to NATO. I greatly valued his support and wisdom throughout a very difficult period.

There had been discussion of also visiting Jordan during this trip. At Lenny's suggestion, he and I had exchanged emails with Bassem Awadallah, the protégé of Jawad Anani who had been so helpful in 1996 when the project was being considered by the Dutch and the Council of Ministers. With the change in government following the death of King Hussein, he had become the chief economic advisor to the Royal Court. Moreover, he was an old and close friend of the new King Abdullah.

Lenny's idea was to propose to Bassem that the King might be interested in promoting a Regional Water Authority along project lines. I sent some materials and received some indication of favorable interest. However, Bassem and the relevant Jordanian water people, including Hazim, were going to be out of the country at the time I could visit, so the meeting had to be postponed—and ended up never taking place.

Palestine

The part of the trip that we spent with the Palestinians had mixed results. Hans had sent repeated faxes and letters and made calls to Ali Sha'ath requesting a meeting, but had received no reply. It turned out Ali was in the Netherlands while we were in the region and could not meet with us. He had still not signed on to the Interim Phase as his dispute with Harvard over money persisted. The Dutch government was going to settle the matter by

---

[*] In fact, Ariel Sharon shut down the Ministry of Regional Cooperation in 2003. Peres became president in 2007, serving in that position until 2014. In 2009, the Ministry of Regional Cooperation was recreated.

paying the Palestinians.

But Ali also felt mistreated by the project, a feeling left over in particular from Shula's insistence he attend certain meetings. As a Deputy Minister— one for whom meetings in East Jerusalem were politically difficult—it had raised suspicions about the project. Of course, the fact that Shula is Israeli and that Lenny and I are Jewish probably never helped, and the fact that Shula was no longer a part of the project didn't seem to matter.

But there was more to it than that. In July 2000, Ali still had not signed, although his signature appeared quite close. He insisted that he was too busy with negotiations over the Port of Gaza to pay attention to it. It was clear, however, that, since much of those negotiations were with the Dutch government, he was using the project to secure concessions. When the agreement over the Port was signed in March, Ali continued to stall as negotiations over details continued. He was a thoroughly political person.

Unable to meet with Ali, we visited the Ministry of Planning and International Cooperation and got a decidedly cool reception. The group we met, led by Luai Sha'at, expressed annoyance that we had come without having communicated with Ali, which was not true on our side. Rather than discuss future plans, they again brought up the Harvard money. Further, they complained that the computers of the Palestinian team had not been turned over to MOPIC at the end of Phase I and that by the time they were, they would be obsolete. That was particularly annoying to us, since the responsibility for securing them rested entirely with the Palestinians. It was a dispiriting meeting, but fortunately was the only meeting like that.

Earlier on the same day, we had met with Nabil Sharif in his office, a meeting attended by the always helpful and clued-in Khairy El-Jamal. That meeting was disappointing only in that Sharif seemed not to remember anything about the project, so Khairy had to explain it to him all over again. Sharif wanted us to begin the Interim Phase by working on Gazan water problems, coordinating our efforts with other teams that were planning various forms of infrastructure. We agreed to do so, although we remained a bit uncomfortable about the constant non-inclusion or lesser inclusion of the West Bank. There was considerable enthusiasm for the project among the relevant Palestinians on the West Bank, and they were anxious to begin.

But, before describing our official meetings, I must describe the meeting that we had on the previous day with my old friend, Jad Isaac.

I had decided that too much time—three years—had transpired since I had last been in contact with Jad, and that we should meet again—Palestinian politics or not. I was nervous that Jad might feel aggrieved at the lack of contact, but it turned out such worries were for naught. Jad was enthusiastically welcoming, and our meeting—including dinner and a nighttime walk in Bethlehem—was perfectly delightful.

We caught up on our professional activities. Jad asked me how I knew that he had ended his battle with MOPIC. I replied that I had not known. By coincidence, just at the time in December that I had emailed Jad to ask for a meeting, he had seen Samih El-Abed, the Deputy Minister of Planning in Ramallah, who had said that the time had come for reconciliation. That, indeed, had happened.

Jad also said he would suggest to Samih that Samih take over responsibility for our project, mentioning Ali Sha'ath being too wound up in the Port of Gaza projects to pay any attention to us. He also said he would express to Samih his own eagerness to return to working on it. That, of course, would have been wonderful, and a great thing for the Palestinians. But when we met with Samih a day or so later, he didn't bring up the subject, and it seemed inappropriate for us to do so.

Samih was, however, highly enthusiastic about starting the Interim Phase in the West Bank. Also present were Anan Jayyoussi and Karen Assaf. She was even more enthusiastic. She had been thinking about the project and had concluded that it was really exciting—although it was, as she said, "scary, at first." We went to her office at the Water Commission and had a lengthier discussion.

Among other things, Karen was a Palestinian representative to the Multilaterals on water. She brought up the idea of having the Multilaterals use the project—something Uri had thought would be a good idea and politically safe. Karen said it would only happen if the Dutch pushed for it. But Louise Anten, who remained very interested, opined later that we would have much less freedom in the Multilaterals context than we would outside it. I decided to follow Louise's advice, as I always did, and not try to move further in that direction.

By the way, Jad mentioned the project was very well known and highly controversial in the Arab countries. He had seen a pamphlet in Lebanon denouncing it as a Zionist plot. We were not always favorably regarded in the

Israeli press, either. A few weeks after the mission, Ze'ev Schiff—a highly regarded *Haaretz* columnist Shula had always been trying to convince to write supportively about the project—wrote a piece criticizing my stand on the value of water. Jad and others called it to my attention. I wrote to Schiff and offered to discuss matters with him, but he did not reply.

A Slow Period

Upon our return to Cambridge, the project entered a slow period. As we continued to wait for Ali Sha'ath to sign on to the Interim Phase, Annette and I continued to work on a multi-year version of the model. We concentrated on the demand and infrastructure side, and succeeded in producing a version that would find the optimal schedule of infrastructure construction from a predetermined menu of projects, given a sequence of demand conditions. That version would also deal with optimal inter-year storage and with the optimal rate of depletion of fossil aquifers. Naturally, convergence time was substantially longer than for the single-year model, but it was still under control, and we had not yet attempted to reduce it directly. We believed that if and when the project really got going again, we would have made considerable strides. Ultimately, that came true.

On a different front, Leah Kronavetter, a student of Uri's, had been working on how the model could be used in negotiations. A number of things remained to be worked out, primarily because I hadn't had the opportunity to sit down with Leah and Uri and thrash things out so we all understood each other. But it seemed clear that her work would constitute a real contribution.

One issue we had not settled had to do with the question of whether or not *levels* as opposed to *changes* in social welfare were relevant to negotiations. My position was, and remains, that they were not, in part because the demand curves were arbitrarily cut off at high prices. That was done early in the project, because we were using constant-elasticity demand curves with elasticities less than unity in absolute value (which made the integral required for social welfare calculations always infinite if we started from zero—and hence we adopted arbitrary ranges over which to work). It was also in part because we had no information about demand at very high prices; use of the area under the full demand curve to measure consumer benefits yielded unreliable results. Looking at changes in that integral typically involved only

374

regions where we were surer of the position of the demand curve.*

Leah and Uri shared the position that levels might matter in negotiations because parties may compare their own benefits to those of others and then argue on equity grounds. Of course, the fact is that, whatever the shape of the demand curves, such inter-personal or inter-party comparisons are illegitimate as a matter of theory. But I knew that would be hard enough to explain to Leah and Uri, neither an economist, and far harder to explain to negotiating parties.

Research also continued on agriculture and AGSM. Ilan proposed and received funding for a joint project that would look at long-term effects and the possibility of using agriculture as a buffer in dry years. Further, there was work being done in Jordan. Amer Salman and Emad Al-Karablieh, two young Jordanian agricultural economists who had worked on Phase I of the project, had begun to write papers and then send them to me for comments and further work. This had already happened twice: once with a model for predicting wheat yields in Jordan given information only on rainfall in the early months of the rainy season; and once with a use of AGSM for the Jordan Valley that explicitly provides for inter-seasonal and inter-district transfers of water. They planned to continue. When they sent me the first paper and asked me to be a co-author, I replied that I would give them as much help as I could but that I did not believe it appropriate (as it evidently is in the natural and biological sciences) to put my name on their work just because it was done in my laboratory, so to speak. I had, however, put in more than enough work on both papers to make co-authorship acceptable.

Amer and Emad were smart and eager. They were also young, somewhat insecure, and not very well trained or experienced. They had a tendency to throw in every technique they'd studied, whether appropriate or not, and a tendency to make large, and sometimes incorrect, claims.

---

* Certainty here is a relative thing. In the summer of 1999, when we were doing the runs in Delft for the Minister of Foreign Affairs, Hans convinced me that we should increase the (absolute value of the) elasticity of Palestinian household demand from 0.2 to 0.7 because the lower value implied very implausible spending on water. A review of the literature by Irena Goldenberger, then my undergraduate student, showed that nearly everyone agrees that household demand is inelastic but that there are few numerical results. Indeed, much of the literature cited project papers for that proposition.

In the case of the AGSM extension, I was extremely busy following my first set of comments. This caused me to delay commenting on their response for several weeks and then to be very anxious to provide a fast turnaround. As a consequence, I failed to realize that we should have sent the paper to Ilan at an early stage. When we did send it, Ilan, who tended to be quite sensitive, complained that all the new developments were already in AGSM. His tone offended Amer and Emad. I made peace by pointing out that those developments were not explicit in AGSM, even though Ilan could easily see that the AGSM methods would permit them, and that we should moderate the tone of the paper so as not to claim that we were correcting defects in the AGSM model.

During this period, I was glad to have my own research money thanks to the Carlton Chair. I used it to pay Annette and for travel expenses.

Another project was also a possible source of funds. For a few years, Carl Steinitz, a professor of landscape architecture and planning at the Harvard Graduate School of Design, had been doing a project on the upper Jordan Valley, involving Jordan, Palestine, and Israel. The project concerned land use and Steinitz had brought in Annette and had used our project's models, principally AGSM. Steinitz then proposed to widen his project to include Lebanon and Syria. Unfortunately, he had no entrée in those countries, but had sent proposals to the MacArthur and Ford Foundations that would have included a large role for us had they been accepted. Eventually, though, the plan was dropped altogether.

Meeting with Larry Summers

In December, Zvi Eckstein suggested that I take advantage of a network of particularly well-known and influential economists, all of whom were old friends. I consulted Elhanan Helpman, a greatly admired Israeli economist associated with Harvard and Tel Aviv universities. But the main suggestion was that I should speak with the man Zvi called "the world's most powerful economist"—Lawrence Summers, an old friend, former student and colleague, and then the U.S. Secretary of the Treasury.

It took some time to arrange an appointment, but in early March, I went to Washington for a meeting at Larry's office that also included Adnan Kifayat, Treasury's Middle East person who set up the meeting, and David Fischer, a son of my old friends Stanley and Rhoda Fischer, who was working for Larry. I have known David from birth.

I had sent the relevant papers and made two points at the meeting. First, it appeared water would be a big issue in the stalled Israeli-Syrian negotiations as well as in the Israeli-Palestinian ones. The day might come when the parties asked the United States to buy them out of the problem by financing large desalination plants. When that happened, the United States, and especially the Treasury Department, would have an interest in being sure there had been rational investigation of all other alternatives.

That point met with easy agreement and, indeed, activity along those lines began shortly thereafter. In May, it was reported that President Clinton, in a meeting with Prime Minister Barak, had indicated a particular U.S. interest in water. Barak responded by asking for desalination plants at a cost of well over $1 billion. (I did not know how far that got or whether Treasury played any role.) At Lenny Hausman's suggestion, I asked whether the president might be interested in promoting global, not just Middle East, water cooperation. Larry told me he didn't think that would happen.

The atmosphere of the meeting was extremely friendly, and we discussed various ways in which Larry and Treasury might be helpful. Larry said he would speak favorably of the project to his colleagues—particularly Dennis Ross. I commented that Ross, who wanted to get *some* agreement without much regard for its content, had never seemed very interested. Larry said he hoped I wouldn't be offended by his response.

"Ross," he said, "cannot tell the difference between you and Jay Forrester. But he knows he can't and he knows I can."

The reference to Forrester was quite amusing. Then a retired MIT professor who had gained considerable influence, he had invented the magnetic core memory for computers, and was the father of "system dynamics," a form of computer modeling that he extended into "industrial dynamics," "urban dynamics," and beyond. His modeling technique (to give a somewhat unfair description) involves building non-linear models with feedback loops and then seeing whether the resulting dynamic behavior somehow fits the data. And therein lies the "joke" part of Larry's response. Forrester's technique involved little serious fitting to data, and many academic economists regarded him as something of a charlatan.

We also discussed working with the World Bank. I said that I had had some communication with the World Bank over the years, but that it never seemed to go anywhere—perhaps because I had been forced to keep a low profile during the Netanyahu years. Larry promised to intervene. Eventually, probably thanks to Larry, I had a promising conversation with Jeff Goldstein, the Bank's managing director, which appeared to produce interest in cooperation from John Briscoe, the water person at whose conference I had spoken in 1995. But Briscoe and I had yet to meet again. I planned to tell him that this couldn't wait for a time when I happened to be in Washington, as he suggested, and that I was willing to make a special trip.

Second World Water Forum and the Multilaterals

One immediate result of the Summers meeting occurred at the end of March 2000. I was in a small session at the Second World Water Forum in The Hague. The session was essentially focused on the project, with papers by Annette, Anan, and me. There, I said hello to Chuck Lawson, the scientific head of the U.S. multilateral delegation on water. I had seen him twice before: once in 1994, in his State Department office, when I gave him and some colleagues a general report; and once at the conference at the Technion in 1997. He had listened politely on the first occasion and had nothing to say to me on the second. Now he greeted me with the remark that Treasury, which was seriously interested in my project, had called him.

The occasion of The Hague conference was also noteworthy for a few other reasons. It provided an occasion to see Louise Anten, who remained devoted to the project even if it was no longer her responsibility. And it afforded an opportunity for Hans, Annette, and I to meet at the Foreign Ministry with Norbert Brackhuis, who had replaced Gerben De Jong as the deputy head of the appropriate section, and with Cees Smits Sibinga, who had replaced Louise as the officer in charge of us. He was very pleased when I greeted him in a few words of Dutch, left over from my year in Holland in the early 1960s.

I reported on the various events in January and on my meeting with Summers. Brackhuis appeared very energetic, and I hoped he would continue to be proactive. He was hopeful that Ali Sha'ath would sign off on the Interim Phase within the next week when the negotiations over the Port of Gaza were completed, but that turned out to be overly optimistic.

The remaining events in The Hague centered around the Multilaterals session, at which the various representatives put a good face on the fact that the Multilaterals had done essentially nothing. When the time came for questions, the first speaker was Hillel Shuval, who had spoken at MIT a week before. Though typically unrestrained, he stated that I had presented an important paper and model the day before and demanded to know why the Multilaterals were not using it. He was followed by Anan, then on the Palestinian Multilateral team, and then by a Jordanian, Oded Jayyoussi (a cousin of Anan) who made the same point in a more restrained way.

However flattering that was, none of it seemed to me to be politically wise. I explained both to Chuck Lawson and to Ram Aviram, the Israeli speaker, that I had not prompted any of those spear-carriers.

Incidentally, Oded Jayyoussi had recently asked for a version of the model for academic purposes. When we discussed his request at the Conference, he explained, understandably, that he wanted a version with the Jordanian data. I said we could not provide such a version without the permission of Hazim, who did not attend the Conference. An exchange of emails with Hazim did not secure such permission, Hazim claiming that even he did not yet have a working version for Jordan—something I found difficult to understand. There had been substantial delays and a repeated failure of Salem Hamati to install a working version on Hazim's machine, which seemed suspicious, but Annette had visited Jordan in January and had installed it herself. In any event, I had to abide by Hazim's wishes.

After the session, Uri made a point of introducing me to Noah Kinarti. I did not handle that particularly well.

"You look all right," Kinarti said to me. I now realize he probably meant I didn't have horns and a tail, but at the time I didn't recognize his mild peace gesture and simply responded, "You look all right, too." And when he said, "No, I mean it," I stumbled over my own words. Generally, I found our very brief comments uncomfortable and perhaps unfriendly.

Uri said that he thought the exchange was all right and that Kinarti would not have talked with me at all had I requested a meeting. He also said I was unlikely to be regarded with affection by someone who thought (quite correctly, I may add) that I was out to get him dismissed from his job.

The Conference also provided another opportunity to talk with Karen Assaf. Later, in May, Aliza Mazor at the New Israel Fund forwarded to me an email from the Edberg Foundation in Sweden asking for nominations for the Edberg Fellowship, of which I had not been aware. They were interested that year in water and the environment and were looking for younger female candidates from the Middle East or North Africa. After corresponding with Karen, I strongly nominated her, pointing out that she was such a great candidate that they should interpret the "younger" requirement very broadly. (Karen was probably in her fifties). I was delighted when she told me that, were she to get the fellowship, she would like to put in her fellowship year working on the project.

It was doubly rewarding to nominate someone who really was a superb candidate and with whom I would earn some additional good will. She appeared to make the short list; I never heard anything further.

One thing I had hoped would happen at the conference did not: a meeting with Bassem Awadallah. Lenny and I had been attempting to meet with him to discuss the question of King Abdullah of Jordan's possible sponsorship of a Regional Water Authority. He hadn't been able to see me when I was in the region in January, and I thought he might attend the conference. But he did not. He then offered to meet me in late May when he would be coming to the United States with the King. We scheduled a meeting in Boston, but he did not show.

When we finally made contact, he explained that it was only after he boarded the plane that the Jordanian mission was informed that the trip would not include Boston, for security reasons. By then, it was too late to call me. He was very apologetic. We next tentatively sought to meet a week later in Washington, but, as Bassem had feared, he did not stay long enough for that to happen.

I then proposed coming to Amman during the last week of June, taking a day from my planned attendance at the semi-annual board meeting of the New Israel Fund in Jerusalem. It seemed Lenny would also be in the region, and I thought it important that he attend the meeting with Bassem, since he knew Jordan better than I and really wanted to push for action there. Bassem agreed, and we made an appointment for Tuesday, June 27, at 9:00 a.m. Then ensued a series of events, large and small, that interfered.

The week before the meeting, we learned that Lenny's plane to Tel Aviv would not arrive until late in the morning on our meeting day, so we switched the time to late in the afternoon. Then Lenny's travel agent moved him to a plane arriving at 5:00 p.m., and so we proposed to Bassem that we come to Amman that night and meet the following morning, Wednesday, June 28. It would be difficult, because we had a meeting back in Tel Aviv at 2:00 p.m., but Bassem seemed agreeable. However, he said he could not confirm the meeting until Monday, June 26, as he would be out of the country until then. He also indicated that he wanted to consult the new Water Minister (the Jordanian government having just changed, as it does periodically) as well as Hazim El-Naser.

By Monday, I was in Israel. Lenny sought confirmation from Bassem for our rescheduled meeting, but he simply did not return Lenny's calls—something that surprised Bassem's secretary. By Tuesday, Lenny was en route, so I had to keep calling Jordan while traveling with a New Israel Fund tour. First, I was first informed that Bassem was at a meeting at the Foreign Ministry and then, eventually, that our meeting could not take place because he would be out of the country until the following Sunday, by which time, as we both knew, I would be back in the United States.

I am not sure what happened. The simplest interpretation is that the change in the Jordanian government put serious demands on Bassem's time and schedule, making the meeting truly impossible at the last moment. An alternate view is that the change in government and the appointment of a new Water Minister who (so far as I knew) knew nothing about the project made this a bad time to meet. If so, my experience with both Jordanians and Palestinians is consistent with the view that Bassem may not have wished simply to cancel, so instead put us off with "scheduling problems." Of course, our own repeated schedule changes did not make things any easier. I emailed him stating my hope that we could meet at a future, possibly more appropriate time. I trusted he could read between the lines.*

Trilateral Gains From Trade

In preparing for the meeting, however, Annette and I discovered a new phenomenon. That often happened when we did runs testing something we had not previously investigated. Those were typically bugs in the model or other difficulties, but on this occasion we discovered something we should have expected—and that was of some potential importance.

---

* In early July, I saw a copy of an email from Bassem to Lenny apologizing, asking that his apology be conveyed to me, and stating that he hoped to meet, perhaps in the United States, that September. That did not happen, either.

I had thought it would be useful to be able to give some illustrative examples of what Jordan could expect to gain from cooperation in water and trade in water permits (that is, permits to use another country's water). So, we did some runs involving cooperation between Jordan and Israel. Moreover, since the purpose of the meeting was to push for Jordanian support for a Regional Water Authority, we did some runs involving trilateral cooperation among Israel, Jordan, and Palestine. It was the first time that we had done such runs, always before concentrating on bilateral cooperation between Israel and Palestine.

We did runs for 2010 assuming (for illustrative purposes only) that each party owned more or less what they had at the time. The results showed very high gains from bilateral cooperation between either Jordan and Israel or Palestine and Israel. In both cases, the Arab partner became a buyer of water permits and got most (but, of course, not all) of the gains. When we did a run with trilateral cooperation, the *total* gains were, of course, even higher. But, as compared with bilateral cooperation, Israel gained from trilateral cooperation, but both Jordan and Palestine lost slightly.

A little thought showed that this was to be expected. In these runs, Israel had water permits to sell, while each of the Arab countries wanted to buy. When both such countries were included, there was competition for water-permit purchases from Israel, as it were, and the shadow prices at which trade took place were higher. This made the seller better off and each of the buyers worse off than in a bilateral arrangement. (Of course, there could be a parallel situation with two sellers and a single buyer.)

This meant that when a Regional Water Authority with n participants is considering the admission of an $n+1^{st}$, with $n > 1$, it need not be the case that all existing participants find that admission in their immediate interest. However, the fact that both the new participant and the existing participants considered as a group gain from the admission means that there will always be a way to redistribute the gains so as to remove the opposition of the (otherwise) losing incumbents.

In the case examined, the new entrant would gain so much and the losing incumbent lose so little that it would be trivial for the new entrant to pay a comparatively small fee to accomplish that end. In principle, however, the gaining incumbents might also have to contribute to the compensation of the losing ones.[1]

More on My June Visit to Israel

Having failed to meet with Bassem, I focused during my trip to Israel in the last week of June on the two primary reasons for the visit: to attend the semi-annual meeting of the board of the New Israel Fund; and to speak at an antitrust conference arranged jointly by Applied Economics, Zvi Eckstein's consulting firm, and my own firm Charles River Associates, which were forging a relationship. While in Israel, I had two other meetings, both promising, and one potentially very important.

On Monday evening, June 26, I had a long talk with Alon Liel, who was then the Chief Scientist of the Peres Ministry. He also claimed to be the originator of the idea of importing water from Turkey, an idea then under very active consideration. When I spoke of my (then projected) visit to Jordan, he asked me to inquire whether the Jordanians would like to cooperate in such imports. He was well aware that the efficient thing to do was not to pipe the imported water to Jordan from the seacoast but rather to sell water from the Sea of Galilee to the Jordanians, replacing it with imported water. We discussed that the model could reveal the efficient price at which to sell (or buy).

Alon seemed confident that, as the example suggests, a number of practical uses of WAS to guide cooperation in water could be found, provided the Interim Phase and the project generally started up again.

"We need meat," he said. He also observed that he thought we could "get around Kinarti."

Earlier that day, Eliezer Yaari, the Executive Director in Israel of the New Israel Fund, had told the NIF Board that he had seen Shlomo Ben-Ami at breakfast. Shlomo, a former NIF Board member, had become the Minister of Justice and the leader of the secret negotiations with the Palestinians. He had told Eliezer, "We have a deal."

When I mentioned this to Alon, he observed that the difficulty was that "Shlomo runs very fast, but the question is whether the rest of the government would catch up." Indeed, not long thereafter, it appeared that there was no deal, and the Barak coalition began to fall apart, with Barak departing for Camp David on July 10.

I also pointed out to Alon that water had disappeared as an issue mentioned in the media for the negotiations. He said he did not believe that there was an Israeli-Palestinian agreement on water.

My other meeting was with Uri Sagie, on Wednesday, June 28. He was an ex-general who was in charge of the Syrian negotiations. Lenny and Baruch Levy, who had arranged the meeting and had told us Sagie wanted to see me, were also present, as was Nachum Mintzker, a water expert known to and well regarded by Shaul Arlosoroff. We met for an hour in the lobby of the Dan Hotel. I found Sagie to be open-minded and intelligent, and I do not think that I held that opinion merely because of the outcome of the meeting.

It began with me offering an explanation of the project and its implications for water negotiations with the Syrians. Sagie then asked Mintzker for his comments, which were critical but were also of such a nature as to reveal that I had, indeed, over-simplified. Issues brought up included the need for a multi-year analysis, the fact that a single model for the entire country or region would not suffice, the necessity for dealing with water quality and hydrological uncertainty, and, especially, the need to be able to examine changes in conditions, for example, in population size.

I pointed out that we had either handled or had plans to handle each of these, and the discussion turned quite friendly. Sagie very quickly suggested that each negotiating party could use the tool (a better word than "model," which tends to frighten non-economists—as I should have remembered) to examine the consequences of proposed agreements— "what-if" games. That, of course, was one of the basic ways to use the project in negotiations.

Sagie was concerned, of course, with issues of pollution under Syrian control of water. He also pointed out that the Banias, which I had used as an example, was not the only issue; the Syrians also wanted to discuss the other sources of the Jordan, the Hasbani, which rises in Lebanon, and even the Dan, which rises in pre-1967 Israel but emerges from an aquifer that begins in Lebanon. He also observed that the other side often seemed more interested in stating ideological positions than in rational discussion. The death of Hafez El-Assad, he thought, provided a new opportunity and a breathing space.

Despite this, Sagie plainly thought we could be useful in the way described. He asked me whether I would help and indicated that we would have to receive confidential data. I could barely conceal that I wanted to do this with every fiber of my being, and pointed out that if the data were too confidential to be shared with an American Jew (and I probably did not need to see the data directly, anyway), he could well rely on the Israeli project team, principally Uri Shamir, to make experiments for him.

In doing this, I pointed out that Uri might feel under some constraint in view of Kinarti's position. Sagie smiled and said he was higher up than Kinarti.

"We will make him an offer he can't refuse," he added.

The meeting ended with me giving Mintzker some of the project papers, principally the Fishelson Lecture, and our agreement that they would get in touch with me with questions. I made it very clear that I would come any time, that I don't get paid, and that this is what I do.

It was the closest we had come to the negotiations themselves. I hoped something would come of it.

Alas, the situation was about to deteriorate.

As already discussed, Israel was the first to sign on to the Intermediate Phase, agreeing even before the change in government to that of Ehud Barak. I remained on good terms with Ben-Meir until someone, probably Dreizin, warned him that the project also cared about Arabs. Jordan was next, waiting until after the Barak government had been formed. Last were the Palestinians, who had wanted it most. They negotiated over funds with the Dutch, and finally received from the Dutch the nearly $200,000 they felt owed by Harvard. Then the Palestinians finally signed on.

That signing took place around the beginning of September 2000, just in time for the Second *Intifada*.

VI. 2001 to August 2016

27

Moving On

With only one exception—our opportunity to engage the World Bank—the project came to a virtual standstill for months. Anyone who remembers what happened in the Middle East beginning at the end of September 2000 will understand much of the reason. The collapse of peace negotiations, the return to violence in the form of the Second *Intifada*, and the election of the Sharon government were all causes for despair.

But I was still alive, and so was the project—at least in some form. If no one else thought my dark joke during this period was funny, at least Uri did: "You see. We were right all along. The next war was not about water."

I was determined not to give in to despair, however, and I cherished the heartening moments during this period when my friends and collaborators, Israeli and Palestinian as well as the Dutch government, displayed continuing interest. The writing of a book about the project was progressing, too, with very interesting additional results. And then there was the very, very pleasant fact that, thanks to the efforts of Eytan Sheshinski, I was to receive an honorary doctorate from the Hebrew University on June 3, 2001—and the citation would mention the water project. I looked forward to that; it would be a great occasion.

Something that occurred in February 2001 also improved my mood. I was grandchild-sitting the two sons and two daughters of my son Abraham and his wife Colleen. At the time, they ranged in age from two-and-a-half to eleven. It was early morning, and I was responsible for breakfast—the preparation of which is not high in my skill set. So, I took them across the street to Henrietta's Table in the Charles Hotel, Cambridge.

As we waited for the food, I told my grandchildren that the restaurant was where people came for "power breakfasts." Naturally, they asked me to explain what that meant. As I began my attempt to explain, Teddy—then six and a half—asked me to take him to the men's room, which I obliged.

As we returned to our table, I saw coming towards us one of the great "power breakfasters"—Nabil Sha'ath. It was a complete coincidence. Several other Palestinians, whom I assumed to be mostly bodyguards, accompanied him. We greeted each other, and he expressed the hope that we would again get back to dealing with cooperation in water. Nabil then asked to be introduced to my grandson.

When I complied, he bowed very low and shook Teddy's hand.

At this point, Beth, the oldest grandchild, appeared, and so I introduced her to Nabil as well. He bent even lower and kissed her hand. This charmed her beyond belief. When I later explained who Nabil was, she resolved to take her hand to school for "show and tell."

Beth has never forgotten this encounter. Many years later, she asked me again to explain who the man who kissed her hand was.

While his interactions with my grandchildren were charming, more important was the fact that Nabil himself brought up the water project, saying that it was still a matter for discussion at home and that they looked forward to the day when it would be possible again to proceed—a day he thought would come before too long. I reported this to the Dutch, who were heartened, but wanted to get the other countries to agree before moving on.

Regretfully, I have not seen Nabil Sha'ath, still an important Palestinian statesman, since that morning.

Of course, I never heard again from General Sagie. I prompted him in mid-July to no avail. By then, Camp David was the focus for everyone. That and its aftermath swept Syrian negotiations into the back seat.

The World Bank

In mid-September 2000, Annette and I went to the World Bank and made the lead presentation in a workshop on water conflict resolution. The opportunity was thanks to Larry Summers. The seminar was extremely well attended. Among others, Harsh came.

One of the other presenters was Dale Whittington, an economist from the University of North Carolina who had been working on the Bank's Nile Project for two years. He was thoroughly familiar with our work and very appreciative of it. In fact, his work on the Nile Project had been along similar lines, although with less powerful tools than we had developed.

The other important players at this meeting were John Briscoe, then the Bank's chief water person, and David Grey, who was in charge of the Nile Project and whom I first met in 1994. At that time, he was with the United Nations Development Programme and had attended the Cyprus conference. While it appeared unlikely that we would work on the Nile Project—there was no real discussion of that as such—joint work with Whittington certainly seemed a possibility at some point.

Most of the discussion as to next steps was of a wider nature. There was general agreement that work on international waters lacked defining methodological principles and that our work could provide them. The Bank group wanted to do this internally in a training program for their people, especially project managers, as well as externally. For the latter, the talk ran in terms of paralleling the Dublin Conference by convening a group of 30–40 important people and exposing them to a discussion of ways of thinking about water primarily along the lines of our project work. The idea was to make our approach just as generally known as people know that water is an economic good.

It was decided that the Bank would promote our way of looking at water conflicts as a fundamental way to think about water. They would train their people and push this at international meetings. David Grey was tasked with writing a memorandum memorializing that decision and suggesting next steps. While it remained to be seen what would actually happen, Annette and I went away very pleased.

The memorandum, however, never appeared. In November, an exchange of emails between Grey (who was often away) and me brought word that the Bank viewed the occasion to act as being the "Albright Initiative" conference on international water conflicts, to be held in Germany in March. In fact, both Grey and Briscoe later acted as though they had never agreed to anything with us.

Meanwhile, I proposed to the Dutch government that they should push and take credit for our way of resolving water conflicts, expanding their focus beyond the Middle East. This led in February 2001 to an exchange with Bert

Diphoorn, whom I had apparently met previously, and who was one of the organizers of the previous year's Hague World Water Conference. Bert was very receptive and proposed to bring this up at the Albright conference on March 12.

Bert also he mentioned Aaron Wolf, a geography professor at Oregon State University, who also consulted to the World Bank on the interaction between water science and policy. At Annette's suggestion, I asked Uri to speak with Wolf; serendipitously, they were meeting that very evening.

Later, Bert reported that the discussions at the conference were at too high a level of abstraction for the subject to arise. I suggested again to him that the Dutch should move forward, but received no response.

As for the World Bank, I emailed both Briscoe and Grey asking what, if anything, they now proposed to do. I received no answer.

At the beginning of May 2001, however, I received an invitation to a May 15 conference on Middle East water conflicts jointly sponsored by the World Bank and the Center for Middle East Peace and Economic Cooperation, an organization with which we had dealt over the years but that never seemed very interested in us in a substantive way. Speakers were to include Munther Haddadin, Aaron Wolf, Sharif El-Musa (the Palestinian-American poet and environmental scholar), and John Briscoe himself. I tried reaching Briscoe to ask what was going on, but he was away.

All in all, this was strange behavior on the part of the Bank's people. My only hypothesis was that they originally agreed to meet because of pressure from above, starting with Larry Summers. With Larry gone from the government, that pressure was off, and they didn't want to do anything. But I do not really know whether that was true. Their enthusiasm when we had met in September certainly seemed genuine.

There was certainly enthusiasm elsewhere. Hossein Askari, a professor at George Washington University, had been a thesis student of mine some thirty years earlier. We had remained in touch, although not often. He had come to the October 1999 conference and dinner that the MIT Economics department had given in honor of my 65th birthday, and was the discussant for Uri's paper describing the project—which he called my "most important legacy." That enthusiasm persisted.

In December 2000, Hossein—who was born in Iran—sent me a paper, done jointly with a student, that described the water situation in the larger

Middle East and urged the World Bank to move forward on a regional water cooperative system along our project's lines. He then combined a much-shortened version of his own paper with material from our principal project paper (essentially the Fishelson Lecture) and submitted it under both our names to an International Monetary Fund journal, where it was accepted.[1] The paper also urged the World Bank to move forward.

Hossein was active in other dimensions as well. He corresponded with a Lebanese minister, spoke with representatives of the Iranian government, and had contacts at Bechtel, the huge international engineering firm. He and I also applied jointly for a "development grant" from the World Bank that would finance his and my activities going forward. He was a tireless source of energy, determined to make things happen.

Bilal Zia, from Pakistan, was another enthusiastic new player. A master's candidate in Urban Studies at MIT, he came upon the project through MIT, became actively interested in using and spreading the methods, and began to work for me as a research assistant. Bilal wanted to build a model to examine issues involved in the construction and operation of the Kalabagh dam in Pakistan. Essentially, he used a WAS model to estimate the losses to water users due to the dam and then asked whether the dam was worth building and, if so, how much of the profits from electricity generation should be used to compensate the losers. In 2001, Bilal won MIT's Carrol Wilson Research Award and used the $5,000 grant to go to Pakistan that summer and gather data.

Meanwhile, Annette continued to work on the project part time, with both her and Bilal financed with funds from my chair. She did find full-time employment, joining the Tellus Institute—which worked on similar problems, sometimes with some of the same players as we. Later, she moved to freelance consulting and then to the position of senior scientist at the Stockholm Environment Institute (SEI), as well as other positions with SEI. In later years, she continued to work on the project and became both an invaluable colleague and, generally, my co-author.

Bilal's official status at MIT was something of a problem. He wanted to become a Ph.D. candidate in the joint program the Economics Department has with Urban Studies. I was informed that admission to that program, so far as the Economics Department was concerned, was typically entirely up to the chair of our Admissions Committee—who happened to be me. But I was unable to broker a deal in which Urban Studies and Economics would share fellowship costs for Bilal, and when Urban Studies refused even to admit Bilal to the joint program because was "too interested" in economics, I arranged for him to do his Ph.D. in the Economics Department. After graduation, Bilal went on to work in the World Bank's Development Research Group.

Modeling Prospects Elsewhere

We had so far been unsuccessful in marketing WAS models elsewhere in the world or making our ways of conflict resolution standard practice. Despite an agreement to work together, neither Charles River Associates nor Delft Hydraulics were very proactive in this regard. Delft Hydraulics, in particular, appeared to have done nothing and to be very reluctant to share in any marketing expenses. Given that situation, it is probably no surprise that all activity came from Lenny Hausman.

First there was Singapore and environs. Lenny had spoken S.K. Chak, a Singaporean businessman Lenny employed as he was building a firm to do university education on the Internet, primarily for developing world customers. He proposed that we retain him as an agent to sell projects in his area. Jim Burrows, the CRA president, Brian Palmer, the CRA vice president in charge of water activities, and I had discussed this at an earlier time and agreed to retain Chak for six months, partly on contingency, but through some slip-up, Chak never received our emails—a fact we only discovered somewhat later.

Lenny also suggested that I ask Lester Thurow of the MIT Sloan School of Management whether he could arrange for me to lecture on water in Singapore; he had given a named lecture there at some time in the past. Lester obliged, but the upshot was an email stating that Singapore already had a contract with MIT to study its water issues and that I could work through that.

That news, of course, surprised me. About two years earlier, Professor David Marks of what was then called the MIT Center for Environmental Initiatives, and Professor Fred Moavenzadeh of MIT's Department of Civil and Environmental Engineering had come to me and asked me to participate in a project on water they were proposing to the Singapore government. I was later told that the Singaporeans had decided to wait on it. Now I was learning that it had gone forward without me. Despite two very polite requests to discuss the matter so that I wouldn't trip over them again, neither Marks nor Moavenzadeh responded after their first email informing me generally of the situation.

I was beginning to feel like the guest no one wanted to have. It appeared easier for water people to ignore my existence than to deal with it.

Lenny was even more active in Kuwait. Beginning in late January, he had been speaking with Sheikh Ahmed Al-Sabah, a man in his late twenties whose father—according to Lenny—was effectively the head of the Kuwaiti government. Sheikh Ahmed was investing in Lenny's Internet education company, but also wished to endow, with $25 million, a center for the study of the problems of small countries. At first, Sheikh Ahmed had thought of placing his center at Harvard, but Lenny convinced him to do it at MIT. Lenny also pointed out that water was one of the big problems and advertised me as the one who could solve it.

Apparently, Sheikh Ahmed was very responsive. He invited Lenny, Phil Khoury, then MIT's Dean of the School of Humanities and Social Sciences, and me, on very short notice, to join Lenny and come to Kuwait to attend the celebration of the tenth anniversary of the country's liberation from Iraqi occupation. Neither Phil nor I could make the trip, but Lenny went. There was then discussion of a trip for me in middle- or late March to meet with water people in Kuwait. But the visa never came, and the trip was postponed at the last minute.

The Sheikh then proposed to come to Cambridge in early April to meet with Phil, Lenny, and me, with the idea that we would take a trip to Kuwait very soon thereafter. The date of the Sheikh's arrival in Cambridge, however, kept moving. He came on April 19, by which time Lenny and I had agreed on May 6 to go to Kuwait. But that day the Sheikh broke appointments with both Phil and Dwight Perkins, a Harvard economics professor, and then failed to show up for lunch with Lenny and me the next day, without even a phone call. When Lenny finally reached him the next day, Sheikh Ahmed explained that he and his wife had spent two anxious days at Massachusetts General Hospital where she was tested for a brain tumor (which, fortunately, she did not have). Lenny told me the Sheikh would call me the following Monday to apologize—which he didn't.

I canceled the May trip. The Sheikh was said to be very serious about MIT and would return in June for serious discussions, but I decided I would believe it when I saw it. Meanwhile, I had several times that spring maintained an ages-long tradition of my family: I had not gone to Kuwait.

Getting Public Attention

As all of this unfolded, and largely didn't go anywhere, Lenny Hausman continued to push me to get more public attention for the water project. He has always been totally committed to the considerably exaggerated view that I have the world's solution to all its water problems. Believing that implies, quite incorrectly, that *I* alone am the project. For years, he urged me to go public in the popular press and perhaps elsewhere, believing it would catch the eyes of politicians. Doing so would have been a disaster during the period when we had to keep the project partly secret, but we had reached a point where such a constraint was no longer required. So, I began to consider that Lenny might have a point, if things were handled properly.

In early 2000, when it appeared that water might be an issue in the Israeli-Syrian negotiations, Lenny suggested I get in touch with Michael Weinstein of the *New York Times*. Mike, a former MIT graduate student, was an excellent writer of economics-based editorials, and we had talked many times over the years, usually on "background," about antitrust issues. Mike was quite interested in writing something about the project, but he was unable to interest his superiors in the editorial office or anyone at the *New York Times Sunday Magazine*.

Lenny's second specific proposal was that I should write an article for *Foreign Affairs*, something non-technical that focused on conflict resolution and the use of the tools for negotiations. It was an eminently sensible idea, but one I kept not getting around to it. When I finally did, the editors returned it without reading, stating they already had too many pieces in the pipeline.

Finally, Lenny consulted Peter Rosenthal of Rubenstein Associates, a major public relations firm. (I believe Rosenthal had been a member of the ISEPME board.) I traveled to New York City and met with Rosenthal and another officer of his firm on November 9, and I was extremely impressed. They believed they could place articles, get me invited to the annual World Economic Forum in Davos, Switzerland, and other such events—and without me feeling as an "also-came"—as I had at the Casablanca Conference in 1994—and generally make our way of dealing with water extremely prominent. In December, they proposed a six-month contract at $8,000 per month to begin the work.

I took the matter to CRA. The company was not having a very good quarter or so, and Jim Burrows, the president, was understandably reluctant to put up a lot of money with no direct commercial prospects. Delft Hydraulics absolutely refused. Since the Rubenstein activities would have benefits for the conflict-resolution parts of the project as well as for securing of model-building contracts, I offered to pay a substantial part of the costs. We eventually agreed to that, but wanted to see whether part of the fee could be on a contingency basis.

Unfortunately, we were unable to discuss this with Rosenthal, who was on medical leave for some months. We were told he was working from home. Repeated messages asking him to call elicited no response.

Publications

In addition to the possibility of an article in *Foreign Affairs*, I pursued several other prospective project-related publications. The most important of these was *Liquid Assets: An Economic Approach for Water Management and Conflict Resolution in the Middle East and Beyond*—a full-length, edited book.[2] After a long hiatus, I returned to my work on it, focusing on the Israeli chapter. Hans, who had left Delft Hydraulics to work at the Dutch Ministry of Transportation, wrote a draft of the Jordanian chapter, and Annette was to write a Palestinian chapter.

The work on the Israeli chapter was fascinating, but also difficult for two reasons. First, we discovered that the Israeli report showed there were some sources of non-potable water used in agriculture that had not been included in the runs the Israeli team made with the model. It took us several weeks to ensure that we had the issue and the data straightened out to correct that.

Second, the Israeli team had done an "administrative run" for 1995 in which they gave the actual water consumption of the various sectors in the various districts that were induced by the Israeli system of fixed-price policies. I decided to evaluate the social benefits lost by such policies compared to a system of pricing at shadow values. That was not easy, because the model was not set up to impose fixed water amounts. Further, the social accounting part of the program required some adjustment.

The experience revealed how difficult it is to specify every piece of a complex program exactly, but also showed how thinking about matters in a particular framework can lead to error if one is not careful regarding how that framework is used.

It was natural to think of social welfare outcomes in the model in the standard framework of consumer (or buyer) surplus, producer surplus or profits, and deadweight loss. The problem is that doing so in a simple way can lead to anomalies. When the capacity constraints on infrastructure (conveyance links, recycling plants, desalination facilities) are binding, the shadow value of the water produced or transported by those facilities will include the shadow value of such capacity constraints. If consumers buy at the shadow values of water, then their water payments will include charges for the shadow value of infrastructure capacity. That will naturally reduce their surplus.

But one must realize that, in the model, such payments were not automatically received by the sellers of water as an addition to their profits. Hence, in doing the social welfare accounting, one must remember that there are payments, as it were, to the owners of infrastructure. If one drops the notion of "profits" and thinks in terms of scarcity rents, this becomes very clear and the whole matter falls neatly into place. This was particularly true when we adopted the extremely useful device of printing out both the maximized value of the objective function and the sum of the social welfare calculations over the three parties. If these did not match, then we knew there was something wrong, and thinking about things in terms of the objective function would reveal how to fix it.

We had failed to notice that issue (except for conveyance links, where we had noticed it about a year or so previously) because we tended to run the model with no infrastructure capacity constraints. That can be a useful procedure, but close analysis of whether to build infrastructure (in the case of the Israeli chapter, whether to build an additional conveyance link to Jerusalem) required paying attention to capacity.

It is worth mentioning that the finding going back to 1994, that profits from recycling get allocated back to the producers of effluent, is not correct when one considers capacity constraints. If a constraint on the capacity of a recycling plant is binding, then what appear to be profits on recycled water will include the scarcity rent of that capacity, and that part of payments get allocated to the owners of the recycling plant. What does get allocated to the producers of effluent is the shadow value of the effluent production itself.

This is positive where the shadow value of recycled water is above its marginal cost of production (for agriculture) and the amount of effluent available to be processed, rather than the capacity of the plant, is the binding constraint in producing recycled water.

In any event, I was very happy with the draft of the Israeli chapter. We found that it cost Israel about $40 million per year in 1995 to retain the fixed-price system rather than charging (and subsidizing agriculture) in a different way. The beneficiaries of that system were mostly farmers in Israel's south.

Further, our results showed that, in years of normal hydrology, desalination on the Mediterranean coast, even at $0.60 per cubic meter, would not be needed even in 2020. It would take a substantial drought to make that technology efficient. In general, new sources of water would not be required, despite the popular view at the time that was influenced by years of low rainfall. The need for more water in drought was influenced by the inefficient subsidy of southern agriculture.

Incidentally, someone told me around the time I was working on the chapter that Israel's "failure" to build desalination plants was attributed to me. I supposed that was possible, but it might also have just been an "amusing" remark circulating among economists.

In addition to working on the book chapter, I revised the Fishelson Lecture for publication in Autumn 2000 and—largely on Uri's advice—submitted it to *Water Resources Research* (*WRR*), a premier water journal. I do not know why it had taken me so long to get around to doing that.

In the interim, a version of the paper had been published in Hebrew in the Israeli *Economic Review* in March 2000, with the shorter authorship list.[3] I had assented to its publication immediately after the Fishelson Lecture, without thinking. In honesty, I told the editor of *WRR* about this. He was quite concerned, even though I pointed out to him that the Hebrew piece would be quite inaccessible to most of his readers and that even most Israeli readers would not have seen it. He had me supply the Hebrew version and said he would arrange for a translation, after which he would ask the translator whether *WRR* should publish both versions. I pointed out to him that there was no doubt that the reply would be in the negative and that a better test would be to inform the referees about the Hebrew publication and ask them whether *WRR* should nevertheless publish the paper. Of course, I believed and hoped (and Uri agreed) that the referees would find the paper so important that they would answer affirmatively.

WRR had a peculiar custom as to referees. They permitted me to suggest up to five; presumably, they would also use others of their own choosing. After some consultation, I did this. I first thought of suggesting Munther Haddadin.

Meanwhile, before the actual submission of the English-language version of the piece, the issue of authorship arose again. When given in January 2000, the paper had listed Annette, Hans, Zvi, Shaul, Uri, and Anan Jayyoussi as authors in addition to me. Before the Hague Conference, I had asked Hazim El-Naser to decide about Jordanian authorship, but he had indicated he could not sign such a paper, presumably for political reasons. By then, the *Al-Aqsa Intifada*, the *Second Intifada*, had begun, and I was not surprised by his refusal once again. He made it clear, however, that I was free to ask other Jordanians, and so I asked Munther Haddadin (which, of course, would make him ineligible to be a referee), making it clear that I knew it might be difficult politically.

While I had no details, it appeared that Munther's political situation, even as a private citizen, was shaky, in part because he was identified with the peace camp. At the time, Munther was refusing even to be interviewed by Israeli newspapers because of the political situation. He told me that his home had been bombed with a rocket at Christmas, and that the police were refusing to investigate. But the article, he said, was a "scientific paper," and so he agreed to be named as one of the authors.

Anan also agreed, as did his colleague at An-Najah National University, Amar Jarrar, whom I added at Anan's suggestion. Munther also suggested adding Salem Hamati, who accepted.

The political situation, of course, affected everything. Anan emailed me, saying he was up for promotion to associate professor and wanted to know whether the paper had been published or at least accepted by *WRR*, as that would be important in his promotion case. I replied that the decision was still pending, and I told him about the publication of the Hebrew version. I asked him whether that would do him any good. He replied that, of course, it would not. "Can you imagine," he asked, "giving a Hebrew paper to a Syrian judge [referee]?"

I offered to write a strong letter on his behalf and Anan responded with his unequivocal hope that our joint work would go forward.

I ended up suggesting as referees Yoav Kislev (who then held the Sir Henry d'Avigdor-Goldsmid Chair in Agricultural Economics at Hebrew University); Jad Isaac; and David Grey. John Briscoe, and Dale Whittington, who had been at the World Bank meeting reacted positively to the project.

There were other joint papers as well, related to AGSM. Ilan and I had already published two papers in *Agricultural Systems*, the original AGSM paper and one using AGSM to show the pitfalls of using two policy instruments—prices and quantity allocations—to accomplish a single end, namely the restriction of agricultural water consumption.[4] In both, I supplied the economics after Ilan did all the work. It was surprising how matters that seemed completely straightforward to an economist came as a surprise to experts in other fields. I suppose that's true the other way around.

I followed my standard policy in agreeing to co-author these papers with Ilan. I told him I would advise and assist, but I did not subscribe to the practice that the senior professor automatically gets co-authorship of papers written in his lab, so to speak. Once the paper was revised and ready to be sent out, we could then decide whether to list me. In the case of the papers mentioned above, I did enough—or in the case of one of them, just enough—to be listed, although I insisted on being listed last.

The same was true of other papers I co-authored around the same time or later with Amer Salman and Emad Al-Karablieh, two Jordanian academics who worked on the agricultural part of the project.[5] One was a study of the ability to use rainfall figures early in the season to predict the ultimate crop of wheat and so assist with inventory and import decisions. Another was an extension of AGSM, or at least a use of AGSM, that permitted inter-district and inter-seasonal transfers of water.

In a second paper, there *was* an unnecessary quarrel to be avoided. Amer and Emad were young. As often happens, they were quite aggressive in characterizing their paper as an improvement on AGSM, correcting flaws in the original. Because I was fearsomely busy when reviewing the paper, I failed to tone down the language and also failed to require that we send it to Ilan before submitting it for publication. When Ilan finally saw the paper, he was quite offended, and this, in turn, offended Amer and Emad. I successfully mediated this affair by suggesting appropriate language.

Honorary Degree

I went to Israel at the end of May 2001 in advance of what I considered a great occasion: the Hebrew University was giving me an honorary degree on Sunday, June 3. The high point was not the degree ceremony itself, but the luncheon given by the Rector that day—at which I had been asked to speak on behalf of all the honorees. While the speech was not about water, the degree was given with my water work as a principal reason.

The preceding Thursday, May 31, Ellen and I had attended the wedding of Noa Sheshinski, Eytan and Ruthie's youngest daughter, at the Dolphinarium, a nightclub on the Tel Aviv beach. The next night, it was bombed—and 21 Israelis, 16 of them teenagers, were killed. Such bombings were rarer then than they are today. My speech—which I reproduce here—thus came at a particularly sensitive time.

It was a very emotional moment both for me and for the audience. For the next few days, people came up to me and said how much they had been moved, or told me how much they regretted missing the speech, about which they had since been told. Here is the speech that I gave:

> I have been asked to speak briefly on behalf of the honorees, and I am honored to do so. Please bear with me, for I shall speak in a way that appears to be highly personal. I assure you that it is, in fact, highly relevant and, as I hope, will strike themes that go beyond my personal experience.
>
> The Hebrew University is about to bestow a great honor on me and the others here, and we are very grateful. Indeed, in my case, for reasons that I am about to explain to you, I would rather have an honorary degree from the Hebrew University than from any other institution in the world.
>
> I have had a long and close association with this country, and with this university. Further, my wife, Ellen and I have made many very close friends here, including two former Hebrew University presidents—Don Patinkin and Yoram Ben-Porath—who are, sadly, no longer with us. But there are other reasons as well.
>
> Now I am going to do something which none of you expect and which will leave Ellen flabbergasted. I am going to talk about my father. For the honors that we are being given by Israel's first general university not only permit all of us to connect with the land of our ancestors in general, but in my case, with my father in particular.
>
> My father, like my mother, was very proud of me. But he was never quite sure what I did, exactly—possibly a very sensible attitude. On this occasion, however, he would be in no doubt as to how much to *kvell*, for

he would have seen me get a degree from an institution that he considered to be "his university."

Indeed, I recall that when my parents visited Israel in, I think, the late fifties my father went to the Hebrew University library and was very excited to discover that it had the issue of *The Quarterly Journal of Economics* with my first article in it. And make no mistake: the thrill was not at finding the Hebrew University's library to be so complete; it was at finding that his son had written something worthy of being found in the library of "his university."

Why did my father think of the Hebrew University in this way? He had six degrees, but they were all from American institutions. He never studied here. Indeed, he only visited here twice. But the feeling was there, nevertheless.

In his later years, my father was fond of proudly saying that he had been a Zionist before the Balfour Declaration. My father was a child prodigy, including apublic speaking, and I am sure that this claim was true, although I have been unable to find any official record showing that Dad's public pronouncements at the age of 14 influenced the British foreign office.

In any event, my father remained an early and staunch supporter of the Zionist cause and then of the State of Israel all his life. Both he and my mother were very active in that way before the establishment of the state and during the War of Independence. But the reason that I think of him tonight is because of an article that he published in 1932, when he was 29. That article appeared in a volume of essays in honor of Justice Louis Brandeis put out by *Avukah*, a Zionist organization. (I have some fondness for *Avukah*, for it was at an *Avukah* summer camp also in 1932 that my father and mother first met.)

The article is called "For Love of Alma Mater" and describes my father's undergraduate experience at New York University—NYU—in the early 1920s. In it he tells how his enthusiasm for NYU progressively died.

There are two major episodes that the article describes. The first concerns the NYU debating team. My father was a star of the team, but, when the team was invited to a debating tour of the United Kingdom, there were serious doubts as to whether it was appropriate for a Jew to participate as a representative of the university. Oxford University, in particular, had problems, and NYU itself was not about to rise to the defense of Jewish rights. Indeed, the NYU coach and some members of the faculty tried to remove my father from the team.

In any event, my father went, but was restricted in his debating role at

Oxford, where he was not housed with the other team members. I am happy to say that when the Jewish communities of Sheffield and Edinburgh discovered his religion, he received an excited and very warm welcome—and probably far better food than the other team members ever got.

The second episode involved the junior prom, a dance that was a major social event. The prom was not open to Jews. Indeed, it was put on by the fraternities, and the Jewish fraternities were not even informed of the date. My father set out to correct this situation, writing to the college newspaper in a letter that he had to struggle to get published.

For this, my father was publicly vilified; indeed, he was physically threatened. Two Jewish members of the football team slept outside his door to protect him. Most shocking, he was called in by one of the deans of the university who said to him: "my fathers came here on the *Mayflower*. They came here for religious freedom. Your people came here to make money." My father then informed the dean about the effects of the Russian May laws. But the prom was only opened to Jews years after my father graduated, and he went—the only occasion on which I know he was actually proud to dance.

And so? This is all very interesting, but why am I telling you about it? Certainly, my father's academic experience in the United States is not reflected by my own. Although I did briefly encounter some anti-Semitism as an undergraduate at Harvard, it had no effect on my career. One can hardly say in these times that being Jewish has been a bar to rising to prominence as an academic economist, either for me or for many, many others. Moreover, the American Jewish community is alive, vibrant, well, and in no danger.

I have told you this story because, despite these differences, I understand very well what my father meant when he wrote the last two sentences of his article. What he wrote now, I shall repeat. It expresses why we are proud to be with you today, here, at this university, on this mountain, at this moment in Israel's history.

Lifnay shishim v'teishah shanim, abba sheli katav b'anglit, aval, hayom, ani omer b'ivrit: "yesh li raq universitah achat.... Hee al har hatsofim."

Sixty-nine years ago, my father wrote in English; but, today, I say in Hebrew: "I have but one university. ... It is on the hills of Scopus."

It was a speech that I had, in truth, begun to prepare long before knowing I would actually be asked to make it.

The days and nights that followed the Hebrew University ceremony were ones of considerable anxiety, during which one was careful about where one went in Jerusalem. On Saturday evening, June 2, we had plans to dine with Rachel and Alon Liel, who were always more than pleasant to be with and always very supportive. When they picked us up at our hotel, they said that their children had said they should not go to downtown Jerusalem, just as the children had been forbidden. That was fine, since we had no intention of going downtown. The Liels took us to a restaurant in Abu Gosh, an Arab town outside Jerusalem that had remained loyal to Israel since 1948. Even there, the restaurant was fairly empty.

One day, I had lunch with Maarten Gischler, then serving at the Netherlands Representative Office in Ramallah and really just beginning to learn about the project. His interest had been aroused, and I believed he would become actively helpful. He confirmed that the Netherlands Ministry of Foreign Affairs planned to send a mission to the region in September to inquire as to the interest of the parties in continuing in some form.

While in Israel, I was also invited to something else to take place the following September. Alon Liel suggested my name to the organizers of a conference on the question of whether autarchy or cooperation was the way to think about water. Joyce Starr, who for years had been a leading proponent of the view that the next war (indeed, the next wars) would be about water, was one of the organizers. I prepared a paper on the question of water as a *casus belli*, or a source of cooperation, and planned to attend the conference, despite that my teaching schedule would mean having to fly in, go straight to bed, give the paper the next day, and fly right back.

A Flurry of Conferences

I had been scheduled to return to Israel at the end of June 2001, for the semi-annual board meeting of the New Israel Fund. But in late May I received a water-related invitation that conflicted with that meeting, and one that I decided took precedence. It was to attend a one-day water conference in Paris being arranged by the Friends of the Technion, featuring Uri's Water Research Institute.

The Paris conference was a pleasant occasion, and I was glad to see Munther. But two things were annoying. First, so far as the presentations were concerned, our water project might as well never have existed. One woman from the World Bank spoke about water shortage and stress in the old-fashioned, quantity-driven way, and predicted conflict over water. At least Kenley Brunsdale, of the Center for Middle East Peace and Economic Cooperation, who explained desalination, set her straight.

The second annoyance was seeing John Briscoe and asking him what had happened to the great plans for the World Bank. I got the runaround. He told me that my memory of what had been agreed was just wrong, and that people at the Bank were saying that I was somebody with a solution looking for a problem—adding that he had told those people I was broader than that and they should try to involve me more generally. I am still waiting for the call.

Paris was not the only conference during this period. I was invited to speak at several, often with frustratingly little effect. In November, the Middle East Center at the University of Oklahoma held a conference on water. The invitation to me first suggested a short talk, and I said I would not go for that. I ended with a featured dinner presentation, and I spoke on cooperation versus conflict and presented new results featuring graphs showing the benefits of cooperation and the value of shifts in ownership at different ownership allocations of the Mountain Aquifer. The results were visually quite striking, showing how cooperation was more valuable than ownership shifts and how cooperation greatly reduced the importance of ownership allocations. The proceedings were later published.[6]

The conference had invited a large number of the right people, including Munther and Jad. Unfortunately, no Palestinians actually attended, and Munther was called to accompany Prince Hassan to a trade meeting in Dubai. That was a good sign, suggesting that both Munther and Hassan were once again active in Jordanian affairs.

From my point of view, the most important attendee was Yoav Kislev, at whose suggestion I had been invited. He had become a strong supporter of our methods, and, indeed, had written to Ben-Meir some time earlier urging him—unsuccessfully, of course—to make use of our tools.

I also met Eran Feitelson for the first time. I had known him as active in studying and promoting joint aquifer management.

The Oklahoma Center proposed to form two working groups on water. One would include Israel, Jordan, and Palestine; the second would include Jordan, Syria, Lebanon, and Turkey. (I forget whether Egypt and Iraq were to be invited.) The idea was to involve Syria and Lebanon in a forum not involving Israel but with crossover because of Jordan's joint membership. I do not know if anything ever came of that.

In a session at which ideas and suggestions were discussed, Kislev proposed the use of WAS. I said that the Dutch would have to be asked, but that WAS, at the least, could be used as an organizing tool with which to evaluate various proposals.

Another invitation came from Bob Lerman, an MIT economics Ph.D. of many years earlier and a friend of Lenny's who was at the American University in Washington. He asked me to present the project at a session on water to be held at an upcoming conference of the Association for Israel Studies at the end of May.

Most interesting of all, I received an invitation to be the keynote speaker at a *general* conference on water organized by Ellen Wiegandt and others to be held in Switzerland in mid-October. Wiegandt was an internationally famous anthropologist who had become interested in environmental issues in the late 1980s and had done important work on water resources. I presented both a keynote talk with a general presentation of our project methods, and a research paper with detailed results for Israel and the Middle East.[7] It was a breakthrough. The organizers had apparently done some work in the Middle East, which is how they had heard of the project, and they had apparently realized that the tools had wider applicability.

During this period as well, the American Academy of Arts and Sciences was starting a working group on trans-boundary water disputes and invited me to be a member.

Meanwhile, not all was bleak on the World Bank front. In January 2002, I received an invitation to speak at a meeting of Friends of Ben Gurion University in New York. At first, I indicated reluctance to *shlep* to New York on short notice to speak for fifteen minutes, but Shawki Barghouti, a Jordanian at the World Bank, called and urged me to come.

Avishai Braverman, the president of the Ben Gurion University, was at the meeting. The university was opening a new water center under director Eilon Adar, who was also on the program. I had met him at the Oklahoma conference.

Apparently, I had met Barghouti on a previous occasion—probably when I spoke at the World Bank in 1995. He asked what had become of the project. He was a friendly guy and, when I mentioned my difficulties with the Bank, he commented that it was useless to deal with Briscoe, who he said would never accept anything not invented in his shop. Barghouti added that he was working in a different part of the Bank, under Ariel Dinar, whom I had met when Uri was visiting Harvard in 1998. They would arrange for me to come and teach for half a day about the project's methods—but that did not happen.

On the publications front, there was finally a decision regarding the paper submitted to *Water Resources Research* a year earlier. After considerable deliberation, William Gray, the *WRR* editor, decided that the original version of the paper that had appeared in Hebrew should not be published in *WRR*, but that he would publish a version with new material. So, we added results for Israel and for cooperation, as in the Oklahoma paper. Gray also wisely suggested that I seek the advice of Dennis McLaughlin of the MIT Civil Engineering Department, who was an associate editor of *WRR* and was someone I knew a bit. He was more than helpful as the paper went through two revisions, ultimately becoming much longer and far better than the original.[8]

28

Not Forgotten

The breakdown of the peace process and the outbreak of violence that continued from the end of September made it impossible to continue in the region. Even though each of the parties wanted the Interim Phase, none was willing to proceed—especially the Palestinians. Negotiations between Ali Sha'ath and the Dutch Foreign Ministry over acceptance of the Interim Phase, particularly Ali's demand for another $50,000, had finally made some progress just as the *Intifada* began.

But there were definite signs that we were not forgotten. When Hans Wesseling left Delft Hydraulics in late November, the Foreign Ministry sent him a letter (and copied it to me) in which he was thanked for his work on the project and was asked to assure me that when the time again came that it would be appropriate to talk with the parties in the region, the ministry would be there in support of the project. Confidence that such a time would come was also expressed.

Some reinforcement for that confidence came along in January, when Itzhak Levanon, the Israeli Consul-General in Boston, telephoned me. He had just spoken with Alon Liel, and was passing on Alon's message that he had just met with Herman Froger, the new Dutch ambassador to Israel, and had emphasized to him that Israel very much wanted to continue. It was the moment at which my entrée into the Israeli government was best.

Still, it was the case that water was far from anyone's thoughts about international issues those days in Israel, although it remained a constant domestic problem.

After my February chance meeting with Nabil Sha'ath in a Cambridge restaurant (described in the preceding chapter), I hoped he was right about to the possibility of continuing. But I was not content with mere hoping. I had emailed Cees Smit Sibinga and Norbert Brackhuis at the Dutch Foreign Ministry, reporting the encounter. Cees told me there was to be a meeting of EXACT, the technical group of the water Multilaterals, in Europe in early March, although the parties did not want it known it was happening. In view of my information, Cees proposed to use that as an occasion to sound out the parties regarding continuation of at least the Interim Phase of the project, which we were calling the Middle East Water Project (MEWP).

The response he got was favorable. The Israeli representative stated Israel's positive view, and, while the Jordanian representative knew nothing about the project, more than one of the Palestinians stated how eager they were to proceed. They were, of course, from the Water Authority, and I was quite certain one of them was Karen Assaf. Cees pointed out the difficulty the Dutch had experienced in negotiating with MOPIC, and the Palestinians promised to try to get around that. I would have felt more hopeful if they hadn't told me much the same months before.

Based on this meeting, the Dutch decided to move further. During my next trip to Israel, scheduled for June, I was to meet Ambassador Froger and with Maarten Gischler, the water expert in the Dutch Representative's office in Ramallah. Cees also told me the Dutch were planning to send a mission to the region in September to discuss all their water-related projects and that the future of the Middle East Water Project would be high on the agenda.

"From his and my mouths to God's ear," I thought.

The Renewal of Dutch Interest

I was scheduled to leave for Israel on September 11, 2001, for a very quick trip to the conference on autarchy or cooperation as the way to think about water. Of course, the attacks in the United States that morning made it impossible. I told my Israeli friends that I had decided to stay in a country safe from terrorism—a joke that was feeble then and seems more so now. Shaul spoke in my stead.

I assumed that the attacks, as well as what seemed like a major eruption of violence between Israel and Palestine, would also stymie the Dutch mission to the region in September to inquire about support for the project. But it only seemed to delay the effort, because on the morning of November 9, I had a major surprise as I sat down, still in my pajamas, to read my emails after breakfast. There was a note from Maarten Gischler.

Somewhat embarrassed, I opened the MEWP file and found the documents you sent me in June ... unread. This week a Dutch mission visits the Middle East, discussing prospects to revive old and develop new water projects. Good opportunity for me to dive into your documents!

The parties are interested to revive the MEWP. The Jordanians see WAS as a very useful tool for national (or regional) water planning but they had a blunt question: "what happens if Frank Fisher is run over by a train (or other). Is there an organization behind the WAS package? Or is there just Frank Fisher?" Without the backing of an organization, they are reluctant to use WAS as a tool for sector planning, which I can understand.

Can you help out?

I replied instantly.

YES!!! This is very, very good news.

The question the Jordanians ask is a very good one, even though I have no current plans to be run over by a train. There are several answers.

1. While I think it is currently the case that I am the person most highly skilled at interpreting and using WAS, there are others quite capable of doing so. In particular, this is true of Annette Huber-Lee (American) and Hans Wesseling (Dutch). Were there not political problems, I would surely add Uri Shamir (Israeli). Others who worked on the project are also capable, but not as much so.

2. My consulting firm, Charles River Associates, has economists quite able to do this after instruction.

3. MOST IMPORTANT (and this should be emphasized to the Jordanians), a basic point (possibly, *the* basic point) of the proposed Interim Phase is to transfer such ability to the participating governments. We would work to train Jordanian water experts in the skill required.

I am flattered by the Jordanian view that I am indispensable. This naturally matches my own view of myself. But it isn't really true now, and the idea is to make it much less so. The Jordanians need to "own" their WAS model themselves, and that has always been the plan.

I am excited that the mission has gone and that the Dutch government

411

both thinks it useful to send one at this time and still stands behind this project—as do I and, I trust, my colleagues and possible successors.

This was followed by message from with Cees Smit Sibinga:

Things seem to come together a bit at this moment. Not only did the mission discuss the MEWP. I just returned today from an EXACT meeting (I think a while ago I informed you about EXACT). During that meeting the MEWP was subject of discussion. PWA expressed the wish that we would look into the possibility of "reviving the interim phase." Israeli representatives gave a careful reaction, mostly because the "new" Ram Aviram, Yaacov Keidar (director multilateral peace talks water issues), was not well informed. I had met Keidar before, in September, and I had talked to him about the MEWP, but he admitted to not having had the time to look into the issue since then. I will make sure he will have to soon! Let me make clear that at this point the parties would not be willing to go any further than introducing the WAS model for "national" water issues only. Another point that was made is that when 're-introducing' the interim phase, this should be done in a flexible manner, that is, the project proposal of two years ago should be "negotiable" for changes. The idea is to possibly integrate the project within the EXACT framework. Parties agreed there is room for doing this. I favor this approach if only because it would give us a possibility to redefine responsibilities on the side of the core parties.

At the meeting it was agreed that the Netherlands would send a letter to the parties suggesting that the core parties consider taking up the MEWP under EXACT. I intend to send this letter shortly, but after the return of the mission, to be sure to take into account their input. Be assured that when sending such letter, I will copy you.

I replied and emailed the good news to Hans and Annette, although I assumed it would be some time before Annette received the message, since she was flying over the Pacific to Sri Lanka. I then went to take my morning shower. I took off my pajamas, but then realized I had failed to inform Cees that I would be in Holland in mid-December; I was going to attend the thesis defense of my student/protégé/friend Maarten Peter Schinkel at the University of Maastricht, and Ellen was coming along. We had already made a dinner date with Louise and Hans and their spouses.

I had grown more forgetful with the years, and feared I would forget to tell Kees about that—and suggest a meeting—if I waited until after my shower. So, I hastened back to my study, buck naked, to email him again. In the few intervening minutes, both Hans *and* Annette had replied to my good-news email.

I wrote them back.

> Good news has wings. I instantly replied to Maarten Gischler, having received his email just before getting dressed. On returning to my computer to tell him additionally that I will be in Holland in December should they want me, I found emails from each of you.
>
> Please wait before replying. I really need to take a shower now. Given the news, probably it should be a cold shower.
>
> God moves in mysterious ways. Given the nature of Middle East politics, that is probably necessary.

It was all very exciting. I sought permission from the Dutch to tell my regional colleagues. In doing so, I was sensitive to the fact that Munther's political position had been shaky. But the Dutch agreed, and I spread the news appropriately.

In late November and early December, the Dutch mission went to the region and obtained fairly positive replies. Then, on December 12, I went to the Dutch Ministry of Foreign Affairs and met with Cees Smit Sibinga, Kees Bons, my contact at Delft Hydraulics, and two other people, one of whom had been on the mission. We agreed that the Dutch would send an official letter to the parties asking for their participation in the renewal of the project and spelling out some particulars.

It was made clear that such a renewal would be a version of what we had been calling the Interim Phase, in which each party would learn and seriously consider adopting the technology for domestic purposes. Although the project would probably come under EXACT, the parties had made it clear to the Dutch mission that they wanted to emphasize that it would not be a joint project, but rather that there would be three parallel projects, each a bilateral venture between the central team and one of the parties. To emphasize that, the parties had asked that the name Middle East Water Project be changed to eliminate any reference to the region. After some discussion, we agreed to change the name to Water Economics Project (WEP). I had thought of Water Economic Evaluation Project, or WEEP, but that seemed too pessimistic.

Alas, it wasn't. The Dutch letter was supposed to go out in late December 2001 or early January 2002, with an eye toward the next scheduled EXACT meeting in April. But a greater round of violence had broken out, and the Israelis were attempting to quell it on the West Bank. That cycle continued into April, when, after a wave of suicide bombings by Palestinians culminating in a slaughter at a Passover *Seder* in Netanya, the Israeli army launched a full-scale invasion of the West Bank.

That violence had already broken out while Ellen and I were in the Netherlands. We had dinner with Louise and Hans and their spouses, as well as Kees Bons, which was extremely pleasant. But during the course of the evening, both Hans and especially Louise expressed their strong aversion to what Israel was doing to the Palestinians. Ellen and I, of course, are long-time supporters of the Israeli peace movement, and we too were very unhappy with Israel's actions. But I was surprised that neither of them said anything about the suicide bombings. There was no blame for the Palestinians. It was a disturbing attitude from people I regarded and still regard as extremely fair and well-intentioned.

I encountered the same attitude, widespread among Europeans, during the April incursion. It revealed a one-sided view, probably induced by some one-sided reporting. There was a flurry of emails at the time regarding a proposal that European academics boycott meetings with Israeli academics. I weighed in against that, pointing out that punishing the supporters of the peace movement hardly seemed a way to have the desired effect. I sent that to a lot of people, including Hans and Louise. Hans responded with a pretty one-sided email; a later note in that exchange seemed quite balanced.

Munther Haddadin also entered the exchange, as did Zvi Eckstein. Not surprisingly, Munther's views were entirely pro-Palestinian. He even went so far as to write in a way that could be construed as denying the Holocaust. I replied—in the friendliest manner I could—that while I agreed with much of what he had to say, I found that part extremely offensive. He then said that he never meant to do that, and we appeared all to be friends again. But passions obviously, and understandably, ran high.

A New Year

Finally, in late March 2002, Cees Smit Sibinga told me that the letter from the Dutch was being sent. The three parties' positions on the eve of the sending varied. The Palestinians remained very eager to go forward—something Karen Assaf had confirmed to me when I emailed her inquiring after her safety and well-being during the fighting in Ramallah. The Dutch view was that the Palestinians needed no further lobbying.

The Jordanians had already asked what would happen were I to be hit by a train, and had been assured that the whole purpose of the renewed project was to teach them how to use the tools without my assistance. Also, Hazim El-Naser, who had become the Water Minister, had told me in late November that the project was no longer a high priority item for Jordanians. So, the Dutch instructed me that they would deal with the Jordanians and I should stay away.

I couldn't quite do that, at least not altogether. Lenny had been speaking with Imad Fakhoury, Najeeb and Jacqueline's son whom I had met briefly years before when he was an attaché at the Jordanian Embassy in Israel. He had since become one of the nine "fair-haired young men" who were close to King Abdullah—a group that included Bassem Awadallah, who had become the Minister of Planning. Fakhoury and Lenny had been discussing plans for the development of the Aqaba area, which was likely to require cooperation with Israel. Among other things, that meant water, and Fakhoury had suggested he could get me a meeting with the King—a meeting Lenny thought would be a long, relaxed one at the King's residence in Aqaba. The Aqaba development project was very dear to the King, and that would seem to be a time to bring up Jordan's ongoing participation in the project and, perhaps most important, the idea of Jordan leading a movement for regional cooperation in water as the beginning of the way back towards peace. The Dutch approved of pursuing the possibility of such a meeting.

As for the Israelis, Water Commissioner Shimon Tal seemed at best uncertain about the project. We decided I should lobby them. I asked Uri to talk to him, but as a consultant to Tal, he uncomfortable about pressing him on behalf of a project in which he had been personally involved. I could understand his view, even though I did not agree with it. I had also asked Shaul to talk to Tal, and he did so. He then sent emails about his efforts.

Dear All—Following my previous message from this morning (Since then the Bus exploded in Wadi Arrah), I feel we have a problem with Dreizin, who became very negative' following Ben-Meir's position and instructions.

415

I suggest that Uri should speak to Shimon Tal before anyone else, as he might ask Dreizin for his opinion. Uri—if you wish I shall join you when you meet Shimon.

***

I met Shimon Tal on Pesach eve, (March 27) I am afraid he came specially for that to the office. We had a rather lengthy discussion. Let me try to cut a long story short the situation is as following:

1) The Dutch govt./Embassy had given him a list of projects, around 10, to consider and propose some of them, for possible cooperation. He is collecting views from colleagues, including Yossi. After that, he will reply to the Embassy.

2) He asked why should we invest resources into it: Without knowing the content of the others I tried to express my/our position that our project if completed—(to include the proposals for the missing stage such as—Stochastic rain fall impact, multi-annual analysis etc)—will give him and others an instrument where Engineering aspects and Economic ones, will enable a much more educated and sophisticated dialogue with the Gov. Knesset, etc.

His 2nd point was that as long as they are not involved with the project, if these are games that Fisher, Uri, Shaul, etc. are playing—it does not "disturb me" Shimon or the gov. However, when we are involved, we cannot allow others dealing with our models, producing results which can be misused, misinterpreted, etc." We hear all kind of presentations in the world of this model with deductions and conclusions that we disagree with. We cannot control the use of it, etc., etc." Hinting that his colleagues consider our group as uncontrollable, and they are unhappy about it. I think Yossi had already expressed his views to him. Yossi would like to be the source of any analysis concerning the Macro and Micro results of the water resources in Israel and of course when they know that the project managers can always come to regional conclusions on water trading, etc.

Sorry, it is longer than I wanted it to be—I did not come out very optimistic from our dialogue, especially, as "our" project competes with others that Shimon was not ready, very tactfully, to discuss.

I suggest that Uri will take it from here and continue the dialogue. I am leaving the country and will be here only a couple of days in April.

At Shaul's suggestion, I sent Tal a long email on April 23 asking for a face-to-face meeting. I attached an explanation of the project:

Dear Commissioner Tal:

Shaul Arlosoroff has suggested that I write to you directly concerning the question of Israel's continued participation in the Water Economics Project (WEP) proposed by the government of The Netherlands. I am happy to do so. Indeed, if you think it would be useful, I would be glad to come to Israel and discuss the subject with you in detail. I am not only personally committed to this project, but I believe that Israel can greatly benefit from continuing in it and acquiring for your office a powerful and useful tool to be used at your discretion.

I emphasize that the tool is to be used at *your* discretion. In its first phase, the Project created a powerful tool, which I describe briefly below. That tool was created with a good deal of Israeli input, but its Israeli participants acted in a private capacity. The purpose of the new phase of the Project is to turn that technology over to the water authorities of the participating parties – *each one for its own domestic purposes*.

Let me be clear as to what this means. Once the turnover-learning process has been accomplished, the tool will be yours to use. There will be no need for you to call on me or other outsiders for help (although you will be welcome to do so). Further, during the process of acquisition, there will be no need for me or anyone else outside your office to see or use confidential data. The function of the Project's participants will be to aid you and your staff in learning how to use the developed technology and, together with suggestions from you staff, to improve that technology so as to make it even more useful than it is at present.

Of course, as I am sure you know, the Project was originally conceived to assist in bringing about resolution of water conflicts between Israel and its neighbors and cooperation in water management. Unfortunately, in the present state of the world, there can be no question of that for the foreseeable future. Hence, although you will see that subject prominently discussed in the existing papers of the Project (including the one that I am sending with this letter), *participation in the proposed continuation now under consideration in no way implies Israel's assent or participation in such cooperative work. While a similar offer is being made to the Palestinians and Jordanians, it is not contemplated that any joint work will take place. Rather, there will be separate participation between each individual regional party and the central team.*

I now turn to a description of the tool that has been developed. (You will find a more detailed discussion in a paper that will appear shortly in *Water Resources Research*, which also gives illustrative examples of the possible use of the tool for Israel. I do not know whether the conclusions reached as to those examples are accurate or not, but they serve to illustrate the power of the tool in potential applications. Uri Shamir can provide you with a copy should you wish to see it.)

There are actually two tools. The major one, called "WAS" for Water Allocation System, operates as follows. Israel is divided into 20 districts. Within each district, water sources are specified with annual renewable quantities and extraction costs per cubic meter. Demand curves for each of three user groups (households, industry, and agriculture) are also specified in each district. It is important to realize that these are not fixed consumption quantities but rather quantities that would be demanded at different prices. Finally, infrastructure (retreatment plants, desalination plants, and conveyance lines) are given, together with capacities and per-cubic-meter operating costs. Such facilities are either actual or potential.

The model user specifies various restrictions and values. For example, prices to consumers can be set with different prices for different classes of users; the use of treated wastewater can be restricted; and water can be required to be set aside for particular purposes.

The user can also easily change the input assumptions. For example, the user can specify a lower-than-normal annual amount from naturally occurring sources, or alter population projections.

The model then takes all this, *including the special restrictions and values imposed by the user*, and assigns the available water flows so as to maximize the total net benefits derived by Israel from water. Such net benefits are defined as gross benefits, measured as the total willingness to pay of consumers, less the costs of producing and delivering the water involved.

The results provide a very powerful tool for such things as the cost-benefit analysis of proposed water infrastructure. The benefits measured are system wide and take account of all effects on consumers as well as on costs – differing in this way from previous cost-minimization models with fixed delivery amounts.

Examples of model use include:

• Analysis of the price that Israel should be willing to pay for desalination plants at different locations and under different assumptions and water policies;

• Analysis of the price that Israel should be willing to pay for water imports from Turkey or elsewhere;

- Cost-benefit analysis of an expanded conveyance line to Jerusalem.

At present, the WAS model is a steady-state annual model (although the conditions of the steady state can be easily varied). We have begun work on a multi-year model that would handle such things as the cost-benefit analysis of inter-year storage facilities, aquifer management, and the timing of infrastructure projects. If thought desirable, that work would continue during the proposed phase of the project.

The second tool produced by the project is called "AGSM" for Agricultural Submodel. Developed originally by Professor Ilan Amir of the Technion, this is an optimizing model of crop choice as a function of water availability and water prices. It has been fitted to Israeli data and provides a means of evaluating the effects of water policy on agriculture.

AGSM is not discussed in the *Water Resources Research* paper mentioned above. If desired, literature can be easily supplied.

I hope you will give this favorable consideration. As already stated, I would be happy to come to discuss these matters with you in person. If that were to happen, I would hope to have enough time to demonstrate the tools in detail.

Again, I believe the project can provide tools of considerable value to your office. Participation in the next phase of the project would require only the devotion of sufficient internal resources to learn how to use and understand the tools. The government of the Netherlands would finance the work of the central team.

I look forward to hearing from you.

Sincerely,
Franklin M. Fisher
Jane Berkowitz Carlton and Dennis William Carlton Professor of Economics
Massachusetts Institute of Technology

Tal agreed to a meeting, and Uri made the arrangements. I was hoping it could be held at a time that he could also attend. I was scheduled to be in Israel from May 9–14 to attend a meeting of the steering committee of the Green Environment Fund (a joint venture of the New Israel Fund and several other philanthropies), as well as an antitrust conference at the Academic College of Tel Aviv–Yaffo. When the violence became extreme, I questioned whether these activities were sufficiently important for me to go to Israel. Even though the risk was probably not all that big, it was certainly there, and

Ellen very strongly did not want me to go. I decided that they were not important enough. A meeting with the Water Commissioner was a different story, however. To her credit, Ellen said instantly that I had to go for that.

Uri was able to join the meeting. The entirely reasonable Tal had unfortunately already told Herman Froger, the new Dutch Ambassador with whom Alon Liel and I lunched earlier, that our project was now low priority for Israel. That may have been due partly to Yossi Dreizin, who Shaul said told Tal we could not be "controlled"— an accurate statement.

It was quite a surprise that the Jordanians also said the project was a low priority—despite Hazim El-Naser being the Water Minister. I do not know what happened, but Faisal Hassan, who had been the other possible choice for Water Minister, implied to me that it was political. Faisal remained a very, very strong supporter of the project, predicting that Jordan would rejoin and that there would eventually be a regional version. He was very helpful later in assisting Annette and me with a necessary revision of the Jordanian chapter of *Liquid Assets*.*

Unlike the Israelis and Jordanians, the Palestinians expressed great eagerness to move forward. The Dutch decided that could be done under a different program: bilateral cooperation between the Dutch and the Palestinians rather than the original "Dutch Initiative." Negotiations to accomplish that had been going on in some form for some time.

Publications and Conferences

Meanwhile, with the project no longer a secret, publications had begun to appear with some frequency. The most important of these was the ten-author article in *Water Resources Research (WRR)*, published electronically in November 2002 and available in hard copy in March 2003. Getting it to that point had been a long and difficult process. The publication of the *WRR*

---

* It was late May 2003 when we had to rework both the Palestinian chapter, which was essentially complete, and the Jordanian chapter, both because the teams had made the same major mistake—a habitual one for water engineers. Despite repeated warnings and explanations, their forecasts of demand at given prices turned out to have been forecasts of *consumption*, taking into account their forecasts of available supply. Faisal generously supplied data for the revision of the Jordanian chapter. It did not help the speed with which we were able to finish the book that Annette had a full-time job outside of the project.

article was plainly attracting attention, but some of that had begun even before the piece was out.

In the summer, an email arrived from an organization called "Global Dimensions" inviting me to a conference on natural resources to be held in Moscow in September 2002. At first, I was inclined to refuse, but three things convinced me to go. First, it turned out that the director of the organization was an old friend of mine from the London School of Economics, Meghnad Desai. He had become Lord Desai of St. Clement Danes. In January 1994, when Ellen and I were passing through London on our way to the Middle East, Meghnad had provided dinner. I had talked then about the incipient water project, and he had remembered.

Second, Meghnad informed me that the Conference was to be about two resources, water and oil, and he wanted me to be the keynote speaker on water. Third, it was to be a small conference for policymakers. I was particularly interested that David Owen, who had been the United Kingdom's ambassador of the United States during the Kennedy administration, would be present.

Ellen and I had a nice trip—but I should not have bothered. Oil and water truly do not mix. Having attended the first session on oil, I then spoke on water, after which the Russians went immediately back to oil. David Owen didn't even stay in the room to hear what I had to say.

Just before attending the conference in Sion, Switzerland in mid-October (see the previous chapter), I was asked to supply a write-up to be published in the journal, *GAIA*, which did it very nicely.[1] In connection with the Switzerland trip, Ellen and I decided to spend a few days in Barcelona, and I emailed Andreu Mas-Colell, a very distinguished economist and an old (if not extremely close) friend. He turned out to have become Minister of Science in the Catalonian government and arranged a lecture for me at the engineering school. There I was fortunate to interest Professor Lucila Candela, who was serving on an international committee dealing with water on the Mediterranean islands. Her responsibility was Majorca. I gave her a WAS license, and there was discussion of developing a WAS tool for Majorca. It turned out that no funds were available, though, and nothing happened.

Before that, there had been discussion of a similar development for infrastructure evaluation in Bogota, Colombia. It followed on my presentation in a week-long program in water pricing that had been held at

MIT in June 2002, and was organized in part by Dale Whittington of the University of North Carolina, whom Annette and I had met at the meeting at the World Bank in 2000.

Ellen forbade me to go to Bogota in person—a decree that acquired even more force (if possible) when we learned that the woman in charge of the water system had 16 full-time bodyguards. But in the end, that opportunity, too, came to nothing. The Colombians sent documents in Spanish to CRA, and we had them translated, but it became clear that they thought we could provide them with a formula (or at least a generalized piece of software) into which they could plug their numbers and get answers without a great deal of work. Moreover, they had rather rigid ideas as to what such a system should tell them.

29

Into 2003 and Beyond

In early March 2003, I received an email from Husam Tubail, an engineer with the Palestine Water Authority who had been assigned the job of making a new program—bilateral cooperation between the Dutch and the Palestinians—move forward. He and I exchanged a number of emails in which he sent me the Palestinian plan for water and I provided a description of what the project could do and how it could assist Palestine. He went to The Hague in late April 2003, and it appeared possible that the work would begin in October.

The most interesting thing about the Palestinian plan was that it clearly contemplated future negotiations, and Tubail was interested in using the project in that regard. I thought I heard the sound of one hand clapping. My suspicion was that Karen Assaf had had a hand in promoting what Tubail was doing, even though she had left the Palestine Water Authority after spending a considerable time in the United States the preceding winter.

There were other developments around the same time. That spring, I wrote to George Shultz, the former U.S. Secretary of State in the Reagan administration. I knew him only slightly; he had been an economics professor at MIT a bit before my time. I sent him material about the project and asked if he could take it up with the George W. Bush administration.

George, being an economist, was extremely enthusiastic. He proposed to carry my material personally to then Secretary of State Colin Powell and speak with him—which he did. Sometime later, George sent me a reply letter from

Powell. In substance, it said that people in the State Department were already familiar with my work and would get in touch with me if they so chose.

It was, of course, a brush-off. It was also quite inaccurate. I had visited the State Department only once, in 1994, and had met with a group headed by Chuck Lawson, then the chief American officer dealing with the water multilaterals. That was when the project was in its infancy. While I had seen Lawson once or twice at conferences, he had never shown any interest. And even though I had met later on with Dennis Ross's number-two, Aaron Miller, I could not accept the idea that the State Department now knew everything there was to know about the project.

Needless to say, I never heard from Powell's State Department. I asked George to pursue it further, but he never replied—perhaps because he'd become preoccupied with Arnold Schwarzenegger's campaign for California governor (Shultz had been named to the campaign's economic council).

Meanwhile, as these efforts played out, an attempt began to resurrect ISEPME or its equivalent and find it a home at MIT. The Water Project was to be the principal activity, and, indeed, donations were made to MIT by former ISEPME board members in the expectation of success. Lenny took the lead, supported by Larry Susskind, a professor in the MIT Urban Studies and Planning Department who also worked with the Harvard Law School-based Negotiations Project. I also put in a large amount of time.

After interviewing everyone we could find at MIT with an interest in the Middle East, I made the mistake of thinking we could house it in the Economics Department, possibly postponing my retirement. I had several discussions with Olivier Blanchard, then department head, and he canvassed the other full professors.

At first, Lenny was going to be the director, but it immediately became apparent that this would not fly. His reputation had been ruined by his dismissal from the Kennedy School. Even after Lenny agreed to play only a minor role, his involvement remained an issue. I thought Olivier had secured the assent of our colleagues before putting it to a vote, but that was not the case, and there were sufficient objections for Olivier to conclude it could not happen. (Ellen's opinion was that Olivier really had little enthusiasm for the work and did not make any serious attempt to interest colleagues in the department, let alone persuade them.)

With the Economics Department out of the question, Lenny, Larry, and

I discussed where else at MIT we might house the project. Lenny and I had extensive talks with Richard Samuels, the director of the Center for International Studies. These seemed very promising, but, in the end, the Center decided that if my own department had taken a pass, then there must be something wrong.

Led by Larry, we then made an attempt through the Urban Studies and Planning Department. At first, we were more successful, winning the approval both of the department and the Dean of the School of Architecture and Planning in which it was situated. But then, to our disappointed surprise, MIT Provost Robert A. Brown, absolutely turned it down. He was concerned that Lenny would have any connection with the proposed center, and, even though we had proposed that I would be the acting director pending completion of a search for a permanent one, he stated that having a retired professor in that position would not do. (I had announced that I would become *emeritus* as of June 30, 2004.)

The attempt ended there. We asked donors who had given money to MIT whether their donations could now be used to support the water project, and a number of them agreed. That was a major help, since the Dutch funds were no longer available, and the project was largely being supported by the $20,000 per year that went with my Chair—which would cease when I retired.

I asked MIT how much was in the fund these donors had created. To my surprise and fury, I found that Larry Susskind had appropriated part of it for his Middle East work—without discussing it with either Lenny or me.*

An Approach to Lebanon

In the fall of 2003, I made contact with a Columbia University professor from India, Upmanu Lallat, who was interested in water issues. I gave a paper at his seminar and met with Jeffrey Sachs and others who thought they could raise money. Nothing came of this directly, but later I was invited to a meeting

---

* I regret to say that this was not the only occasion on which I disapproved of an action by Susskind. In 2012, he and Shafiqul Islam of Tufts University published a book with a section that described attempts to resolve water conflicts in the Middle East.[1] It was seven years after our book, *Liquid Assets,* had come out from the same publisher, RFF Press, and Susskind was obviously well aware of the project, but there was no mention of it whatsoever. I called Susskind and complained, and he was suitably embarrassed and apologetic.

at Utah State University along with Lallat, Annette, and Anan Jayyousi (for whom it would be a return visit to his Ph.D. *alma mater*). That meeting took place in January 2004—which, fortunately, was ski season, so Ellen accompanied me for part of that trip.

The water group at Utah State—part of the Utah Water Research Laboratory that was officially launched in 1965—had done, and continues to do, a good deal of work on the Middle East, particularly Jordan. It included, in particular, Mac McKee and Jagath J. Kaluarachchi, both professors of civil and environmental engineering, and David Rosenberg, a student of Jay Lund at the University of California, Davis.

The meeting eventually became a discussion of how to resume project work in the Middle East. One suggestion was to attempt to restart in Jordan by speaking with Munther Haddadin; a second, put forward by several of those present, was to approach Fadi Comair, the long-time Director General of the Lebanese Ministry of Energy and Water. I learned that Fadi had attended the seminar that Uri Shamir, Howard Raiffa, and I had given at Harvard in spring 1998.

Fadi became our very successful introduction to Lebanon. He set up a workshop in Beirut, scheduled for October 2004. But before that, I was a keynote speaker at a conference in Antalya, Turkey, on Palestinian water organized by the Israel–Palestine Center for Research and Information (IPCRI) and held in Antalya Turkey. It afforded an opportunity for Ellen and me to sightsee in Turkey for the first time.

At the conference, my paper (later published in an edited book[2]) drew considerable attention. Discussing the issue of whether one participant in a WAS-guided treaty would be injured were the other to withdraw, I first pointed out that it would not be in the interest of either to withdraw, since all parties would gain from WAS-guided cooperation. But I then went on to point out that that might not be a sufficient answer—and offered others. In so doing, I told the story of the duck and the scorpion, well known in the Middle East.

> A duck is swimming in the Suez Canal when he is called to one of the banks by a scorpion.
> "Mr. Duck," says the scorpion, "I cannot swim and I very much need to get to the other side of the Canal. Would you please be so kind as to put

me on your back and take me across?"

"Mr. Scorpion," the duck replies, "you must think me an awful fool. You're a scorpion. We'll get halfway across and you'll sting me and I'll die!"

"Oh, Mr. Duck," says the scorpion, "I would never do that, because it would be greatly against my own self-interest. If you were to die, then eventually you would sink and so would I. Then I would die too, since I cannot swim, which is why I need you in the first place."

Says the duck, "You've got a point there, Mr. Scorpion. Hop on."

So the scorpion jumps on to the duck's back. Halfway across the Canal, the scorpion stings the duck.

"Why?!" asks the duck, with his dying breath.

"What did you expect?" replies the scorpion. "This is the Middle East."

There was then question time. Yossi Dreizin, a member of the conference committee, went first. He was an undisguised bigot who knew the project and had done what he could to kill it because it involved the Arabs. He asked an angry question: "Why have you come here to tell us we should give water to our neighbors? Why don't you tell the United States to give water to Mexico?"

As politely as I could, I pointed out to him that in the proposed arrangement, Israel would not "give" water to the Palestinians. Rather, it would *sell* water to the Palestinians *only if it were in Israel's interest to do so.*

When I was done, my dear friend Alon Liel, said to me, "I am so glad you told that story. There stands the scorpion!" He pointed to Dreizin.

Alas, there are too many "scorpions" in the Middle East who would rather hurt the supposed enemy than benefit their own country.

I headed to the Lebanon workshop where WAS would be the principal topic. Annette attended, as did a delegation of Palestinians led by Amjad Aliewi, who I met for the first time, along with Karen Assaf and Amar Jarrar. Cees Smit Sibinga, then with the Dutch Embassy, was also there.

Before the workshop began, Fadi Comair made clear to me that he wanted a WAS model built for Lebanon, both for domestic purposes and "to get ready for negotiations with our neighbors. And I mean *all* our neighbors," he emphasized. And in case I didn't understand the message, he added, "And I include Israel."

That was very good news indeed, and after the workshop Fadi prepared a Lebanese proposal for the Dutch (through Sibinga). At the same time, the

427

Palestinians proposed to the Dutch a renewal of their project.

Unfortunately, and despite the helpful advice and influence of Louise Anten, these proposals came to naught. The Dutch were revising how they dealt with such applications, which took a long time. By the time they were ready to decide, Lebanon had destabilized and would not deal with the Water Minister, who was a member the Islamist political party Hezbollah. Similarly, by the time Palestinians applied, Hamas—another Islamist group with which the Dutch would not deal—had won a plurality in the January 2006 Palestinian parliamentary election.

Developing MYWAS and Completing the Book

During this period, Annette and I had worked to develop MYWAS—the multi-year version of WAS. Those activities continued for some time. At first, there were various snags. I explained earlier how we solved the problem of costing out changes in capacity for a given project. Unfortunately, Annette never transferred that solution to the correct computer, so we had to reconstruct it—which led to another complication.

Annette had found it difficult to construct a user-friendly interface, and so we farmed that out to David Rosenberg, who had since joined the Utah State faculty. But the program he sent back was rather more than building the interface: he had changed the programming so it no longer worked. It took us considerable time to fix.

Annette and colleagues at the Boston-area branch of the Stockholm Environment Institute (SEI) then produced a useful interface using (the Water Environment and Planning (WEAP) system. WEAP is not an optimizing tool, but, with some difficulties, it was adopted for the MYWAS interface. The properties of MYWAS made it an extremely powerful tool for efficient water management, and led to a new round of model building. (The appendix provides a list of MYWAS applications.)

Also in this intervening period, my book with Annette Huber, *Liquid Assets*, finally came out in the fall of 2005.[3] We had worked on it for many years. The delay was not because of Israeli opposition, something implied in a question I was asked later by Tony Allen, an important water expert in London. It just took much longer than we anticipated. Two things had slowed us down in particular. Annette received her degree and found a full-time job, which kept her from working on the project as much as she had previously and would have liked to continue. The most important reason, though, had to do with the demand curves used in our project.

The demand curves we used were generated first by specifying a price for water and then estimating the quantity of water that would be demanded at that price. That—plus a posited demand elasticity—produced the desired demand curves, but it had to be done both for a base year (1995) and for projected conditions in two (then) future years (2010 and 2020). We produced such curves for households, industry, and agriculture in a number of districts within each of the countries involved. The demand curves so generated were then used to calculate (in the relevant range) the benefits to users of receiving the water. But for that to produce usable results, the fixed-price estimates of demand had to be estimates of how much water the users would buy *if there were no constraints on supply*. We emphasized that strongly to those making the estimates.

Unfortunately, that message did not get through in the cases of Jordan and Palestine, where those doing the estimation took "demand" to mean consumption, taking into account the expected constraints on supply. That was (and to some extent still is) common usage among water engineers and was only natural for the Jordanians and Palestinians who were living under severe constraints on water availability, but it meant that the chapters on results for those two countries had to be completely redone (as described in the preceding chapter). That was a major effort and could not be accomplished in a short time even with substantial help from the regional experts. It took roughly a year for each of the two chapters.[*]

Once published, the book became quite well known, and, within a year was being recommended to some water students as a "must-read."

Immediately following the book's release, Alon Liel arranged for the board of IPCRI to nominate the project or me for the 2006 Stockholm Water

---

[*] Another reason for delay was the obstinacy of the production editor. The book was divided into two parts: first theory, and then actual results. At the beginning of the second part, we placed a two-page note entitled "A Note on Some Sensitive Issues" that explained we were handling (as regarded water) the questions of the Israeli settlements and the issue of Jerusalem. The production editor observed that we had no similar note for the first section and insisted there had to be symmetry. I could not see why that was necessary, and I eventually refused to comply. But that contributed to the time it took.

Prize, a prestigioius award based on the Nobel. I insisted it be for the project.

We obtained supporting letters from a variety of important people, including: Como van Hellenberg Hubar, the former Dutch ambassador who had supported the Project in its first phase); Yossi Beilin, my old and distinguished friend; economist Eytan Sheshinski; Henry Rosovsky, the Harvard economic historian; and, of course, water academics such as Jay Lund of the University of California, Davis, who had produced optimizing models of water usage.

Annette and I were convinced, naively, that we could not lose. I even declined to take up a suggestion by Lenny and Musallam Musallam, a very rich and powerful Saudi, to compete for the Saudi Water Prize, because it would make me ineligible for the Stockholm prize. Annette and I were wrong.

Alon went on to renominate the project and me every year for at least a decade after informing me that he would continue for as long as it took.

Later, on the prize front, Lenny had this crazy idea in 2009 that I should be awarded the Nobel Prize in Economics and insisted that I provide him with a paper describing the reasons why that he could circulate to some mutual friends for their opinions. I was grateful for such a huge vote of confidence (others had mentioned the possibility from time to time), but I was rather embarrassed, too. Lenny insisted, however, and I produced what I called "An Immodest Statement."

Lenny gave my paper to Stanley Fisher, Eytan Sheshinski, and I believe some others. I was not surprised (admittedly, I was only a bit disappointed), when they replied that the profession had moved beyond me and that I should concentrate on the Stockholm Water Prize.

The entire incident reminded me of an anecdote from 1970, when Ellen and I took our three children (Abraham, then 9; Abigail, then 6 and a half; and Naomi, then 6 months) on a short vacation in Pennsylvania and Washington, D.C. To amuse the children, I told them about the time when I was in high school and my friend Richard Friedberg and I had talked about creating a product we called "Water Mixo." The idea was it would produce dehydrated water by taking water and evaporating or boiling it, and then the dehydrated water could easily be carried, especially in the desert where water was needed. All one would need to do is add water.

Abe and Abby thought it was a great idea.

On the next day of our vacation, we heard the expected announcement that Paul Samuelson, who the kids knew, had been awarded the Nobel Prize. The children were quite interested, and we overheard them discussing whether their daddy would ever win it. One of them (probably Abe) said that I would have to discover something first, and I had not done that.

"Yes, he has!" interjected Abby. "He invented Water Mixo!"

30

The Model Is Adopted

In 2006, Hillel Shuval brought to my attention a relatively new organization, EWE, based at the London School of Economics (LSE). The "energy, water, and environment communities project" came to play an important role in our project. Hillel had joined EWE's board. Munther Haddadin was also involved.

EWE has some auspicious origins. In 1996, Václav Havel, the first president of the Czech Republic, Japanese philanthropist Yohei Sasakawa, and Nobel Peace Prize Laureate Elie Weisel launched a joint initiative called "Forum 2000" that became an annual conference to "identify the key issues facing civilization and to explore ways to prevent the escalation of conflicts that have religion, culture or ethnicity as their primary components."[1] Over the years, the conferences have drawn a number of prominent participants, including—among many others—Bill Clinton, the Dalai Lama, Shimon Peres, and Prince El Hassan bin Talal, the former Jordanian Crown Prince with whom I had repeatedly sought an audience.

At the suggestion of Prince Hassan, an offshoot body was formed to consider issues of energy, water, and environment in the Middle East and North Africa. EWE's director is Professor David Held of LSE's Centre for the Study of Global Governance (CSGG); his two co-directors are Dr. Michael Mason of LSE's Department of Geography and Environment and Dr. Pavel Seifter, a former Czech diplomat at CSGG who had served as the Czech Ambassador to the Court of St. James, had married an English woman, and now lived in London.

Hillel had me invited to EWE's meeting in Prague in July 2006. After I spoke, he succeeded in getting the board to adopt our project as one of its own two signature projects, which proved to be a major event in renewing our work—despite the fact that EWE had little, if any, money. What the group did have was a great number of connections and contacts to whom our project could be introduced—beginning in Jordan.

Jordan

We made our first connections in meetings in Amman in the fall of 2006. Annette, Pavel, and I visited a number of ministries with a focus on the use of MYWAS for domestic purposes—something that, at the least, would benefit Jordan. These meetings came off rather well. We suggested a two-day workshop the following year to educate the appropriate people regarding use of WAS and MYWAS.

Unfortunately, however, they led me into making a blunder when, on the final day, I met with Prince Hassan himself (along with staff and a number of other people, including Pavel and Munther. I began (as I had several times in the previous days) by explaining the merits of MYWAS as a way in which Jordan could manage its water system, and did not explain at length how it could be used for peace.

But, of course, the Prince (as I should have realized) was primarily interested in promoting peace. He responded with a scolding speech. Fortunately, I had two chances to mollify him.

The first was that he mentioned very favorably the report that had just been issued by Peace Now on the extent of Israeli settlements built on Palestinian private land. He really wanted to see that report and encourage others like it. Fortunately, when I got the chance to comment, I said that the report in question was a product of *Shalom Achshav* (the Hebrew for "Peace Now"), and that I was the board chairman of its sister organization American Friends of Peace Now and had come to Jordan directly from a meeting of the board at which the report in question had been distributed. Indeed, I said I thought I had a copy back at our hotel.

The prince said he would very much like to see it and asked how he could contact "Mr. Achshav." This, of course, made me realize that he did not know the Hebrew name of Peace Now.

The other opportunity was for me to remind him just before leaving that I was a friend of Julia Neuberger. This increased his interest in me. He expressed great regard for Julia, whom he said he would see in a few weeks.

I could not find the report back at the hotel, so I immediately called American Friends of Peace Now in Washington and had copies sent to the prince. Those on the other end of the phone were quite excited.

Moreover, the prince had sent a staff member after me with a request that I vet a letter that the Prince was about to send to someone in the U.S. government. I was able to make a few suggestions.

The prince ended up agreeing to convene the workshop we proposed, to be held in October 2007.

Before that, at the end of September 2007, a conference on Sustainable Development and Management of Water in Palestine was held in Amman. I presented a paper entitled: "Analyzing Future Palestinian Water issues with the WAS Model."[2] The conference was organized by a committee led by Amjad Aliewi and involved Karen Assaf, Anan Jayyousi, and Shadad Attili, who was the chief water negotiator for the Palestinians and who later would become the head of the Palestinian Water Authority. I spent some time with Attili, who had heard about WAS and, although he would not use it for negotiations, considered it as a long-run plan for regional cooperation.

While in Amman, I had a pleasant dinner with Jawad Anani, who has always been a fascinating conversationalist and always worth listening to. He told me a story about going with Crown Prince Hassan to see Bibi Netanyahu during his first term as Israel's prime minister. They brought a major proposal for cooperation between Jordan and Israel in the Rift valley, which includes the Jordan River.

Netanyahu listened. He then noted that there were black flies on either side of the valley, and that Jordan and Israel were independently spraying the flies. He suggested that Israel and Jordan should coordinate the spraying activities.

"Mr. Prime Minister," Jawad replied, "wouldn't it be simpler just to announce a no-fly zone?"

When they left the meeting, Prince Hassan turned to Jawad and said, "If you go home and tell my brother [King Hussein] that you said that instead of me, I shall kill you!"

After that conference, Annette and I returned to Amman for the October workshop, at which we were principal players. Prince Hassan welcomed us, and we had dinner at his home. In some ways, we felt we had come for a great occasion.

The workshop lasted two days; a third day was scheduled for discussions with the appropriate authorities. On the morning of the first day, members of the diplomatic corps were invited, but they largely disappeared once we got down to business in the afternoon. Of course, our friends Jawad and Munther attended. The discussion focused on what WAS and MYWAS could do for Jordan. On the second afternoon, we had a group of Jordanians learning to use the model and evaluate the results. Not surprisingly, they were particularly interested in what the model suggested about the desired Red–Dead Canal.

In our discussions on the third day, officials promised to get back to us in a week or two with proposals about how to get started on the multi-year Jordanian model. That stretched into months. I believe that being promoted by Prince Hassan did not produce enthusiasm on the part of the government bureaucracy.

Finally, the following spring (2008), I went to Jordan again and Munther Haddadin and I succeeded in having a meeting with two relevant secretaries-general at the Water Ministry and Jordan Valley. I drafted the following letter to be signed and sent to Pavel:

> As you will recall, following the very exciting symposium on the WAS and MYWAS models held in Amman at the end of October 2007, a meeting of the principal participants adopted the following conclusions:
>
> "The WAS and MYWAS models are potentially very useful decision-support tools for Jordan, in particular: the planning and evaluation of future infrastructure projects, i.e., given information on a menu of possible future projects, the models can give advice on which ones should be built, when, what order, and to what capacities; the evaluation of the likely impacts of policies towards water—for example, different price policies; the exploration of the impacts on Jordan of different regional water agreements or actions by other parties; and the exploration of the effects of climate change. It should be noted that WAS and MYWAS models take into account social and environmental values, as well as purely economic values.
>
> "On the basis of this, the parties recommend: the WAS model should be considered as one of the tools for water allocation on the national level,

to be used in conjunction with other appropriate tools; the WAS and MYWAS models can and should be adapted to better handle issues of water quality; the further development and adaptation of the Multi-year WAS (MYWAS) will be a joint effort of the parties, including statements about possible policies and social values, and will be applied at both national and regional levels.

"According to WAS (using data from the 1990's), there may be circumstances where it would be inadvisable to build the proposed Disi pipeline. We recommend that the relevant Jordan water authorities run the model with updated data and appropriate social policies, and draw a local conclusion on the Disi project. In this update, we recommend taking into consideration the possibilities of the Red Sea-Dead Sea project.

"Intensive training of Jordanians, with a goal of developing capacity to use these models themselves and to be able to train others."

Those recommendations were to be implemented following a follow-up meeting in the following week. Unfortunately, for administrative reasons having nothing to do with the substance of the conclusions, that meeting was not held, and the matter has remained suspended until this time.

We now propose to proceed with the implementation of the recommendations as quickly as arrangements can be made. In that connection:

A meeting of the Jordanian institutions involved will be held shortly (in June 2008) to discuss the concrete applications of MYWAS in Jordan and to chose the first pilot project for implementation. Administrative arrangements as to which Ministry will take the lead will be worked out.

The Jordanian side will inform you about its decision that the WAS model will be considered as one of the tools for water allocation on the national level, to be used in conjunction with other appropriate tools.

The Jordanian side is prepared to set up a joint Jordanian/MYWAS team for the further development and adaptation of the Multi-year WAS (MYWAS). This will be a joint effort of the parties, including statements about possible policies and social values, and will be applied at both national and regional levels. To that end the Jordanian side will nominate their coordinator for the project.

It must be remembered that the existing WAS model for Jordan was constructed a decade ago. A major part of the project will have to be the updating of the data and projections used.

The Jordanian side is prepared to cooperate with the MYWAS team to organize a program of intensive training of Jordanian experts, with a goal of developing capacity to use these models themselves and to be able to train others. (It may be that the optimal place to begin that program will be in Cambridge Massachusetts with follow-up sessions in Amman, but that remains to be worked out).

Of course, budgets and funding will need to be determined and approved.

We look forward to working with you, EWE, and the WAS and MYWAS team on this important project.

I do not remember whether the letter was sent, but it was certainly the case that the secretaries-general signed such an agreement. Indeed, when others soon replaced the secretaries-general involved, we needed to meet anew and obtain signatures. Munther insisted, quite correctly, that I make a special trip for the meeting, even if only for one day.

At one of those meetings, I regret to say that I made a serious *faux pas*. As the meeting was ending, Munther asked me what the role would be for him and his consulting firm. When I replied that I did not know, Munther interpreted it to mean that I was going to drop him, even after all his work setting up the meeting. He was very unhappy.

Of course, I should have said that we needed to discuss it. Eventually we did have a meeting of minds, but relations were very strained for a while.

With the agreement of the secretaries-general in hand, we now faced the issue of finding the money to support the Jordanian effort. That was not at all easy, but very helpful in the endeavor was H.R.H. Prince Zeid Raad Zied Al Hussein, the Jordanian Ambassador to the United States, who I later realized I had already met at a dinner of the board of American Friends of Peace Now. He was excited by the project and made several suggestions and attempts to obtain financing.

One of the attempts to gain funding had occurred on the last day of the Amman workshop. I became aware that the Dutch ambassador to Jordan was none other than Johanna Van Vliet, whom we had met several times when she was assisting Ambassador Como van Hellenberg Hubar, the Dutch ambassador to Israel. That was very welcome news. She could not attend the workshop, returning only on the closing night, but thanks to an email from Louise Anten, we were able to arrange to have dinner with her.

That could not have been more pleasant. Ambassador Van Vliet, of course, knew all about the project, at least as it had been some years before, and she agreed to see whether the Dutch government would once again finance Jordanian participation. Unfortunately, the answer came back that such assistance would require that the various parties were prepared to cooperate, and that, of course, was not the case.

Munther had also made an important suggestion regarding cooperation: that we should begin by arranging cooperation among Jordan, Palestine, and Lebanon, and then broaden it out with Syria and possibly other Muslim states. At the same time, he hoped to have cooperation among Israel, Palestine, and Jordan, ultimately arriving at regional cooperation. Perhaps one day this will happen.

Palestine

While all this was unfolding on the Jordanian front, we began in 2008 to renew the Palestinian effort and move to MYWAS. Pavel began discussions for funding with the government of the Czech Republic. The Czechs issued a request for three plans, requiring that any plan include me, and also allowing me to indicate which proposal I would favor. I did so, naming the one submitted by Anan Jayyousi and his company, which was successful.

We began with a workshop in Ramallah to describe to the Palestinians the benefits of the plan, if the work succeeded. It was well attended, and, for the most part, enthusiastically received. But there was one dissenter when it came to the proposition that we could solve water disputes through trade in water permits to the benefit of all parties—and I regret to say it was my old friend Jad Isaac.

I had last seen Jad the year before in Bethlehem, when he asked me to join his team that was bidding for the analysis of the Red–Dead canal project. I readily agreed, and I looked forward to being on the same team. Unfortunately, the bid of the consortium that would have included us was not accepted. Now, a year later, Jad strongly took the view that under our system "the rich Israelis would buy all the water!" That was simply wrong. It could not happen unless the Palestinian government wanted to sell all the water, and that would make no sense. We were not speaking of individual trade with poor Palestinians selling all their water for money. Jad knew better, and had for years—but apparently had forgotten.

Fortunately, the other participants did not agree with Jad and clearly regarded him as a cranky troublemaker who should not have been invited to attend. I really regretted that.

For some time afterwards, I visited the Palestinians on several occasions, and once, early in the process, a team of Palestinians visited Cambridge. But the first few years of the Czech support were somewhat difficult: funding had to be renewed annually, and it was only towards the end of a year that a decision about the next year would be made. But finally, in 2011, the Czechs gave a considerably larger grant for work to be done over three years. At that time, the project was transferred to the Palestinian Water Authority (PWA), of which Shadad Attili had become the head. I stopped visiting the Palestinians because the budget could not finance my travel expenses as well as those of Annette, who was irreplaceable.

For a while, things went well, but a number of roadblocks emerged, and it became clear that the PWA was siphoning off money from the project. Further, there were repeated attempts to fire Karen Assaf, who was coordinating the Palestinian efforts. By the end of 2012, all work had come to a halt.

With the intervention of Pavel Seifter, it appeared by mid-2013 that the issues had been resolved, with local Czech representatives securing the return of the misused funds and the project placed in the hands of Annan and Karen. But by late 2013, Pavel reported that the authorities in Prague were objecting, saying that the misused funds had not been very large and that they saw no reason that *any* travel from the United States should be paid for—let alone payments for Annette's time.

The Czech position revealed some serious misunderstanding of what we were doing, MYWAS was not just a software package that could be taken out of a box and used without a good deal of user training and, of course, substantial gathering of data.

As of late 2013, it was not clear what would happen.

More Talks and Papers
Meanwhile, as had been the case for some time, I continued to participate in conferences and write papers.

In early March 2008, I received an invitation I could not refuse. It was the outgrowth of the Annapolis Conference, which had taken place at the U.S. Naval Academy in late November 2007. This was part of an effort to revive the Israeli–Palestinian peace process and implement the "roadmap for peace" that had been proposed several years earlier by the so-called "Quartet

439

on the Middle East," comprising the United States, the European Union, Russia, and the United Nations. A group of people involved with the Geneva Peace Initiative, issued in late 2003 and based in part on the roadmap, had set out to organize a meeting and dialogue about water.

For that meeting, the Palestinians had retained David Brooks (whom I had known since the 1994 Cyprus Conference) and Julie Trottier, at the time a lecturer in politics at the University of Newcastle-upon-Tyne in England and a senior research fellow at Oxford University's Centre for Water Research (who I did not know). They had been retained on behalf of the Palestinians by Friends of the Earth Middle East (FOEME) and had authored a long paper on dealing with water rights. When the Israeli representatives to the meeting—my old friends Hillel Shuval and Shaul Arlosoroff—read that paper, they rejected it outright, and the meeting collapsed.

This was a second attempt. I was asked to attend a meeting near Annapolis in August of 2008. Of course, I accepted.

The people who met really tried. Hillel and Shaul were again the Israeli representatives. Abedrahman Tamimi, director of the Palestinian Hydrology Group for Water and Environment Resources Development, with whom I had worked previously, and Fadia Deibes, a lawyer I had also met earlier, were the principal Palestinians. Samih Al-Abed was also present; I had met him long before when he was a deputy Planning Minister. There were also a number of Americans and some representatives from the Geneva Peace Initiative. At first, the intent was to involve Brooks and Trottier again, but that did not happen. I was one of those who influenced that decision.

The problem was their original paper. It was an interesting discussion and model of how to distribute water rights among very small groups of people—Palestinian farmers—but it could work for allocating water between the two governmental parties. Presumably, Hillel and Shaul had seen that when they rejected it in the first place. David Brooks got in touch with me several times, and I realized that he felt quite hurt about the rejection, but while I could understand his feelings, and was sorry I couldn't see how having his paper a second time would lead to progress.

The principal presentation at the meeting was an eminently fair proposal by Hillel for how water ownership should be divided. While I also made a presentation about our project, Hillel kindly referred to Fishelson and me for the proposition that water could be thought of in terms of money, but did not suggest that the parties should agree to water-permit trading as a beneficial way to get to a "yes-yes" outcome. The Palestinians were simply not ready to agree to the project's mechanism.

After considerable discussion, the task of negotiation along the lines Hillel proposed was turned over to him and Shaul for Israel and Deibes and Tamimi for Palestine. They came very close to an agreement, but the Palestinians wanted more time, presumably to discuss with others.

No resolution was ever reached, and—regrettably—Fadia Deibes was killed in an automobile accident shortly after the meeting.

Also in 2008, I was a participant during Expo Zaragoza in Spain, which had as its theme "Water and Sustainable Development." A number of other water scholars were present, including Jay Lund and, especially, Munther Haddadin, with whom I spent a lot of friendly time.

The week's organizer was Josefina Maestu, who later edited a 2013 book that collected the papers from the conference that took place, including mine.[3] At first, she was sufficiently interested to discuss using the model for one or more river basins in Spain, but MYWAS was not yet ready, and, by the time it was, Josefina no longer held a position in Spain in which it could be used.

In July 2008, I received an invitation from Alessandra Casella, who had gotten her Ph.D. in the MIT economics department and is now a professor at Columbia University. She had been at a party Ellen and I attended at Columbia in mid-2005 to celebrate the 70th birthday of our old friend Jagdish Bhagwati, also a Columbia economist. She had asked me what I was working on at the time and I, of course, told her about water. I said that our book was about to be released, and she asked me to send her a copy—which I did.

Later, after reading the book, she wrote one line to me. It is the nicest compliment I have ever received: "This is why economists were created."[*]

Alessandra sent me an email in July 2008, excerpts of which follow:

> I am spending several weeks in Marseilles, as I have done for the last few years, and have been asked to invite you to give the next "Marcel Boiteux" conference in Marseilles. The conference, considered very

---

[*] Annette, who is not an economist, was somewhat annoyed when I showed her Alessandra's note. It would have been better, she said, if Alessandra had written, "This is why people were created."

prestigious here, is given by an academic economist with experience in the practice of economic policy. Its goal is to highlight uses and limitations of economic theory for policy, or vice versa lessons for theory derived from policy experience. (Marcel Boiteux was for a long-time Chairman of EDF (Electricité de France) and the initiator of France's nuclear energy program. His previous academic career had been successful too, and he had been president of the Econometric Society). In particular, there is a lot of interest here in your work on water in the Middle East.

The email included a list of distinguished previous speakers, and then continued:

> The presentation is usually about 45–60 minutes, followed by open discussion. The conference is open to the public, which is usually quite large, and to the media. The date is up to you, I believe, although it should be during the 2009–10 academic year.
>
> The conference is organized by the IDEP (Institut d'Economie Publique), a group of public economics economists ...

Alessandra provided some names associated with IDEP, indicated who was in charge of dealing with "practical questions and details," identified her own role as "ambassadorial only," and wished me a pleasant summer.

I was delighted to accept, and wrote an email reply to "Ambassador Alessandra."

> This is a considerable honor, and of course I accept. In view of the fact that, after reading my water book, you paid me an enormous compliment, I suspect that the suggestion of inviting me came from you, and I am very grateful.

The conference took place in May 2010, and, while not attended by a large group, it was attended by Marcel Boiteux *lui-même*. Apart from difficulties in travel, we had a very good time. The paper I presented was later published.

Even before that, however, I participated in a session on water sponsored by the MIT Women's League as part of its Catherine N. Stratton Lecture Series on critical issues. This was a large meeting with several speakers, of which I was one. Peter Rogers of the Harvard Applied Sciences Department

442

was the chair. John Briscoe, by then no longer at the World Bank but at Harvard, was also a speaker.

I spoke about the project and the theory behind it, which was very warmly greeted. Briscoe told me he wanted to learn more about what he called an important analysis and asked me to send him a paper—which I did. True to form, however, he never got back to me—just as after our meeting at the World Bank some ten years earlier.

Finally, during this period, there was the Institute conference held at MIT on May 20–21, 2010 on "Rethinking Water: A Critical Resource." Its objective was to bring together all the work on water going on at MIT. I had just returned from the Middle East, where I had been attending the semi-annual meeting of the New Israel Fund, and where I also visited the Arava Institute for Environmental Studies at the invitation of David Lehrer, its director. I had made some suggestions regarding the first day of the conference to Harriet Ritvo, the historian who spoke for MIT's School of Humanities, Arts and Social Sciences, but played no role. On the second day, although very jet-lagged, I did participate in the workshop on "Water Policy, Economics and Business." I was very glad to have done so, because of what happened next.

After the session, a small, older woman came up to me and introduced herself as Miriam Balaban, who had been visiting MIT for the entire academic year and remains very active in water issues, particularly desalination. She had founded the journal *Desalination*, the international journal on the science and technology of water desalting and purification, and served as its editor-in-chief for decades, and has been the Secretary General of the European Desalination Society since 1993.

She wanted to discuss my work. I told her I was very jet-lagged, but she was leaving MIT that evening to return to her base in Europe. I managed to overcome my condition, and we had a long talk. Starting with that, we became friends.

Miriam was arranging a large conference of the European Desalination Society to be held in Tel Aviv during the first week of October 2010. She regarded my work as important and asked me to speak. Of course, I agreed. Indeed, because the paper I prepared was quite long,[5] we later decided to have me speak in two sessions—one on the theory of the project and its use for domestic water management, and the second on the use of the model for regional cooperation. That did not work out well.

The first session was fine, but the second one on cooperation was not. The chairman, who was from the Israeli Water Authority, had convened a luncheon meeting for the speakers and announced that every speaker would have a quite limited time to present. I argued with him to no avail. Finally, well into the evening, the time for my presentation came at last. But then the chairman held the question and answer period before I spoke, despite urgent signals from Miriam. This meant that most of the audience left before I could speak. Miriam was quite apologetic.

It was at this conference that I first met Clive Lipchin, the director of the *Arava Institute* Center for Transboundary Water Management, who had been away when I visited there the previous May. We had dinner together and, at his request, I provided a short explanatory paper—something I had already drafted for a different purpose—that was distributed at the conference and eventually published.[6] Later, in December 2010, Lipchin was one of the main organizers of a conference held in Oxford, England to which I was invited and spoke. The intention was to see how cross-boundary cooperation in water might be attained in the Middle East.

Through all of this, I continued to seek opportunities to interest the parties in the continued work on our water project.

31

Israeli Participation and Other Developments

Through the entire period of the previous chapter, I continued to work on trying to rekindle Israel's interest in our project. Part of my effort was to speak again with Shimon Tal, the Israeli Water Commissioner. He was far more interested than previously, but his primary concern was about security issues in the use of Israeli data. I put it to him that we could work out an arrangement in which I would never see the underlying data. Alternatively, he could just trust me. Further, although I was probably still the person best qualified to run the model, and especially to interpret the results, there were Israelis who could certainly gain the required skills. Tal said he would send someone to Cambridge to learn the system—probably Miki Zaide, who did much of the Israel Water Authority's computer modeling. But his staff members were too busy, and no one ever came.

I did not give up. In the summer of 2005, I was in Israel for the summer board meeting of the New Israel Fund. Lenny Hausman suggested I meet with Shalom Turgeman, an adviser to then Prime Minister Ariel Sharon (and later to his successor, Ehud Olmert). Alon Liel joined me. Turgeman appeared seriously interested, but nothing ever came of it.

The next year, in June, Ellen and I traveled to Israel, again in connection with the summer meeting of the New Israel Fund board. I had lunch while I was there with Zvi Eckstein at Tel Aviv University. During the meal, Zvi received a call telling him that his appointment as Deputy Governor of the Bank of Israel had been approved. I congratulated him.

After lunch, while riding in a cab to meet Ellen, I checked my email and found a message from Daniel Wasserteil, an economist with Applied Economics Ltd., Zvi's consulting firm, from which he was about to resign. Wasserteil and Applied Economics had been working for the Israeli Water Authority and using WAS. He asked whether the Water Authority could have a free license to MYWAS (as they had for WAS), because Miki Zaide was particularly anxious to get one. That was very exciting, and I called Wasserteil immediately.

Of course, though, I could not simply agree to a free license. Annette and I had already decided that we could not give MYWAS licenses away, but had to sell them at some reasonable rate. We had expended considerable time on MYWAS, and I had contributed a good deal of money to keep the project alive. Beyond that, MYWAS was not yet really ready. A lot of work had to be done, especially to provide a user-friendly interface. As it stood, we could show how it worked on some simple, made-up examples, but were not ready for it to be used on real complicated data.

After that, Shlomi Parizat, Zvi's successor as head of Applied, asked whether I could assist the firm in making a presentation of MYWAS to the Water Authority and show how it could be used as they had used WAS. That seemed to be reasonable, so I agreed.

That turned to be an unwise decision. I accompanied Parizat to make a presentation to the Water Authority, in early 2008. The decision would ultimately rest with Professor Uri Shani of the Department of Soil and Water Sciences at Hebrew University's section in Rehovot. He had become head of the Water Authority and was present for a bit of the presentation.

After our presentation, Parizat submitted a written proposal. It was rejected. Uri Shamir, who had been present at the meeting, said in an email that the principal reason was that Applied, with Zvi's departure, lacked the capacity to do the work. I suggested as an alternativee that Annette and I could substitute for some of that capacity, but it did not seem workable.

In Spring 2010, Sharon Freedman—the New England Zone Director for the Jewish National Fund (JNF) and later the National Campaign Director—contacted me. She was responsible for the Parsons Water Fund (PWF), a sub-fund of the JNF created in memory of the late Nathan Parsons, who had served on the JNF board. I learned later that she had met with Nadav Tamir, the outgoing Israeli Consul-General in New England, and asked him for a list

446

of people with whom she should talk who were interested in water. My name was first on his list.

I had gotten to know Nadav pretty well in the four years he had been in Boston. We had spoken about the project several times, and he had arranged for me to talk with two people in the Foreign Ministry during one of my visits to Israel. He was not averse to telling his own government when he thought they were making mistakes in dealing with the United States, and for doing so, he had been summoned to Israel for a dressing down, but then allowed to return to his station. He was strongly supported by leaders of the Boston Jewish community.

When Sharon Freedman came to talk with me, she also invited me to a JNF dinner to honor Nadav on the occasion of his end of term and departure. It was a request with which I really wanted to comply, but I was not anxious to be identified with the JNF, whose actions had not always seemed appropriate. So, I consulted a couple of friends in the Boston Jewish community—primarily Leibel (Leonard) Fein, a friend of many years and who was a leader among American Jews in supporting a two-state solution in Israel–Palestine. Leibel served with the New Israel Fund and Americans for Peace Now. He was a lover of Israel who definitely did not march in step no matter what the Israeli government did. I admired him a great deal. He told me it would be fine for me to discuss water with the JNF and, especially, to go to an event in honor of Nadiv Tamir. So, I did.

That began several fruitless years of trying to get funding from the Parsons Water Fund. After several years of time and effort to get funding from the Jewish National Fund, I went back to looking for a way to get the project restarted in Israel. I stumbled into an excellent solution.

In the process of trying to get JNF funding, I had asked Uri Shamir who might supervise such a project. He suggested Yoav Kislev of Hebrew University. I had known Yoav for a number of years as a really first-rate agricultural and water economist. At various times, he had urged the IWA to use our model, had provided me with data on rainfall, and had even caught a data error in our runs of the WAS model for Israel.

I eagerly contacted Yoav, but he declined on the grounds that he was retired and not up to a new project. But he did recommend an excellent group of three members of his department from Hebrew University, who he said were very good and highly suited for the work: Iddo Kan, Eli Feinerman, and

Israel Finkelshtain. Yoav and I met them all for dinner in Tel Aviv the following February 2012, when I visited for the New Israel Fund's winter meeting, and we discussed matters generally.

I learned in that discussion that they were close to the Ministry of Agriculture, which really wanted such a model. They assured me they could obtain an official letter to that effect—and they did so, very promptly. It seemed clear to me that here was a team that really understood how to do economics modeling for water. They lived up to that view over the following years.

Developments in Palestine and Jordan

There were, however, developments outside of Israel. The most important of these involved Palestine and Jordan.

The Czech Republic had for some time been subsidizing, on a small scale, the building of a Palestinian MYWAS model. Indeed, I used to point out that the Palestinians were well ahead of the others. That seemed especially so when the head of the Palestine Water Authority (PWA), Shadad Attili, agreed to have his group take over the model building itself. That prompted the Czechs to give substantially more funds and time to the effort. However, problems soon arose.

It became clear that attempts were being made at the PWA to dismiss Karen Assaf, who was to have been in charge of overseeing model development. She was indispensable, but had not been receiving any pay in any capacity. Anan Jayyousi, also indispensable, was being pushed out as well. Further, Annette was not getting paid, even for her travel to Palestine. By the end of 2013, all work had stopped. The kindest explanation I could give was that the money for the project was being "diverted." But the staff very much wanted to continue.

Of course, the Czech government wanted the "diverted" money returned and put to proper use. When Shadad Attili resigned from the PWA, it seemed that that would happen—but then he returned. Pavel Seifter then attempted to get the project moved to Anan's control, where it had begun a few years earlier. But as if Attili's return was disastrous enough, the Czechs then replaced Ivo Silhavy as head of the Czech Representative Office in Ramallah. His support had been crucial from the beginning in 2008. His replacement— and, indeed, the Czech government—was wildly out of touch with what had happened and just wanted the funds returned, claiming there was no reason to pay for Annette's travel and seeming not to understand that she also was playing an indispensable role. Even when Pavel pressed them to look at matters correctly, they remained unmoved.

When I saw Anan at a conference in November 2014, he said the PWA was being divided into two groups, and he was hopeful the group without Attili might straighten things out. But by January 2015, I had concluded that there seemed very little hope of proceeding with the Palestinians.

Meanwhile, the Jordanians had finally acquired the funds from the European Union that had been promised for years. Work started on a MYWAS model for at least part of Jordan. Annette took charge of that work, and I was told they were excited and the project was going well.

Another encouraging development was the Israeli acceptance of a proposal from the Jordanians that Israel should buy water from Jordan in the South and sell water to Jordan in the North. An agreement was being signed, although the buying and selling prices had not been set. MYWAS could help that process.

It was a welcome idea that appealed to me in two ways. One was that it meant both countries had come to understand that water could be bought and sold. The other was that the proposal came from Jordanian Water Minister Hazim El-Nasser, who was a serious and helpful member of our water project.

Eco-Peace and Possible Cooperation among Countries

In the summer of 2012, I was invited by the U.S. Department of Agriculture to deliver a lecture on our methods to a small group of Israelis and Palestinians in East Jerusalem. Just as I was preparing to travel, I received an invitation from Gidon Bromberg, the chair of Eco-Peace (the new name of FOEME) in Israel. He was a major mover in a project on restoring the Lower Jordan River, and there was about to be a conference on that in Amman for Israel, Jordan, and Palestine. He very much wanted me to attend, and I was able to fit it into my schedule.

Royal Haskoning DHV, the Dutch consulting team that had been retained, had done a fine job working out how much water would have to be put into the river to reach various levels of restoration. In a presentation, the team also pointed out where that water might come from. When they were finished, Gidon turned to me and asked whether I had any questions. I could think of one.

"I see you've located a menu of places from which water could be taken," I asked, "but have you studied where water can be taken most efficiently?" That, I pointed out, could be done with the help of MYWAS.

When the session was over, I had a good discussion about my question with Jeroen Kool, head of the Dutch consulting team. Kool had read and approved of *Liquid Assets*. I told him about MYWAS, and he agreed that it might be very useful to their project. But MYWAS was not yet ready, so we agreed to talk again when it was. For various reasons, that never happened.

Two summers later, in 2014, I was invited to attend a follow-on conference on the Jordanian side of the Dead Sea during the coming November. I would lead a discussion on economics. Of course, I accepted right away, and I began to consider what I should say about the matter. Ellen and I decided that we would take the opportunity to go to Berlin, where we had never been, before heading to the conference.

Then, while already in Berlin, I received an email from one of the Swedes putting on the conference. There had been a change in the program and economics was no longer to be discussed. I was quite unhappy, having put together some points I thought were of interest, but I decided to attend nonetheless.

With the help of Gidon Bromberg and Anan Jayyousi, I was allowed to speak for a few minutes to about 16 people. The only really useful thing that came out of this were the insights I had put together in expectation of a much larger and attentive audience. My main points were as follows.

Consider first a single country. Then MYWAS will easily find the optimal sources from which to take water to the Lower Jordan. To do that, there will have to be a statement regarding the total amount to be brought to the Lower Jordan. If two or more countries will be providing the water and all are willing to run their MYWAS models together, they will also reach the efficient solution.

However, it is very unlikely that the countries will agree to running their MYWAS models jointly, even though that will be optimal and to the benefit of all. In that case, there will have to be agreements as to how much water each country should contribute.

In the current situation, it appears quite obvious that the efficient way to obtain the necessary water would be to have it come from Israel, who would stop (or at least greatly reduce) the amount of water now taken by the Israel Water Carrier at the Sea of Galilee (also called "Lake Kinneret" by the Israelis and "Lake Tiberias" by the Arabs). The water would be replaced in Israel by the output of desalination plants. But the Israeli government will be very reluctant to do this.

It should also be noted that a similar arrangement would efficiently have the Palestinians pumping much larger amounts of water on the West Bank than are now permitted. There have been offers from Israel to build a pipeline to carry water to the West Bank Palestinians. But that would leave the desalination plants and the pipeline involved under the control of Israel, and, not surprisingly, the Palestinians have refused the offer. It would be far more efficient if the water from the Israeli Water Carrier were to be released, at least in part, to the Palestinians on the West Bank, instead of bringing the water from the Sea of Galilee over the mountains to the coast and pumping the desalinated water back over the mountains.

But, in the current state of mistrust, that appears unthinkable for the present-day government. At least agreements on bringing water to the Lower Jordan are steps in the right direction, even though they are likely to be inefficient.

That is not what happened. But by August 2016, as I brought this writing to a close, it still seemed as if all might not be lost. David Lehrer at the Arava Institute had decided to use WAS and MYWAS in a way no one had ever done. Specifically, he wanted to have all three countries—Israel, Jordan, and Palestine—engaged together in using the model and, by doing so, he would show how everyone could cooperate for mutual benefit.

Meanwhile, I became less able to deal directly with the complex, sometimes almost Kafkaesque issues surrounding funding and cooperation. The project moved to individual groups—Anan Jayyousi and others, too, were continuing to work, and, as always, moving forward with their efforts to get funding—with some help from Annette, but little or none from me. And that is where things stood as I closed my recounting of the project.

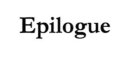

Epilogue

I close first with a *Torah* story and then with two sayings. The story comes from Genesis 21:22–34.

> And it came to pass at that time, that Abimelech and Phicol the captain of his host spoke unto Abraham, saying: "God is with thee in all that thou doest. Now therefore swear unto me here by God that thou wilt not deal falsely with me, nor with my son, nor with my son's son; but according to the kindness that I have done unto thee, thou shalt do unto me, and to the land wherein thou hast sojourned."
>
> And Abraham said: "I will swear."
>
> And Abraham reproved Abimelech because of the well of water, which Abimelech's servants had violently taken away, And Abimelech said: "I know not who has done this thing; neither didst thou tell me, neither yet heard I of it, but today."
>
> And Abraham took sheep and oxen, and gave them to Abimelech; and they two made a covenant.
>
> And Abraham set seven ewe-lambs of the flock by themselves. And Abimelech said unto Abraham: "What mean these seven ewe-lambs which you have set by themselves?"
>
> And he said: "Verily, these seven ewe-lambs shalt thou take of my hand, that it may be a witness unto me, that I have digged this well."
>
> Wherefore that place was called Beer-sheba [Well of the Seven], because there they swore both of them.
>
> And Abraham planted a tamarisk tree in Beer-sheba, and called there on the name of the LORD, the Everlasting God.
>
> And Abraham sojourned in the land of the Philistines many days.

The story from Genesis is the oldest story I know that shows water can be

455

traded for other things. Of course, neither Abraham nor Abimelech knew anything of computers or systems such as MYWAS. But in our work, we liked to think they would have approved of what we were doing.

The following saying has already been quoted in chapter 10. It is also from the *Torah*, Isaiah 12:3.

As it is written: with joy shall you draw water from the wells of salvation.

The second saying also seems appropriate here. It comes from the *Pirkei Avot*, the "Chapters of the Fathers," which is a compilation of ethical teachings and maxims passed down to the rabbis, beginning with Moses. The *Pirkei Avot* is part of the *Mishnah*, the great written collection of Jewish oral traditions first compiled in the third century. The quote is attributed to Rabbi Tarfon, a member of the third generation of Mishnah sages, who is said to have lived during the period between the destruction of the Second Temple in the year 70 and the fall of Betar in the year 135.

It is not incumbent upon you to complete the work, but neither are you at liberty to desist from it.

APPENDIX
The Multi-Year Water Allocation System

The Multi-Year Water Allocation System (MYWAS) deals readily and directly with a number of problems over time by maximizing the present value of net benefits over a number of future years or time periods using a discount rate specified by the user. In all of its applications, as in all WAS applications, system-wide effects and opportunity costs are automatically dealt with, and the user's own decisions and values are implemented.

A partial list of MYWAS applications includes:

- *Timing, Order, and Capacity of Infrastructure Projects*: MYWAS allows the user to specify a menu of possible infrastructure projects, their capital and operating costs, and their useful life. The program then yields the optimal infrastructure plan, specifying which projects should be built, in what order, and to what capacity. It is a major advance.

- *Storage Management*: Most obviously, it is now easy to deal with storage issues, in particular the decisions as to how much water should be stored or released from reservoirs. The decisions involved can be for inter-year or for inter-seasonal storage.

- *Aquifer Management*: Man-made storage is not the only kind. Water can also be transferred between time periods by increasing aquifer pumping when water is relatively abundant and reducing it when water is relatively scarce. It means that the use of aquifers and other natural water sources no longer needs to be restricted to the average yearly replenishable amount in the model (with that average adjustable by the user). Rather, by specifying the effects of withdrawal on the state of the aquifer, the user can obtain a guide on the optimal pattern of aquifer use over time, including guidance as to aquifer recharge.

- *Climatic Uncertainty*: Of course, optimal planning over time will depend on the climate, and climate—especially rainfall—is variable and uncertain. MYWAS enables the systematic study of the effects of such uncertainty on optimal planning by providing the means to examine optimal decisions as a non-linear function of climate variables. Other uncertainties, such as those involved in population forecasts, can also be dealt with.

- *Global Warming*: Of course, the multiyear nature of MYWAS makes it suitable for examining the effect of different global warming scenarios.

- *Fossil Aquifers*: The rate at which a fossil aquifer (i.e., an aquifer that does not have any replenishment of water but is used up over time) should be pumped can also be determined endogenously, through the use of MYWAS rather than being specified exogenously by the user. That rate will generally vary over time as conditions change.

- *Water Quality*: If desired, MYWAS (or WAS) can be adapted to permit a more sophisticated treatment of multidimensional aspects of water quality than available in the original version of WAS.

- *Effect of Discount Rate*: Obviously, MYWAS can be used to examine the effects of the choice of discount rate on all aspects of the optimal solution.

These properties make MYWAS an extremely powerful tool for efficient water management.

References

Preface

1. Frost, "The Death of the Hired Man."

Chapter 1

1. Fischer *et al.*, *Securing Peace in the Middle East.*

Chapter 3

1. Fisher, Huber-Lee, *et al.*, *Liquid Assets.*

Chapter 4

1. Fischer, Rodrik, and Tuma, *The Economics of Middle East Peace.*

Chapter 5

1. Fisher, Huber-Lee, *et al.*, *Liquid Assets.*

Chapter 7

1. Fisher, "The Economics of Water Dispute Resolution ..."

Chapter 10

1. Fisher, "Water and Peace in the Middle East."

Chapter 15

1. Fisher, "Water: *Casus Belli* or Source of Cooperation?"

Chapter 16

1. Fisher, Huber-Lee, *et al.*, *Liquid Assets.*

Chapter 17

1. Longfellow, "The Masque of Pandora."

Chapter 18

1. Fisher, Huber-Lee, *et al.*, *Liquid Assets.*

Chapter 20

1. Zachary, "Water Pressure."

2. Frost, "The Death of the Hired Man."

Chapter 26

1. Fisher and Huber-Lee, "WAS-Guided Cooperation in Water: The Grand Coalition and Sub-coalitions"; Fisher and Huber-Lee, "WAS-Guided Cooperation in Water Management: Coalitions and Gains."

Chapter 27

1. Fisher and Askari, "Optimal Water Management in the Middle East and Other Regions."

2. Fisher, Huber-Lee, *et al.*, *Liquid Assets.*

3. Fisher *et al.*, "Optimal Water Management and Conflict Resolution."

4. Amir and Fisher, "Analyzing Agricultural Demand for Water with an Optimizing Model"; Amir and Fisher, "Response of Near–Optimal Agricultural Production to Water Policies."

5. Salman, Al-Karablieh, and Fisher, "An Inter–seasonal Agricultural Water Allocation System (SAWAS)"; Al-Karablieh, Salman, and Fisher, "Forecasting Wheat Production"; Salman *et al.*, "The Economics of Water in Jordan."

6. Fisher, "Water: *Casus Belli* or Source of Cooperation?"

7. Fisher, "Water Value, Water Management, and Water Conflict."

8. Fisher *et al.*, "Optimal Water Management and Conflict Resolution."

Chapter 28

1. Fisher, "Water Value, Water Management, and Water Conflict."

Chapter 29

1. Islam and Susskind, *Water Diplomacy.*

2. Fisher, "Water Management, Infrastructure, Negotiations and Cooperation."

3. Fisher, Huber-Lee, *et al.*, *Liquid Assets.*

Chapter 30

1. Demas, "Members of Demas—Forum 2000," https://www.demas.cz/en/members-of-demas/.

2. Fisher *et al.*, "Analyzing Future Palestinian Water Issues …"

3. Fisher, "Optimization Modeling in Water Resource Systems and Markets."

4. Fisher and Huber-Lee. "Using Economics for Sustainability, Efficient Management and Conflict Resolution."

5. Fisher and Huber-Lee, "The Value of Water."

6. Fisher and Huber-Lee, "Water for Peace."

Glossary of Economics Terms

Absolute value: an asset's intrinsic value irrespective of comparison to other assets.

Capital costs: fixed, one-time expenses incurred in a purchase of land, buildings, construction, and equipment, used in producing goods or rendering services.

Constant-elasticity demand curve: a demand curve on which the elasticity of demand is the same at every point; a demand curve that is completely flat is perfectly elastic.

Consumer surplus: the difference between the total amount consumers are willing and able to pay for a good or service (indicated by the *demand curve*) and the total amount they actually do pay (i.e., the *market price*).

Deadweight loss: a social cost created by market inefficiency, that is, when supply and demand are not in equilibrium, or caused by a deficiency resulting from the inefficient allocation of resources.

Demand curve: a graphic representation of the quantities demanded of some commodity at various prices.

Demand elasticity: the sensitivity of demand for a good to changes in other economic variables, such as price and consumer income.

Efficiency price: a price that incorporates all available information.

Equilibrium price: the price at which economic agents or aggregates of economic agents (such as markets) have no incentive to change their economic behavior, because supply and demand are precisely equal.

Fixed costs: costs (such as rent) that are constant whatever the quantity of goods or services produced.

Hypersurface (*from mathematics*): a generalization of the concept of an ordinary surface in three-dimensional space to the case of an n-dimensional space.

Inverse demand function: a function that maps the quantity of output demanded to the market price for that output.

Lagrangian: a functional in mathematics whose extrema—that is, maximum or minimum values in a specified "neighborhood"—are to be determined in the calculus of variations.

Marginal costs: the cost added by producing one additional unit of a product or service.

Microeconomics: broadly speaking, economic analyses of the behavior of individual persons or firms, rather than aggregates that are the subject of macroeconomics.

Nominal interest rate: the rate of interest before adjustment for inflation.

Operating costs: expenses related to an enterprise's day-to-day operations.

Producer surplus: the difference between the amount a producer of a good receives and the minimum amount the producer is willing to accept for that same good.

Real interest rate: the rate of interest received (or expected to be received) after allowing for inflation, that is, the nominal interest rate (see above) minus the inflation rate.

Replacement cost (or value): the amount it would cost to replace an asset at current prices.

Resource allocation: the assignment of a society's limited resources—labor, land, and capital—to produce the goods and services it needs.

Scarcity rent: the cost of "using up" a finite resource because the benefits of that resource would be made unavailable to future generations.

Shadow price: the estimated price of a good or service for which no market price exists.

Social surplus: the total of consumer surplus plus producer surplus (see above).

Unity: in the context of elasticity of demand, a percentage change in price and the consequent percentage change in demand being equal.

Bibliography

Al-Karablieh, Emad K., Amer Z. Salman, and Franklin M. Fisher. "Forecasting Wheat Production: The Case of the Irbid Region of Jordan." *Quarterly Journal of International Agriculture* 41, no. 3 (2002): 191–206.

Amir, Ilan, and Franklin M. Fisher. "Analyzing Agricultural Demand for Water with an Optimizing Model." *Agricultural Systems* 61, no. 1 (1999): 45–56.

_____ and Franklin M. Fisher. "Response of Near–Optimal Agricultural Production to Water Policies." *Agricultural Systems* 64, no. 2 (2000): 115–130.

Fischer, Stanley, Leonard J. Hausman, Anna D. Karasik, and Thomas Schelling, eds. *Securing Peace in the Middle East: Project on Economic Transition* (Cambridge, Mass.: The MIT Press, 1994).

_____, Dani Rodrik, and Elias Tuma, eds. *The Economics of Middle East Peace: Views from the Region* (Cambridge, Mass.: The MIT Press, 1993).

Fisher, Franklin M. "The Economics of Water Dispute Resolution, Project Evaluation and Management: An Application to the Middle East." *International Journal of Water Resources Development* 11, no. 4 (1995): 377–390.

_____. "Optimization Modeling in Water Resource Systems and Markets." In *Water Trading and Global Water Scarcity: International Perspectives*. Josefina Maestu, ed. (Oxfordshire, England: Routledge/RFF Press, 2013).

_____. "Water: *Casus Belli* or Source of Cooperation?" In K. David Hambright, F. Jamil Ragep, and Joseph Ginat, eds. *Water in the Middle East: Cooperation and Technological Solutions in the Jordan Valley*. International and Security Affairs Series (Norman, Okla.: University of Oklahoma Press, 2006), pp. 185–199.

_____. "Water and Peace in the Middle East." Memorial lecture for Gideon Fishelson. *The Economic Quarterly* 43, no. 3 (Nov. 1996), pp. 441–446.

_____. "Water Management, Infrastructure, Negotiations and Cooperation: Use of the WAS Model." In *Water Resources in the Middle East: Israel-Palestinian Water Issues*

– *From Conflict to Cooperation*. Hillel Shovel and Hassan Dweik, eds. (New York: Springer, 2007), 119–132.

_____. "Water Value, Water Management, and Water Conflict: A Systematic Approach." *GAIA – Ecological Perspectives for Science and Society* 11, no. 3, (September 2002): 187–190.

_____. "Water Value, Water Management, and Water Conflict: A Systematic Approach." In *Mountains: Sources of Water, Sources of Knowledge*. Ellen Wiegandt, ed. (Dordrecht, Netherlands: Springer, 2008): 123–148.

_____, Shaul Arlosoroff, Zvi Eckstein, Munther Haddadin, Salem G. Hamati, Annette Huber-Lee, Ammar Jarrar, Anan Jayyousi, Uri Shamir, and Hans Wesseling. "Optimal Water Management and Conflict Resolution: The Middle East Water Project." *Water Resources Research* 38, no. 11 (2002).

_____, Shaul Arlosoroff, Zvi Eckstein, Annette Huber-Lee, Anan Jayyousi, Uri Shamir, and Hans Wesseling. "Optimal Water Management and Conflict Resolution: The Middle East Water Project." *Economic Review* (2000). In Hebrew.

_____ and Hossein Askari. "Optimal Water Management in the Middle East and Other Regions." *Finance and Development* (2001): 52–6.

_____ and Annette T. Huber-Lee. "Using Economics for Sustainability, Efficient Management and Conflict Reolution in Water." *Économie Publique (Public Economics)* no. 26–27 (2011/12): 11–49.

_____ and Annette T. Huber-Lee. "The Value of Water: Optimizing Models for Sustainable Management, Infrastructure Planning and Conflict Resolution." *Desalination and Water Treatment* 31 (July 2011): 1–23.

_____ and Annette Huber-Lee. "WAS-Guided Cooperation in Water: The Grand Coalition and Sub-coalitions" in *Environment and Development Economics* 14, Special Issue 01 (February 2009): 89–115.

_____ and Annette T. Huber-Lee. "Water for Peace: A Game Changer for Israel, Palestine, and the Middle East." *Water Biz* 2, no. 3 (2010): 112–114.

_____ and Annette Huber-Lee, "WAS-Guided Cooperation in Water Management: Coalitions and Gains." In Ariel Dinar, José Albiac and Joaquín Sánchez-Soriano, eds. *Game Theory and Policy Making in Natural Resources and the Environment* (London: Routledge, 2008).

_____, Annette Huber-Lee, Ilan Amir, Shaul Arlosoroff, Zvi Eckstein, Munther J. Haddadin, Salem G. Hamati, Ammar M. Jarrar, Anan F. Jayyousi, Uri Sharmi, Hans Wesseling. *Liquid Assets: An Economic Approach for Water Management and Conflict Resolution in the Middle East and Beyond* (Washington, D.C.: Resources for the Future, 2005).

_____, Annette Huber-Lee, Karen Assaf, Ammar M. Jarrar, and Anan F. Jayyousi. "Analyzing Future Palestinian Water Issues with the WAS Model." Unpublished.

Frost, Robert, "The Death of the Hired Man," in *North of Boston* (New York: Henry Holt and Company, 1915).

Islam, Shafiqul, and Lawrence E. Susskind. *Water Diplomacy: A Negotiated Approach to Managing Complex Water Networks* (Washington, D.C.: RFF Press, 2012).

Longfellow, Henry Wadsworth. "The Masque of Pandora" (1875).

Salman, Amer Z., Emad K. Al-Karablieh, and Franklin M. Fisher. "An Inter-seasonal Agricultural Water Allocation System (SAWAS)." *Agricultural Systems* 68, no. 3 (2001): 233–252.

———, Emad K. Al-Karablieh, Heinz-Peter Wolff, Franklin M. Fisher, and Munther Haddadin. "The Economics of Water in Jordan." *Evolving Policies for Development, the Environment, and Conflict Resolution. Resources for the Future,* Washington, D.C. (2006): 116–149.

Zachary, G. Pascal. "Water Pressure: Nations Scramble To Defuse Fights Over Supplies." *Wall Street Journal,* December 4, 1997.

About the Author

Franklin M. Fisher, *Dennis Carlton and Jane Berkowitz Carlton Professor of Microeconomics* at *M.I.T,* was the leader of a group of Israelis, Jordanians, Palestinians and Americans, who together created **MYWAS**, the multi-year **W**ater **A**llocation **S**ystem for managing water systems in the Middle East. This over twenty-five-year long project created a model usable by each participating country for evaluating improvements in each water system. At the time of his death in 2019, Fisher, a man of peace, hoped that someday, the system could also be a tool for facilitating cooperation between the nations. In telling the story of this project he hoped to show how cooperation yields gains greater than the value of the disputed water.

CPSIA information can be obtained
at www.ICGtesting.com
Printed in the USA
FSHW021230080220
66732FS